Real-time PCR

Advanced Technologies and Applications

Edited by

Nick A. Saunders

HPA Microbiology Service Division
Colindale
Health Protection Agency
London
UK

and

Martin A. Lee

Porton Consulting Research Ltd
Salisbury
UK

Caister Academic Press

Caister Academic Press
Norfolk, UK

www.caister.com

British Library Cataloguing-in-Publication Data
A catalogue record for this book is available from the British Library

ISBN: 978-1-908230-22-5

Cover image courtesy of Martin A. Lee.

Printed and bound in Great Britain

Contents

Contributors

Mark A. Behlke
Integrated DNA Technologies, Inc.
Coralville, IA
USA

mbehlke@idtdna.com

Tom Brown
School of Chemistry
University of Southampton
Southampton
UK

tb@soton.ac.uk

Stephen A. Bustin
Postgraduate Medical Institute
Anglia Ruskin University
Chelmsford
UK

stephen.bustin@anglia.ac.uk

Joseph R. Dobosy
Integrated DNA Technologies, Inc.
Coralville, IA
USA

jdobosy@idtdna.com

Christina Egan
Wadsworth Center
New York State Department of Health
Albany, NY
USA

eganc@wadsworth.org

Jim Huggett
Molecular and Cell Biology
LGC
Teddington
UK

jim.huggett@lgc.co.uk

Waqar Hussain
Enigma Diagnostics Ltd
Dstl
Porton Down
Salisbury
UK

waqar.hussain.email@gmail.com

Cassandra D. Kelly-Cirino
DNA Genotek
Ottawa, ON
Canada

cassandra.kelly-cirino@dnagenotek.com

Martin A. Lee
Porton Consulting Research Ltd
Salisbury
UK

martin@portonconsultingresearch.co.uk

Dario L. Leslie
Dstl
Porton Down
Salisbury
UK

darioleslie@hotmail.com

Alan McNally
Pathogen Research Group
Nottingham Trent University
Nottingham
UK

alan.mcnally@ntu.ac.uk

Jacob Moran-Gilad
National Microbiology Focal Point
Public Health Services
Ministry of Health
Jerusalem
Israel

giladko@014.net.il

Tania Nolan
Sigma-Aldrich
Cambridge
UK

tania.nolan@sial.com

Richard Owczarzy
Integrated DNA Technologies, Inc.
Coralville, IA
USA

rowczarzy@idtdna.com

Scott D. Rose
Integrated DNA Technologies, Inc.
Coralville, IA
USA

srose@idtdna.com

Andrew D. Sails
Health Protection Agency Public Health Laboratory Newcastle
The Medical School
Royal Victoria Infirmary
Newcastle upon Tyne
UK

andrew.sails@hpa.org.uk

Chaminda Salgado
NDA Analytics
Alconbury
UK

chaminda.salgado@nda-analytics.com

Nick A. Saunders
HPA Microbiology Service Division
Colindale
Health Protection Agency
London
UK

nick.saunders@hpa.org.uk

Melvyn Smith
King's College Hospital
London
UK

melvyn.smith@nhs.net

David J. Squirrell
Enigma Diagnostics Ltd
Dstl
Porton Down
Salisbury
UK

david.squirrell@enigmadiagnostics.com

Preface

The polymerase chain reaction (PCR) is a technique so commonplace in the modern-day laboratory that it is easy to forget its revolutionary impact. Real-time PCR has removed many of the limitations of standard end-point PCR, and since its introduction in the mid-1990s there has been an explosion both in available instrumentation and in the number of publications describing real-time PCR applications across many disciplines. The technology is developing from an established powerful research tool into mainstream testing in the regulated markets such as food, veterinary and human *in vitro* diagnostics. The recent trend in these sectors is to adopt the most appropriate technical approach rather than to follow those better known methods of lead marketers.

This book aims to provide both the novice and experienced user with an invaluable point of reference to all of this technology. It separates commercial practice and branding, instead providing the detailed technical insight into underlying principles and methods to help the reader. The initial chapters cover the important aspects of real-time PCR from choosing an instrument and probe system to set-up, nucleic acid synthesis, sample extraction controls, validation and data analysis. It then goes on to provide a comprehensive overview of important real-time methodologies such as quantification, expression analysis and mutation detection. This is complemented by the final chapters, which address the application of real-time PCR to the diagnosis of infectious diseases, biodefence, veterinary science, food authenticity and molecular haplotyping. This essential manual should serve both as a basic introduction to real-time PCR and a source of current trends and applications for those already familiar with the technology. We also hope this text will stimulate readers of all levels to develop their own innovative approaches to real-time PCR.

Martin A. Lee
Nick A. Saunders

Current Books of Interest

Genome Analysis: Current Procedures and Applications	2014
Bacterial Toxins: Genetics, Cellular Biology and Practical Applications	2013
Cold-Adapted Microorganisms	2013
Fusarium: Genomics, Molecular and Cellular Biology	2013
Prions: Current Progress in Advanced Research	2013
RNA Editing: Current Research and Future Trends	2013
Microbial Efflux Pumps: Current Research	2013
Cytomegaloviruses: From Molecular Pathogenesis to Intervention	2013
Oral Microbial Ecology: Current Research and New Perspectives	2013
Bionanotechnology: Biological Self-assembly and its Applications	2013
Real-Time PCR in Food Science: Current Technology and Applications	2013
Bacterial Gene Regulation and Transcriptional Networks	2013
Bioremediation of Mercury: Current Research and Industrial Applications	2013
Neurospora: Genomics and Molecular Biology	2013
Rhabdoviruses	2012
Horizontal Gene Transfer in Microorganisms	2012
Microbial Ecological Theory: Current Perspectives	2012
Two-Component Systems in Bacteria	2012
Malaria Parasites: Comparative Genomics, Evolution and Molecular Biology	2013
Foodborne and Waterborne Bacterial Pathogens	2012
Yersinia: Systems Biology and Control	2012
Stress Response in Microbiology	2012
Bacterial Regulatory Networks	2012
Systems Microbiology: Current Topics and Applications	2012
Quantitative Real-time PCR in Applied Microbiology	2012
Bacterial Spores: Current Research and Applications	2012
Small DNA Tumour Viruses	2012
Extremophiles: Microbiology and Biotechnology	2012
Bacillus: Cellular and Molecular Biology (Second edition)	2012
Microbial Biofilms: Current Research and Applications	2012
Bacterial Glycomics: Current Research, Technology and Applications	2012
Non-coding RNAs and Epigenetic Regulation of Gene Expression	2012

Homogeneous Fluorescent Chemistries for Real-time PCR

1

Martin A. Lee, David J. Squirrell, Dario L. Leslie and Tom Brown

Abstract

The development of fluorescent methods for the closed tube polymerase chain reaction has greatly simplified the process of quantification. Current approaches use fluorescent probes that interact with the amplification products during the PCR to allow kinetic measurements of product accumulation. These probe methods include generic approaches to DNA quantification such as fluorescent DNA binding dyes. There are also a number of strand-specific probes that use the phenomenon of fluorescent energy transfer. In this chapter we describe these methods in detail, outline the principles of each process, and describe published examples. This text has been written to provide an impartial overview of the utility of different assays and to show how they may be used on various commercially available thermal cyclers.

Introduction

The fluorescent real-time polymerase chain reaction (PCR) (Saiki *et al.*, 1985; Mullis *et al.*, 1987) can provide both qualitative and quantitative analysis in various applications. Real-time PCR differs from earlier methods of analysis in that additional components are required to carry out the process. These components include an optical system integrated into the thermal cycler, and a probe that reports amplification during the course of the PCR process. The real-time PCR thermal cycler is discussed in Logan and Edwards (2009a) in detail. In this chapter we discuss the probe technologies or 'reporting chemistries.' Since the first edition of this text new methods have been reported in the literature and have become commercial products. This edition expands to include these developments.

In order to give the reader a complete understanding of current technology, the principles of real-time analysis will initially be presented outside the context of any specific commercial platform. The number of these is increasing and it is certain that the capabilities on offer will undergo continued improvement and allow new fluorescent approaches to be realized. At this point it is important to highlight to new users the relationship between the choice of probe system and the instrument. These are inextricably linked: the optical specification and the analysis tools on any given platform greatly influence the applicability and utility of different probe systems. Whilst the main factors for the choice of instrument are often driven by throughput requirements and the initial purchase cost, careful consideration of the reporting chemistry for the required application should be made before purchase since the operation of a particular chemistry may be greatly compromised on some platforms.

Equally important is the technical support provided by suppliers. This may be limited for non-supported chemistries and so called 'open platforms' from manufacturers that do not support any specific chemistry, or who may not be able to provide technical advice for their chemistry of choice. In this chapter we break down probe technology into a number of components to enable the reader to understand how the different assay systems may be used on current and future fluorimetric thermal cyclers. For new users this will facilitate the implementation of various assays on commercial platforms.

We first present a background to the limitations of PCR without real-time detection and introduce the two main classes of real-time chemistries. A second section on the basics of fluorescent resonance energy transfer (FRET) will allow the reader to understand how dye technologies may be applied in strand-specific applications. An outline of the function and utility of real-time PCR probe methods will provide the reader with an understanding of how such technology may be best applied. Finally, experimental considerations and specific examples will be provided although the reader should refer to the relevant chapters in this text for detailed information.

Background to real-time PCR

Prior to the introduction of the first commercial real-time PCR instruments in 1996 the utility of the PCR was limited. It could only be easily applied as a qualitative method. Analysis of amplification products was carried out at the end of thermal cycling using techniques such as gel electrophoresis or PCR ELISA. The dynamic range for quantitative PCR using these approaches was limited because the accumulation of specific product tends to plateau when the reaction is allowed to progress to completion. At high cycle numbers the amount of product is often unrelated to the initial amount of target nucleic acid. Factors that contribute to the plateau effect include, amongst others, substrate depletion (nucleotides, amplimers, etc.), specific inhibition (competitive binding of products-to-products rather than amplimer-to-product) and non-specific product inhibition (amplimer artefact accumulation, mis-priming and the accumulation of pyrophosphate). Quantitative analysis, with a limited dynamic range, could only be achieved by stopping thermal cycling before this plateau was achieved. Competitive PCR (utilizing molecular mimics that compete for amplimers) could be used to achieve a wider dynamic range. However, this approach is less accurate and more time consuming than the real-time methods described here.

The fluorescent real-time approach requires a thermal cycler that can interrogate the sample throughout the course of amplification. These instruments are discussed in Logan and Edwards (2009a) and the basic kinetic theory of real-time PCR is in Logan and Edwards (2009b). The signal collected during the exponential stage of amplification is of most use for the determination of the number of initial target nucleic acid molecules. At this stage the reaction efficiency is so high that the number of amplicon molecules effectively doubles every cycle. Amplification is observed as a two-fold increase in signal as it rises above that of the background noise. The amount of signal noise in a real-time assay is a function of the type of chemistry utilized, the optical and thermal (since fluorescence has a reciprocal relationship with temperature) performance of the instrument as well as other experimental considerations such as volume effects and the degree of mixing of reaction cocktails.

There are two main classes of real-time fluorescent chemistries: 'generic' and 'strand-specific' methods. Generic methods use probes that bind non-specifically to DNA and include the intercalators and other DNA binding dyes. Strand-specific methods use nucleic acid probes that target the amplicon (product) between the amplimer binding regions. It should be noted that most of the systems described utilize DNA for the probe. The use of modified bases was proposed in 1991 (Holland *et al.*, 1991). Methods for peptide nucleic acid (PNA) probes utilizing FRET have been reported (Ortiz *et al.*, 1998) and described for application in real-time PCR (Svanik *et al.*, 2000a,b), and for those probes that use hybridization alone (rather than hydrolysis) it should be possible to utilize PNA and other analogues such as locked nucleic acid (LNA) (Koshkin *et al.*, 1998) in the chemistries. Nucleotide analogues increase the melting temperature (TM) of the probe improving signalling in most methods, in particular where the target is AT rich. The strand-specific methods have the advantage that amplification of reaction artefacts such as amplimer-dimers do not contribute to the observed signal. Generally they provide higher specificity and better signal–noise ratios than the generic methods that are described next. The type and mode of action of different probe systems is illustrated in Table 1.1.

Generic detection using DNA binding dyes and melting point analysis

DNA-binding dyes have been used extensively in molecular biology research for the direct analysis and quantification of nucleic acids. These dyes bind to dsDNA with enhanced fluorescence. Ethidium bromide, which is used conventionally for staining agarose gels, has been used in real-time PCR (Higuchi *et al.*, 1992, 1993). Other dyes, such as those from Molecular Probes (YoPro® and YoYo®) (Ogura *et al.*, 1994), have also been reported. The fluorescent signal is usually monitored towards the end of the extension step in a 3step PCR.

Of these probes, by far the most reported are the minor groove-binding dyes SYBR®Green-1 (Becker *et al.*, 1996) and SYBR®Gold. These dyes typically exhibit a 20- to 100-fold fluorescence enhancement on binding and are commonly used because their emission maxima closely match that of fluorescein and the optics in most commercial real-time instruments are set to detect in this (circa 520 nm) wavelength range. The SYBR® dyes are popular for real-time detection because they are readily available from PCR reagent suppliers and their use requires little additional experimental design.

The optimum concentration for these dyes for a number of instrument platforms is published in the open literature. For SYBR® dyes this is typically 1:30,000 to 1:100,000 dilution of the reference solution supplied by the manufacturer, although the concentration is dependent on the tube format (e.g. composite glass or native polypropylene) and the optical efficiency of the instrument. Ready-to-go cocktails are available from suppliers with the dye already in the reagent mix.

Binding dye chemistries are excellent for assay optimization. The signal is proportional to the total nucleic acid concentration and therefore directly related to the PCR process. In an optimized assay, they may thus be used to determine accurately the reaction efficiency. It should be noted that this is not always the case when using labelled nucleic acid probes, where the signal generated is related to both the assay and the 'probe efficiency'. In some assay systems the probe efficiency can have a major effect on the signal obtained and this is discussed later for each assay type. In both strand specific and generic RT-PCR, the signal obtained is additionally dependent on the efficiency of the reverse transcriptase step.

Melting point analysis may be carried out on most commercial real-time instruments, usually at the end of the PCR amplification (Fig. 1.1). Temperature-dependent fluorescence measurements are made whilst slowly increasing the temperature of the reaction products

Table 1.1 Summary of fluorogenic reporting chemistries for real-time PCR. The schematics show the probe structure and mode of action. A colour version of this table is available in the plate section at the back of the book.

Method	Principle	Self-probing embodiment	Principle: Probe structure	Principle: Nucleic acid interaction
DNA binding agent	Enhancement of fluorescence DNA binding into DNA duplex	N/A	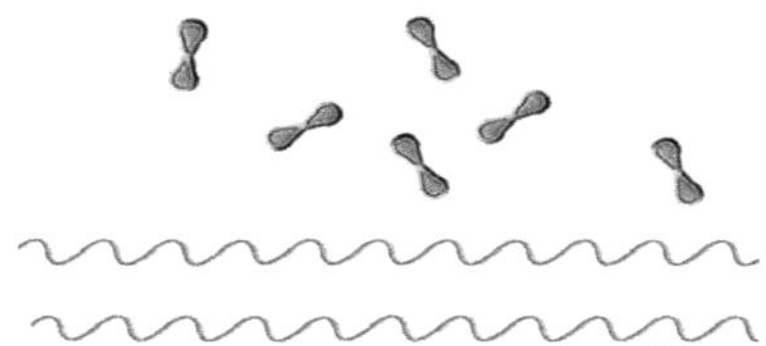	
5′ nuclease assay	Hydrolysis of probe by concomitant hydrolysis of probe during primer extension	Intrataq		
Dual hybridization probe (*cis*)	Hybridization of two linear probes to nascent PCR strand	Embodiment of Scorpions®	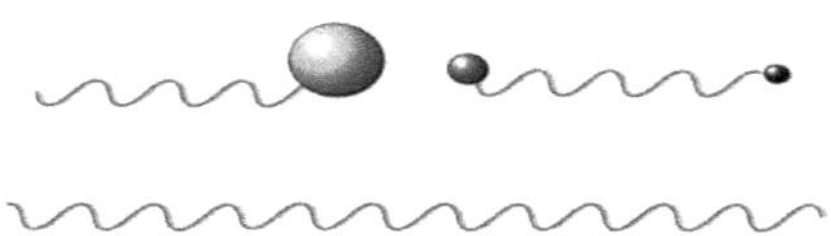	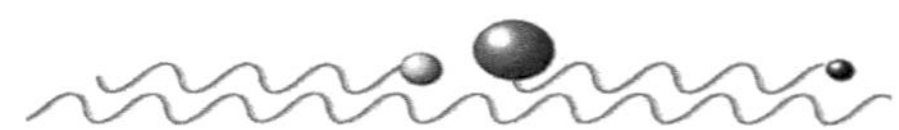
Dual hybridization probe (*trans*)	Hybridization of linear probe to nascent PCR strand generated by a labelled primer	Not reported	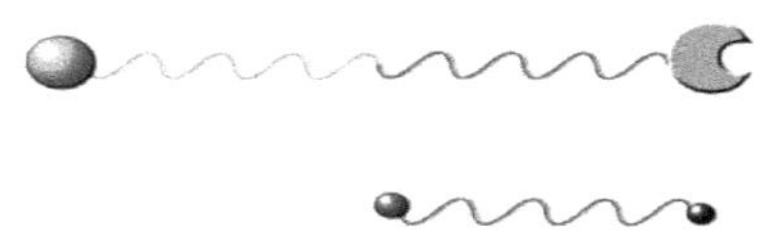	

Method	Principle	Self-probing embodiment	Principle	
			Probe structure	Nucleic acid interaction
ResonSense®	Hybridization of linear probe to nascent PCR strand. Energy is transferred through a DNA binding agent	Angler®	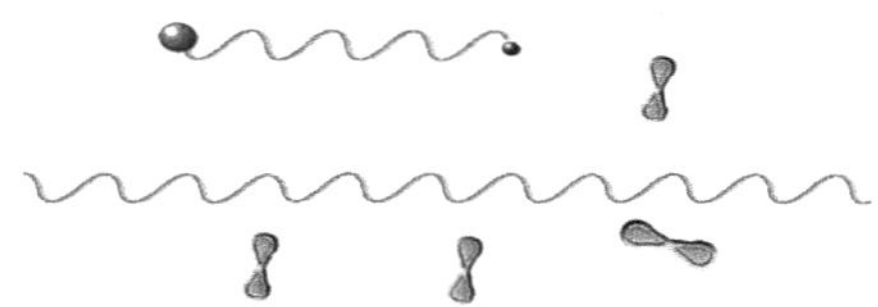	
Hybeacons®	Hybridization of linear probes.	Not reported	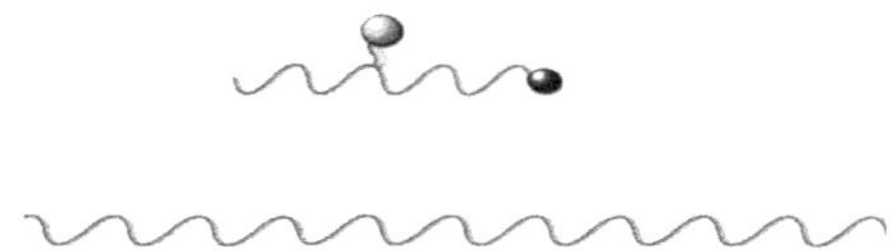	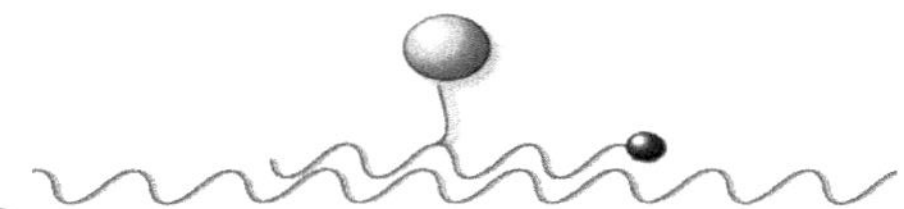
Eclipse®	Linear conformational change. Short probe is stabilized to target using a minor groove binding agent	Not reported	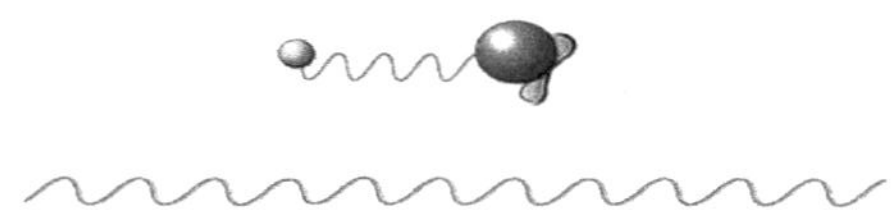	
Light-Up®	Fluorescent enhancement upon hybridization with DNA-binding agent covalently linked to probe	Not reported	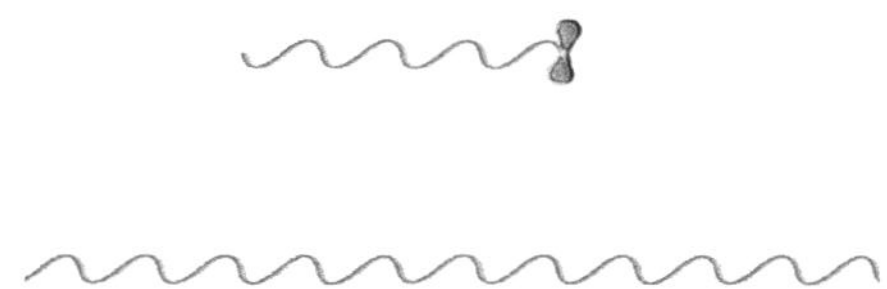	

Table 1.1 (Continued)

Method	Principle	Self-probing embodiment	Principle: Probe structure	Principle: Nucleic acid interaction
Molecular Beacons	Conformational change. Native probe stabilized with hairpin structure	Scorpions®	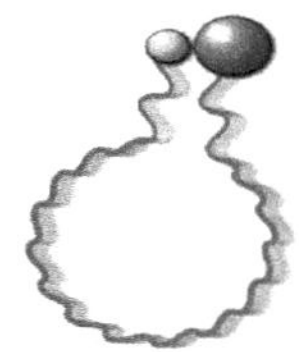	
Ying–Yang®	Competitive binding of linear probe	Duplex Scorpions®	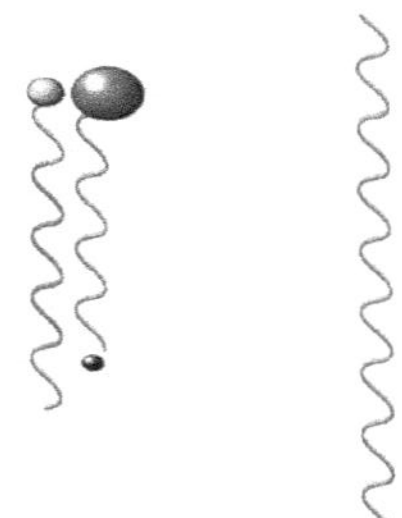	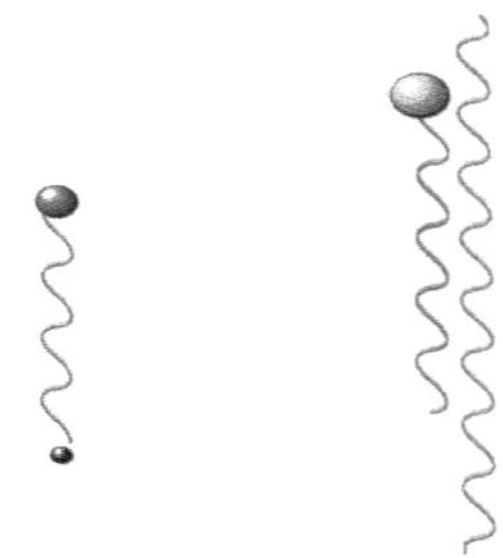

Key: DNA , donor dye , acceptor dye , DNA binding agent , minor groove binder , polymerase , 3' blocker .

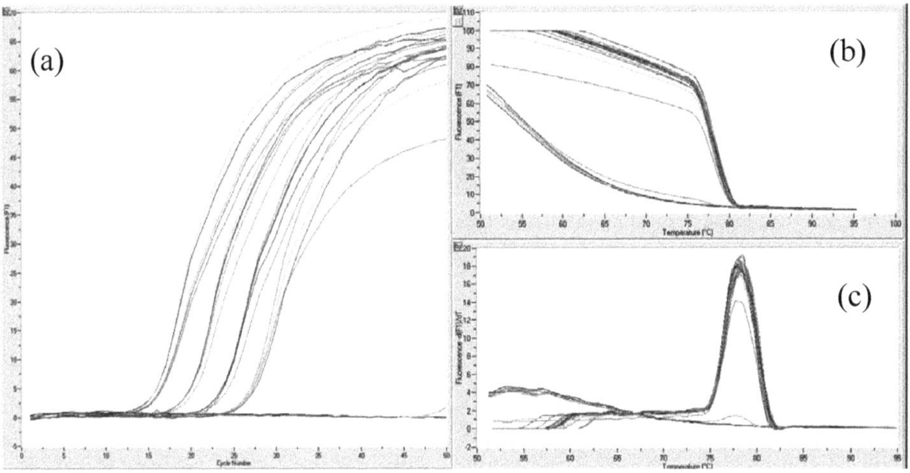

Figure 1.1 Amplification using generic binding dyes (minor-groove binding dye SYBR®Gold) on the LightCycler®. (a) The amplification plot shows 50 cycles (to completion) of amplification of four dilutions (four replicates) of a 10-fold dilution series and four no-template controls, one of which produces a low amplification signal. The reaction utilised anti-Taq antibody hot start and UNG carry-over protection and the data utilise the background subtraction algorithm. (b) Melting point analysis of the amplified products, and (c) the first negative differential of the fluorescence, with respect to Temperature, plotted against Temperature. From this plot it can observed that the signal in the positive no-template control has the same melting point as the specific amplification in positive samples, and therefore amplification is most likely a results of cross-contamination. A colour version of this figure is available in the plate section at the back of the book.

from around 50°C to around 95°C. As DNA duplexes melt the fluorescence decreases as the bound dye is released. Most instruments provide an analysis of this data by plotting the first negative differential of the fluorescence signal with respect to temperature, against temperature. This plot appears as one or more peaks representing the point(s) at which the maximum rate(s) of change in fluorescence occurs corresponding to a particular dsDNA product. Specific reaction products generally melt at a higher temperature than artefacts such as amplimer dimers and misprimed products. Melting temperature is a function of GC content and to a lesser extent product length, which allows the artefact to be observed to melt at a higher temperature than specific products in some reactions, although this is a rare observation.

The melting peak method is analogous to that of agarose gel electrophoresis in that it does not unequivocally determine the presence of the correct sequence. Products with similar molecular masses or melting peaks may not be resolved by either technique. Gel electrophoresis has a higher resolution than that of melting point analysis and the quality of the melting point data collected varies greatly between hardware platforms. Those instruments where the temperature is accurately controlled and recorded, and where the correlation between temperature and fluorescence is good, will produce the highest resolution melting point data.

The major limitation with binding dye chemistries is that they bind to total nucleic acid, and in a PCR of suboptimal efficiency a significant amount of signal will be derived from reaction artefacts. This leads to a dramatic loss in the dynamic range of quantification of samples and standards with low (< 1000) initial target copies. However, a melting point analysis allows non-specific artefacts to be discriminated, to some degree, from specific amplicons (Ririe *et al.*, 1997). Many dyes become limiting at the concentrations they are used in the reaction. At higher concentrations they may inhibit the PCR. When dyes become limiting, melting analysis loses resolution because of competition between products for free dye. New dyes such as LC-green can be used at higher concentration and they have facilitated high resolution melting on specialist instruments; the analysis of polymorphisms within products without the use of probes.

The melting point analysis adds significantly to the utility of fluorescent DNA-binding dyes since the data can be used to determine a number of important variables. Amplimer complements can be synthesized and used in melting point experiments to empirically determine annealing temperatures. The melting temperature of the product when determined may be used to improve the reaction in two respects on some instruments. Firstly, sequentially decreasing the denaturing step to a few degrees above the product melting temperature ensures that high molecular mass sample DNA is not denatured in later cycles which reduces the risks of mis-priming events in later cycles. Secondly, this lowering of the denaturing temperature allows a shortening of the time taken for thermal cycling. If the melting temperature of reaction artefacts is significantly lower than that of the amplicon then it is possible to monitor fluorescence at a temperature just below the melting point of the amplicon, but higher than that of artefacts (Morrison *et al.*, 1998). This serves to reduce the artefact contribution to the specific amplification signal and may improve the dynamic range of quantification for a given assay.

An important consideration when using such dyes is the incorporation of a 'hot-start' to reduce the formation of artefacts and so improve the dynamic range of quantification (Morrison *et al.*, 1998; Lee *et al.*, 1999). It should be noted that most of these dyes are both toxic and mutagenic to varying degrees, due to their ability to bind DNA. Storage conditions for these reagents are critical since they may be photo-bleached by both ambient and artificial light. The use of light-tight containers such as darkened microfuge tubes will reduce this, but most are labile on long-term storage.

When optimized, a SYBR®Green-1 or SYBR®Gold assay can be used for most applications (such as expression studies). The ability to multiplex these probes is limited because the binding, which generates the fluorescence, is non-specific. However, it is possible to multiplex by melting point using binding dyes, providing that the melting peak of each species can be resolved. Whilst allelic variation and mutation detection can be carried out, for example, using the amplification refractory mutation system (ARMS) approach (Newton *et al.*, 1989), this requires assays for both variants to be carried out in separate tubes. Because of these limitations assays that require high-confidence results are best addressed using the strand specific approaches that are discussed next.

Strand-specific detection using labelled nucleic acid probes

The strand-specific methods employ fluorophore-coupled nucleic acids that interact with reaction products, probing accumulating PCR products for the presence of the target sequence. Such PCRs exhibit greater specificity than the respective assay without the probe component. However, the probe targets only a portion of the amplicon sequence and therefore the signal for the product only reports the presence of sequences that are complementary to the probe rather than the entire amplicon. The basis of detection of the product can be classed by the mode of action of the probe.

The efficiency of the probe is important if it is to produce a signal that correlates closely with the concentration of amplicon in the exponential stage of amplification, and thus is used to determine the number of initial target copies. With an inefficient probe the signal may not increase exponentially and not correlate with initial target copy number. Inefficient probes will produce an inaccurate value for the PCR efficiency. The signal generated by a probe is dependent on, for example, sufficient hybridization and/or cleavage as is later described. The target sequence is the main factor affecting probe efficiency because of different sequence-dependent factors for a given probe type. Therefore the benefits of selecting a probe type should be carefully considered when designing an experiment and also when purchasing an instrument, since the preferred assay system may not easily be applied on every instrument.

The optimum composition of the core chemistry may vary with different platforms and this is discussed in Logan and Edwards (2009b). The optical specifications of platforms differ such that they will be optimized for the specific fluorophores. The concentration of probes may also vary between platforms but typically probes are used at a final concentration of around 0.1–1 μM.

In this text probe types are grouped for discussion by mode of action. The evolution of real-time methods is not presented chronologically, although reference is made to how and when they were implemented. Most of these methods use the phenomenon of fluorescence energy transfer (FET). More specifically, many of these methods use fluorescence resonance energy transfer (FRET), in which energy is transferred between two fluorophores, rather than, for example, between a fluorophore and a non-fluorescent quencher. This phenomenon will be discussed first and then the application and experimental considerations of the various methods will be outlined.

Fluorescent resonance energy transfer probes (FRET) and the passive reference

FRET or Förster energy transfer involves the non-radiative (no photon emitted) transfer of energy between fluorophores and has been used for a number of biological assays. There are excellent reviews of principles and applications (Cardullo *et al.*, 1988; Selvin, 1995; Clegg, 1992), so there is no need to include an in depth description of the process in this text. Here we present an overview and the two main approaches to detection and multiplexing using FRET as used in real-time PCR experiments. Reference will be made to specific instruments to illustrate the process.

For FRET to occur there must be at least two fluorophores in close proximity. One fluorophore, called the 'donor,' is excited by an incident light source and emits light at a longer wavelength that falls within the excitation spectrum of the second fluorophore, called the 'acceptor.' When the donor and acceptor are on the same or neighbouring molecule(s) then the emission energy of the excited donor may be transferred through a number of distance and orientation dependent effects including dipole–dipole interactions between the donor and acceptor. Variations in the intensity of emissions of fluorophores on nucleic acids resulting from spatial changes during the course of amplification provide the basis for detecting amplicon accumulation (Fig. 1.2). The choice of donor-acceptor dye pairs

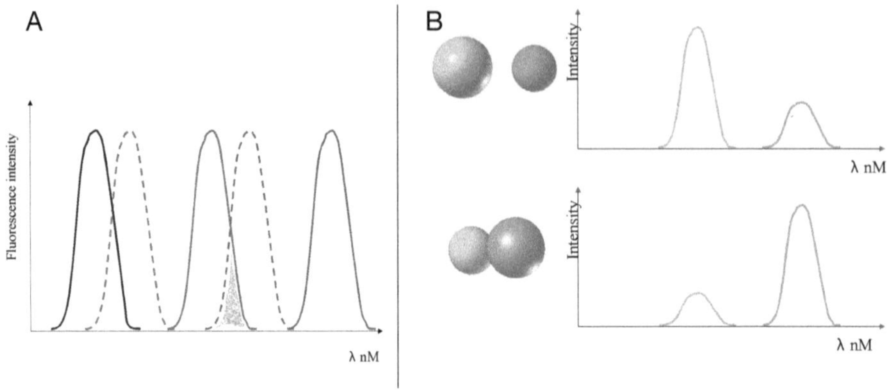

Figure 1.2 Fluorescence energy resonance transfer. (A) The emissions of a light source (blue line) falls within the excitation spectrum (green line dashed) of a fluorophore and cause this dye to fluoresce. The emissions (green solid) of this fluorophore fall within the excitation spectrum (red dashed line) of a second fluorophore that emits fluorescence at longer wavelengths (red solid). When the second fluorophore is close to the first on the same or neighbouring molecule, the energy that would be emitted by the first fluorophore (top of B) can be transferred to the second through a number of energy transfer mechanisms such that the second fluorophore emits this energy at its own emission wavelengths (bottom of B). A colour version of this figure is available in the plate section at the back of the book.

is important in achieving a good signal: noise ratio. The Förster distance (the distance between the dye pair at which the efficiency of energy transfer is 50%) varies for different dye combinations.

Analysis utilizing FRET involves monitoring either a rise or drop in fluorescence of either fluorophore depending on the assay type and instrument. On some platforms it is possible to present this data as a change in the quotient of the emissions of the two dyes. For example, if the basis of detection is the donor fluorophore being liberated spatially from the acceptor then one could express the data as donor/acceptor. If the converse is true the basis of detection is the donor being brought in close proximity to the acceptor one could express the data as acceptor/donor. This simple type of analysis is useful in a non-multiplexed real-time PCR context since subtle changes in sample volume due to pipetting errors and/or changes in sample volume due to evaporation can be normalized using this approach.

Of importance to the signal–noise ratio of assay systems utilizing FRET is the ratio of dyes in a FRET pair. In particular this is important where two dyes may be used on the same probe (dual labelled) as described later. These probes are more complicated in terms of manufacture since they require incorporation of both dyes into the probe molecule. These types of probes are therefore more expensive to synthesize. These oligonucleotide probes are usually made by one of two common methods. The first dye, if this is at the 3′-end of the probe, is part of the support matrix on which the probe is synthesized by sequential addition of each base in the sequence. The second dye is either directly incorporated by the use of fluorophore-labelled monomer, or by the inclusion of a monomer with a reactive group, and the dye is subsequently covalently cross-linked after synthesis. In the latter case, any free dye needs to be removed by a separation step, and this is usually carried out using high-performance liquid chromatography (HPLC). In cases where neither dye is at the

3′-end of the probe (e.g. Scorpions), both dyes are incorporated as fluorophore labelled monomers. In dual-labelled probe types the donor is liberated from the acceptor. If the stoichiometry of the donor and acceptor fluorophore incorporation is not balanced (i.e. the probe is impure) then a low signal and/or a high background may be obtained. The quality of probes can vary greatly between suppliers and even between scales of synthesis. Quality control of such probes is important to avoid wasting time trying to optimize assays with poor quality dual-labelled probes. The authors would recommend that probes should be checked following manufacture by any one or more of a number of techniques to ensure the integrity of each probe. These include mass spectrometry, HPLC separation using analysis of the fractions at the relevant wavelengths. Degradation and subsequent analysis utilizing a spectrofluorimeter is an ideal method of determining dye stoichiometry. Some suppliers offer one or more methods as a service or as part of their quality control system.

Another experimental consideration that is important in improving the signal–noise ratio of assays is the sequence dependent quenching of fluorescence. In particular, a fluorescein dye may be partially quenched by a neighbouring guanine base (G) in the sequence, which should be avoided. Whilst both dyes in a FRET pair may be fluorescent it is becoming increasingly common to find the use of non-fluorescent acceptors (quenchers). These are often termed 'dark quenchers' and can provide collisional or Dexter quenching. No light is emitted from these molecules, rather the energy is dissipated to solvent. Their use is now more common because a non-fluorescent moiety on a dual-labelled probe frees the instrument's bandwidth, increasing both the utility of instrument for multiplexing, and the dynamic range of a given detector where fluorescent emissions may have otherwise overlapped. Such quenching moieties are now commonly employed on dual-labelled probes. One limitation of this approach we have observed is that it is difficult for the end user to determine the efficiency of the incorporation of these quenchers into dual-labelled probes. In practice this may lead to a higher noise for the reasons described above (Zahng *et al.*, 2001). The use of gold nanoparticles as efficient quenching moieties has also been reported (Dubertret *et al.*, 2001); however, this has yet to be demonstrated in a homogenous PCR.

Since most instruments can excite and detect the emissions of fluorescein, a combination of fluorescein and any dye that can act as an acceptor may be used to report amplification. On most instruments the rise or drop in either fluorescein or the acceptor would be suitable to report amplification in various assay types. However, when a multiplexed assay is required, the instrument specification narrows the choice of dyes and assay types that can be used. Two principal approaches are used by different manufacturers to allow multiplexing. Here we describe these as the universal donor method and the universal acceptor method (see Fig. 1.3):

- *Universal donor.* In this arrangement the system is set up to multiplex assays by monitoring the rise in acceptor emissions. Different acceptors are used to report amplification from two or more amplifications in the same tube. Typically the machine has one excitation source and detectors at longer wavelengths that are used when multiplexing, e.g. the LightCycler®.
- *Universal acceptor.* In this arrangement the system is set up to multiplex by monitoring the rise in donor emissions that are being quenched by a longer wavelength or non-fluorescent acceptor. Different donors are used to report amplification from two or more amplifications in the same tube. Typically the systems detect donors that have short

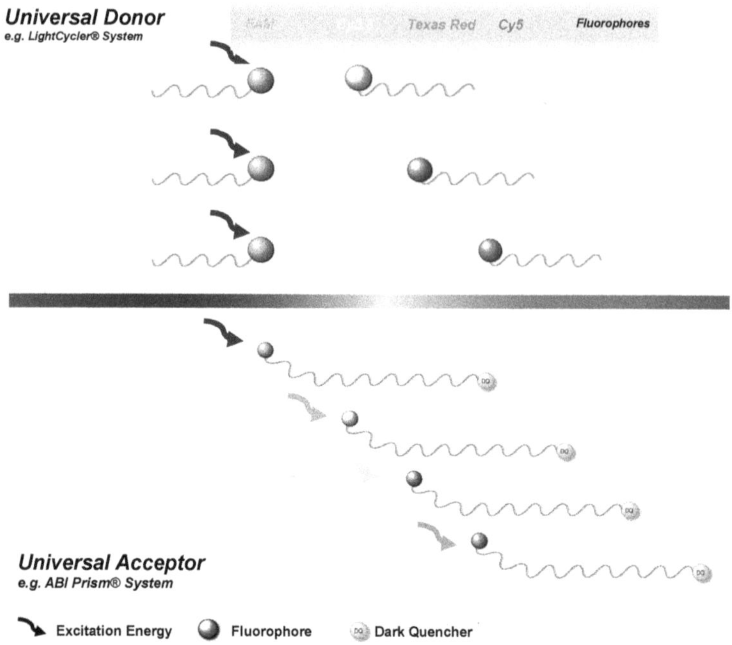

Figure 1.3 Schematic showing the two principal methods for multiplexing using FRET: (top) the universal donor on the LightCycler® and (bottom) the universal acceptor on the ABI 7700. A colour version of this figure is available in the plate section at the back of the book.

wavelength emission and are all excited by the one excitation source, e.g. The Applied Biosystems 7700, 7000. Alternatively, dyes that are spectrally well separated with longer wavelengths may be used on some instruments that have the ability to both excite at more than one wavelength, and measure emission at more than one wavelength, e.g. BioRad iCycler® and Cepheid SmartCycler®.

In theory, most fluorimetric instruments can be used to detect either universal donor or universal acceptor dye combinations, but the software on most platforms is not flexible enough to allow practical application and analysis using both approaches without first exporting the data. The universal acceptor arrangement is the most common because of its effectiveness with quenched dual-labelled probes (5′-nuclease probes, molecular beacons etc.) that are widely used. However, instruments such as the Corbett RotaGene® are flexible enough in optical design and analysis to allow the use of most, if not all fluorescent methods. It should be noted that generally the signal obtained when measuring acceptor fluorescence in universal donor methods, is of a lower value (because of energy transfer inefficiencies) and may inherently generate more noise than the universal acceptor methods, where donor fluorescence is measured. That is not to say that a comparative signal–noise is not achievable with universal donor methods. In particular this is possible where the probe and FRET efficiencies are good and the dyes are spectrally well separated, such that the donor moieties do not significantly contribute to the high background when detecting the acceptor emissions, as is achieved with the LightCycler® dye combinations (fluorescein, LC640 and LC705).

One approach to allow the universal acceptor method to be used on a system optimized for the universal donor method uses the 'big dye' approach to transfer the energy of the excitation source of the detector to the reporting dye (Solinas *et al.*, 2001). This approach has been previously applied in DNA sequencing chemistries and instruments (Ju *et al.*, 1995). For example, on the LightCycler® the dyes for the longer wavelength detectors are only minimally excited by the blue LED. A fluorescein molecule that is chemically coupled to such a dye may now be detected in the longer wavelength detectors since the energy will be transferred by the fluorescein, in turn this energy may be transferred to a longer wavelength or dark acceptor dye. Although less efficient, the method does work and these probes are available from a few suppliers. However, the trend with new instruments is to move away from supporting universal donor chemistries alone, for example the LightCycler® 480 instrument now has multiple excitation wavelengths facilitating multiplex universal acceptor methods and negating the necessity for dye deconvolution algorithms for some applications. Table 1.2 provides a summary of commonly used dye pairs.

A passive reference is used in some systems. This dye is detected and its signal is used principally to normalize for optical variations in the instrument (such as edge effects), but also normalizes for differences in volume that occur between wells and subtle differences in the optical properties of tubes and wells. These differences may be due to either volumetric errors in pipetting during preparation, and/or changes that occur during the amplification as evaporation occurs, causing a drift in the background signal. The reference dye takes no part in the reaction itself. On Applied Biosystems and similar instruments ROX (carboxy-X-rhodamine) dye is commonly used for this purpose. On instruments where there is little surface area for evaporation to occur, such as capillary systems like the Idaho & Roche LightCycler®, and in centrifugal systems where the volume and concentration is maintained in the reaction well, such as the Corbett RotaGene®, a passive reference is less important.

Colour correction algorithms

A prerequisite of multiplexing using fluorescence is discriminating the emission signals of different fluorescent dyes at different wavelengths. The requirement for an algorithm is dependent largely upon the fluorescent principles for which the instrument platform is optimized. For example, on platforms which use the universal donor system a colour compensation algorithm is required to analyse all dyes in a given multiplex assay. This is because there is usually only a single excitation source. When, for example, these systems are applied to universal acceptor chemistries they still require a colour compensation algorithm because the resulting measured energies will be a composite of all fluorophore emissions within a given multiplex. Those systems designed for using the universal acceptor system which have excitation sources that excite only specific fluorophores in a multiplex do not require a colour compensation algorithm. The authors would note that this is technically difficult to achieve. In practice 'cross talk' is usually addressed algorithmically on platforms intended for high-confidence applications such as diagnostics. Therefore, the provision of colour correction processes artificially segments the platform market into instruments intended for applications other than research and 'research-only' instruments.

Table 1.2 Some of the common fluorophore combinations that can be used effectively in energy transfer pairs

Fluorophore	Excitation maximum (nm)	Emission maximum (nm)	As a DONOR when used with	As an acceptor when used with
Cascade Blue	396	410	DABCYL	Fluorescein, SYBR®Gold, SYBR®Green-1
Cy3	552	570	DABCYL, BHQ-2, ElleQuencher, Eclipse	Fluorescein, SYBR®Gold, SYBR®Green-1
Cy3.5	581	596	DABCYL, ElleQuencher	Fluorescein, SYBR®Gold, SYBR®Green-1
Cy5	643	667	DABCYL, BHQ-2, ElleQuencher, Eclipse	Fluorescein, SYBR®Gold, SYBR®Green-1
Cy5.5	675	694	DABCYL, BHQ-2, ElleQuencher	N/A
EDANS(5-((2-aminoethyl) amino)naphthalene-1-sulfonic acid)	335	493	DABCYL	N/A
FAM (carboxyfluorescein)	494	518	TAMRA, ROX, Cy3, Cy3.5, Cy5, Cy5.5, DABCYL, BHQ-1/2, Eclipse	
Fluorescein	492	520	TAMRA, ROX, Cy3, Cy3.5, Cy5, Cy5.5, DABCYL, BHQ-1/2, Eclipse	
HEX (carboxy-2′,4,4′,5′,7,7′-hexachlorofluorescein)	535	556	TAMRA, DABCL, BHQ-1/2, Eclipse	N/A
JOE (carboxy-4′, 5′-dichloro-2′,7′-dimethoxyfluorescein	520	548	TAMRA, DABCL, BHQ-1, ElleQuencher	N/A
LC640	625	640	N/A	Fluorescein
LC705	685	705	N/A	Fluorescein
ROX (carboxy-X-rhodamine)	585	605	DABCYL, BHQ-2, ElleQuencher	Fluorescein
SYBR®Gold	495	537	Cy5, Cy5.5, LC 640, LC705	N/A
SYBR®Green-1	494	521	Cy5, Cy5.5, LC 640, LC705	N/A
TAMRA (carboxytetramethylrhodamine)	564	605	DABCYL, BHQ-2, ElleQuencher	Fluorescein
TET (carboxy-2′,4,7,7′-tetrachlorofluorescein)	521	544	TAMRA, DABCL, BHQ-1/2, Eclipse	N/A
Texas Red	583	603	BHQ-2	N/A
VIC	538	554	TAMRA	N/A
Yakima Yellow	526	549	Eclipse	N/A

DABCYL (((4-(dimethylamino)phenyl)azo)benzoic acid), Black Hole Quencher 1 and 2 (BHQ), ElleQuencher(TM) and Eclipse ™ are non-fluorescent 'dark' quenching moieties.

The authors find that those new to the field who require flexibility in their PCR platform, for example an instrument that may be required as a departmental resource, often make a poorly informed first purchase. The trend in new platforms is to increase the choice of excitation and emission filters which may be used together in different combinations. On the basis of the optical specification alone one might expect that such systems would have excellent flexibility. However, whilst they almost certainly will be able to analyse singleplex assays in all fluorescent configurations (universal donor and acceptor) they may be limited in their flexibility without a colour compensation algorithm. Systems without a colour compensation algorithm will almost certainly not be able to analyse multiplex universal donor configurations. To some degree this makes the flexibility of these 'open' systems rather troublesome rather than offering real utility. However, some third-generation instruments (such as the LightCycler® 480 family) are truly flexible in their ability to multiplex in all fluorophore configurations; having multiple excitation and emission filters and a colour compensation algorithm.

A second important aspect of a colour compensation algorithm is its ability to account for the temperature dependent effects of fluorophore emissions. It will be perhaps familiar to most readers that the spectral nature of fluorophores means there is a requirement to numerically deconvolute a spectrum when carrying out multiplex analyses. What is perhaps less apparent is that most fluorophore emissions are reciprocal with respect to temperature. This is largely to do with the ionic environment of the PCR buffer. For example, Tris-based buffers drop in pH as temperature increases and therefore the resulting fluorophore emission decreases. It is noted here that the ionic composition of the calibrator is an important consideration in the calibration experiment as discussed below. Some colour compensation algorithms also numerically account for these variations. The key benefit is that one may multiplex using melting point analysis combined with strand specific probes as well as and/or multiplex by fluorophore wavelength.

Different approaches are available to achieve this colour compensation:

1. A fluorospectrometer that can measure a given area of the visible spectrum and use a calibration algorithm and software to deconvolute the different dye components. For example Applied Biosystems 7700 and 7900.
2. A fluorimeter with discrete optical detectors using a calibration algorithm and software to deconvolute the different dye components. For example Applied Biosystems 7000, Roche LightCycler®.
3. A fluorimeter with discrete optical detectors and utilizing a 'factory set' calibration to deconvolute the different dye components. For example Cepheid SmartCycler®.
4. A fluorimeter with discrete optical detectors with sufficient bandwidth between dyes and narrow cut off filter to minimize cross-talk. For example Corbett/Qiagen RotorGene®.

All approaches are equally applicable but we would recommend, for high-confidence applications, the use of an algorithm to minimize the risk of false-positives due to erroneous data interpretation.

A good algorithm is one that completely compensates for the contribution of other dyes at the wavelength of the dye being monitored. This is usually achieved using a calibration with pure dyes. For systems that are designed to provide multiplexing using probes and melting point analysis, this includes heating each pure dye through the PCR operating range. The colour compensation process could be broken down and described as:

1 *Optical calibration.* Using a defined preparation of fluorophore that is optimum for the assay to be analysed, 'a calibrator.' This may include several replicate wells of the dye to produce an arithmetic mean of the resulting spectra. Empty or background noise samples also normally form part of the reagents required.
2 *Numerical calibration.* The software of the machine produces a numerical transformation of the fluorophore and background (calibrator) sample data to produce a set of coefficients that characterize each fluorophore's emission energy across all channels of the instrument.
3 *Dye deconvolution.* The coefficients are applied to real data from PCR experiments in a transformation which is an inversion of the calibration process. These data may then be used for secondary analysis such as amplification and melting point analysis.

The flexibility of many commercial systems means that users may deviate from those fluorophore combinations recommended by the system provider. The key benefit in doing so is a greater choice of fluorophores, lower oligonucleotide synthesis costs from a greater range of suppliers, and the application of a growing number of commercial kits often with bespoke dye combinations. As systems have become more complex in terms of their optical configurations the authors find that the successful application of these bespoke colour compensations are too difficult for many users

There are some important practical aspects that one should adhere to in order to ensure that the colour compensation processes works correctly. To a great extent this may be largely negated by the use of system provider calibration kits. A poor calibration will result in either under or over correction which manifests itself as a false-positive signal in an intended negative channel. This will be as a result of 'spill' or 'bleed' over of optical signal from a neighbouring positive channel. These are sometimes evident as inverted amplification and/or melt curves when colour correction is poor. If these signals are above the threshold levels of one's secondary analysis (amplification or melt analysis) then the colour compensation process requires refinement. The important experimental considerations are:

1 *Fluorophore concentration.* Some fluorophores are not efficiently excited by the excitation source of the chosen platform. This is implicit in their intended use within a universal donor configuration. All dyes must be applied in a calibration at a concentration to which they are sufficiently excited as to be within the working dynamic range of the detector (linear range) in the channel to be used for analysis. The same applies to those additional channels to which they are known to significantly overlap. In practice this means that one must manually adjust dye concentrations until they satisfy these criteria. Trial and error using the analysis of raw data is the best method to ensure success. Too high (detector saturation) or too low (below background noise) a signal will lead to an incorrect numerical process and poor secondary data analysis (amplification or melting point). For those systems that provide for melting point analysis with probes, this is more complex since the dye concentration must also satisfy this criterion across the temperature range of the analysis. It should be noted that the solubility of dyes (due to increasing molecular mass) generally decreases with increasing wavelength and

therefore there may be a practical limit to achieving the ideal fluorophore signal on some platforms.

2 *Fluorophore and the ionic environment.* The fluorophore emission spectra may change with respect to wavelength and/or temperature depending upon the ionic environment. The pure dye calibrator should ideally be made using the same solutes as the PCR for which its calibration coefficients are to be applied. The buffer match is crucial since formulations change frequently with different suppliers and assay requirements. For example, *Taq*, *Pfu* and similar DNA-dependent DNA polymerases function in Tris-based buffer systems. RNA/DNA-dependent DNA polymerases such as *Tth* and r*Tth* will be supplied with bicine or tricine buffers and manganese rather than magnesium as their preferred bivalent ion source. Other solutes that are commonly used as PCR adjuncts, for example co-solvents such as DMF and DMSO may also produce significant changes in spectral and thermal emission responses. To a lesser extent other co-solutes and adjuncts such as salts (potassium chloride, ammonium sulphate), sugars (sucrose, glycerol) can also have some effect. There is an increasing trend towards stabilized PCR mixes which may have their own functional components such as freeze drying excipients, including sugars and structural polymers (oligosaccharides and polyphenols). Suffice to say that PCR formulations from different suppliers (and even different batches) may differ and therefore may perform differently with respect to the colour compensation. Therefore, it is important to have an appreciation of the content of such mixes if they are to be applied in experiments that require colour compensation, one may then prepare suitable calibrators.

3 *Oligonucleotides.* Some fluorophore labels are influenced by the primary sequence to which they are attached. This is often termed 'proximal quenching' and is exploited in some fluorogenic chemistries. In some experiments this proximal affect is so pronounced that it will perturb the colour compensation process. It may require that the pure dye calibration be carried out using the oligonucleotide utilized in the desired PCR experiment. If the donor probe in a multiplex experiment is affected then this can cause significant problems. The authors would recommend redesign of the oligonucleotide and/or assay in question. Pooling of different donor oligonucleotides is often recommended but will produce a poor result. It should also be noted that multiplexing of proximal probes using a single excitation source and a colour compensation process is extremely difficult since the calibration can never satisfy the two spectral states of each fluorophore that may change as a result of amplification.

A number of factors can affect the requirement for re-calibration; these include the drift in detector sensitivities and excitation sources. Solid state devices such as LEDs and photodiodes are more robust with time than lamps and photon multiplier tubes. The calibration is also highly dependent on the optical geometry, so any servicing of the instrument may initiate the requirement for recalibration.

The 5′-nuclease assay

The principle of a DNA synthesis-dependent 5′–3′ exonuclease process using the enzyme *Taq* polymerase was first described in 1989 (Gelfand, 1989). The application of this process for analysis of PCR amplification using isotopic labelling and a separation analysis was later reported (Holland *et al.*, 1991). A fluorogenic approach for monitoring the hydrolysis of HIV-1 peptides in real time (Matayoshi *et al.*, 1990) had been described and this principle was subsequently applied to the 5′-nuclease assay. A fluorogenic 5′-nuclease PCR assay was first reported in 1993 (Lee *et al.*, 1993).

The 5′-nuclease assay is now by far the most commonly used strand specific probe method. This is partly because this assay was the principal detection chemistry for the first commercial real-time PCR instrument, the Applied Biosystems 7700 launched in 1996 (Bassam *et al.*, 1996). Applied Biosystems market the 5′-nuclease assay for research applications under the registered TaqMan® trademark of Roche Molecular Systems Inc.

The probe is dual labelled. Typically, one dye of a FRET pair is located at the 5′-end of the probe, usually the donor, and the second is labelled at or near the 3′-end, usually the acceptor. However, the reverse is acceptable and more efficient quenching may be obtained by locating the acceptor closer to the donor. When dyes are coupled in close proximity the signal–noise ratio of the assay improves compared to terminally labelled probes. The optimum distance of the two dyes on the probe is dependent upon the properties of the dyes used in the FRET pair. Typically a spacing of six nucleotides is optimal. If the dyes are too close then there may be inefficient transfer of energy (Heller *et al.*, 1987) and they risk being cleaved together on the same fragment. The probe is protected at the 3′-end to prevent it being extended as an amplimer by the *Taq* polymerase. A phosphate group is normally employed for this but other blocker-groups, e.g. propandiol, work equally well.

The 5′-nuclease assay uses a probe that is designed to be cleaved by the concomitant 5′–3′ exonuclease activity of the *Taq* polymerase (Holland *et al.*, 1991). This occurs whilst the *Taq* enzyme is extending the nascent 3′-end of a neighbouring amplimer bound to the complementary strand. The products of hydrolysis have been shown to include one or two nucleotide products and it is proposed that the polymerase displaces the first one or two base pairs it encounters before beginning cleavage (Holland *et al.*, 1991). The 5′-nuclease assay is fundamentally different to the other strand specific probe assays because the signal is generated by the 5′-exonuclease process. Probe hybridization is necessary, but not sufficient, for this to occur. The probe is consumed within the reaction and therefore the method is unsuitable for melting point analysis.

The Applied Biosystems 7700 instrument utilizes dyes that were previously optimized for Applied Biosystems Sequencers, so the first commercially available 5′-nuclease probes used FAM (carboxyfluorescein), JOE (carboxy-4′, 5′-dichloro-2′,7′-dimethoxyfluorescein), TET (carboxy-2′,4,7,7′-tetrachlorofluorescein), and HEX (carboxy-2′,4,4′,5′,7,7′-hexachlorofluorescein) as the donors, and TAMRA (carboxytetramethylrhodamine) as the universal acceptor. The ROX (carboxy-X-rhodamine) sequencing dye was used as a passive reference. In practice a large number of dye combinations are possible and a multiplex of up to six reporter dyes in one reaction has been used for SNP end-point scoring (Lee *et al.*, 1999).

A number of software tools can be used to design 5′-nuclease probes. Similar instruments that followed the 7700 were developed with optics that supported dual probe systems using the a universal acceptor arrangement, and a large number of oligonucleotide synthesis companies supply the dual-labelled probes with the correct dyes to support the 7700 and later instruments. Applied Biosystems, who market TaqMan® probes, recommend that the design of such probes should follow these recommendations for optimum signal generation: the GC content of the probe should be in the 30–80% range, the strand with the greater number of Cs rather than Gs should be selected; runs of identical nucleotides should be avoided; there should be no Gs at the 5′-end (which would otherwise partially quench the fluorophore); the probe should have a melting temperature that is at least 5°C greater than the amplimers; and, finally, the probe sequence should not overlap with the amplimer regions (Gelmini *et al.*, 1997).

The 5′-nuclease assay was the first probe system reported which used dye pairs that are covalently linked prior to molecular cleavage. The spacing of the dyes may be optimized with respect to Förster distance. Following cleavage, separation of donor from acceptor is maximized, although some spatial effects are observable on binding to the complementary strand. Providing the experimental parameters are optimized, the 5′-nuclease system produces a signal–noise ratio that is usually much better than that achievable with other assay methods (Fig. 1.4). The method is very versatile and can be used on a large number of platforms. However, there are limitations and these will be outlined briefly here.

Applicability of the 5′-nuclease assay is limited by the target and probe sequence requirements for efficient hydrolysis. A PCR may be shown to be highly efficient, but the signal from a hydrolysis probe may be low. Therefore design of an efficient probe is the key to an efficient PCR signal. As for most of the described systems the assay is designed around the probe and not the amplimers *per se*. In particular, the use of this assay system for AT-rich DNA can be problematic. To obtain a TM sufficient to allow efficient cleavage of the probe it should be ~ 5–10°C greater than that of the amplimers. For some sequences this would require the synthesis of very long probes that could only be produced in low yields. The development of the minor groove binding (MBG) probe technology that increases the

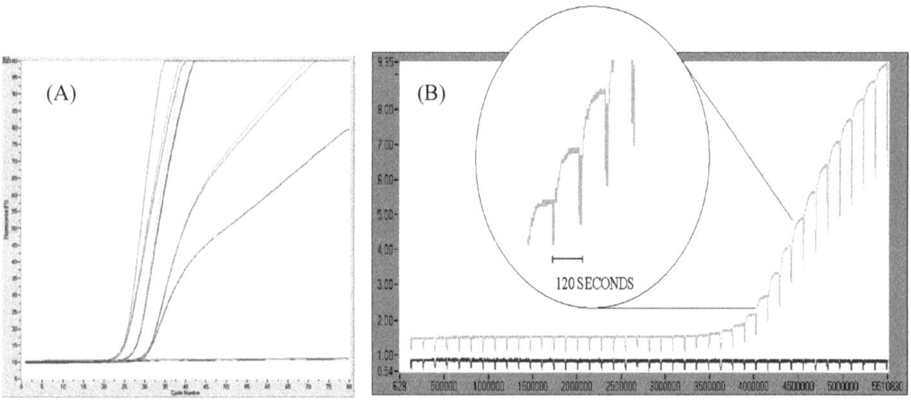

Figure 1.4 Amplification using the 5′ nuclease assay on the LightCycler®: (A) the amplification plot shows 80 cycles (in excess of completion) of thermal cycling of three dilutions (three replicates) of a 10-fold dilution series and three no-template controls on the Roche LightCycler®. The reaction utilised anti-Taq antibody hot start and UNG carry-over protection. The reaction used the primers and probes from the Applied Biosystems Human β-Actin kit which utilises a FAM-TAMRA internally quenched TaqMan® probe. The amplification plot shows raw data (no analysis/ background subtraction) and illustrates the excellent signal–noise ratio achievable with this chemistry. (B) Shows the same reaction amplified on the Idaho LightCycler® carried out by continuous monitoring of a positive and negative sample throughout all stages of amplification. The data shows that ~120 seconds are required each cycle to take the hydrolysis of bound probe to completion (plateau) each cycle. A colour version of this figure is available in the plate section at the back of the book.

TM of such probes (Kutyavin *et al.*, 2000; Salmon *et al.*, 2002) has reduced this problem somewhat but adds to the complexity and cost of the probe. Similarly locked nucleic acids may be used to improve the probe TM (Letertre *et al.*, 2003). The difficulties associated with the synthesis of dual label probes and the signal: noise ratios obtained was discussed in 'Fluorescent resonance energy transfer probes (FRET) and the passive reference', above. We have observed this from almost all suppliers although this is less problematic in small-scale syntheses from reputable companies. The use of an additional complementary probe labelled with a quenching moiety may be useful in improving the signal–noise ratio of problematic probes (Nurmi *et al.*, 2002).

On many platforms the temperature transitions make up a large proportion of total assay time. On instruments with faster temperature transitions such as the Roche/Idaho LightCyler®, Corbett RotorGene® and Cepheid SmartCycler®, the overall PCR assay time may be significantly reduced. Whilst the PCR process has been shown to proceed at considerable speed on fast systems, the signal–noise ratio of the 5′-nuclease assay may be dramatically improved with an increase in hold time at the annealing step. The 5′-nuclease process has been shown to generate signal much slower than the PCR process itself (Whitcombe *et al.*, 1999a; Lee *et al.*, 2002) and this is illustrated in Fig. 1.4.

The 5′-nuclease assay has been reported for a large number of applications and is supported by the majority of oligonucleotide synthesis companies and suppliers of instruments, such that for many new users, the trademark TaqMan® is used as a synonym for real-time PCR. These applications include expression studies, detection of pathogens, allelic discrimination, and genomic applications to name but a few.

Probe library

One of the most recent innovative developments in probe technology is 'Probe Library' (Mouritzen *et al.*, 2005). This is a commercial off-the-shelf array of probes. The probes are not specific; rather each probe is pre-designed, being a 10- to 15-mer sequence that is reproduced frequently throughout a specific genome. The PCR assay design is carried out by the end-user with a web-based design package. A sequence is entered and the software then designs primers around probes which exist within the library. There are probe libraries for different genomes; if a genome is not catered for then a closest match may provide a solution.

The probes themselves are dual-labelled probes. The length of each probe is maintained short such that the sequence occurs with enough frequency to be useful. The TM of the probe is raised using locked nucleic acids such that it hybridizes to the amplicon. The signalling is produced from the 5–3′ hydrolysis of the probe concomitant with primer extension, i.e. the biochemical process is the same as the 5′ to 3′ nuclease assay except that the probe is not designed specifically for the amplicon. The specificity of the reaction is derived only from the amplimers. When the assay is optimized the end-user is able to carry out a large scale synthesis of the sequence or purchase it from the supplier. This should allow the assay to be applied further on a wide variety of instrument platforms. Additional benefits will be realized from the kit-scale synthesis and the associated quality control that is required for such a product. Problems with dye stoichiometry observed in research-scale synthesis of dual-labelled probes will be minimized.

Probe library is useful for the rapid development of quantitative real-time PCR assays.

Market literature suggests an assay can be optimized in several days. The main useful application will probably be in the rapid development of assays for gene expression studies. A useful feature for this application is that the probes are designed to have similar efficiencies. This significantly simplifies the optimization process and eliminates some aspects of experimental error. The authors would not consider this probe system to be suitable for high-confidence applications such as diagnostics. The high background in some samples may lead to a false-positive signal. It may be possible to carry out an allele specific multiplex assay using two or more probes over a polymorphism but this has not been reported as-of-yet.

Direct hybridization methods

Direct hybridization methods use one or more probes that hybridize to the product strands during the low-temperature stages of the PCR each cycle and are released from the product strand during the high temperature stages. Unlike the 5′ nuclease assay the signal is not cumulative, but does increase each cycle as the amount of probe binding rises with product accumulation. Since the signal is derived from a hybridization event it is proportional to the product concentration, but is also a function of the probe efficiency. The signal from hybridization assays may be based on direct binding of product, or on the competitive formation of different hybrids. Generally, where the hybridization event is simple the effect of probe efficiency will decrease. Here we describe the different types of hybridization probe in detail.

Linear hybridization probes

A number of different probe systems fall into this class. These use the direct hybridization of linear oligonucleotides; these are oligonucleotides that are designed to have minimal secondary structure when not bound to target. Here we will describe the various types and use of these probes in more detail.

The most often reported probes of this class are dual hybridization probes that were the strand-specific chemistry developed for PCR applications on the LightCycler® instrument (Cardullo *et al.*, 1988; Wittwer *et al.*, 1997). The term hybridization probe is often used as a synonym for this probe method. This probe system uses two oligonucleotide probes each labelled with one dye of a FRET pair. Typically one dye is located at the 3′-end of one oligonucleotide and the other dye is located at the 5′-end of the second oligonucleotide. The second oligonucleotide is also protected to stop primer extension using a 3′-phosphate or other blocker molecule.

These oligonucleotides are designed to hybridize adjacently on one of the nascent PCR product strands such that the dyes are in a *cis*-configuration with respect to the nucleic acid. Both probes are designed to bind to the non-amplimer region of one of the amplicon strands. As product accumulates both probes bind to the strand and the basis of detection is to monitor the drop in donor fluorescence, or the rise in acceptor fluorescence, or to monitor the quotient of acceptor/donor in non-multiplexed reactions. Typically fluorescence is measured at the end of the annealing step and requires 10–30 seconds to generate sufficient signal for this application.

Dual hybridization probes were first described for use in PCR on the Idaho LightCycler®. The two channel optical systems of these machines used fluorescein as the donor moiety and the Amersham dye Cy5 as the acceptor. The later three channel Roche LightCycler®s use three dyes, two of which are optimized for the LightCycler®, LC640 and LC705, although it is possible to multiplex using other dyes with emission wavelengths that correspond to the discreet optical detectors of the LightCycler™. The assay is applicable to other platforms such as the Corbett RotorGene® in either a universal donor arrangement or a universal acceptor arrangement by using a number of dye combinations and a non-fluorescent quenching moiety.

For applications requiring accurate quantification of initial target copy numbers a requirement is that the probes be designed to bind at approximately the same temperature, but above that of the amplimers so that efficient signal is generated during the annealing step, and melt at a temperature that will allow efficient extension by the *Taq* polymerase. This typically requires the probes to have a melting temperature of 65–75°C. If the probe TM is too low, insufficient signal will be generated each cycle. If the probe TM is too high, the probes may clamp the PCR and reduce amplification efficiency. The probe efficiency (the amount of signal generated by hybridization) can be significantly reduced by probe–probe interactions if there exists any sequence complementary between probes. This is important in multiplexed reactions where the concentration and number of probes increases. Ideally the probes should be designed to bind to one strand, close to, but not overlapping the reverse amplimer region. The TM of probes can be calculated using software tools such as the Idaho Technology applet found on their website, or using other commercial packages such as that supplied by Roche for use on the LightCycler®. If the design of probes is carried out as described then the probe efficiency will be high enough to provide accurate quantification of initial target molecules. A commonly observed effect in this and other hybridization assays is a decrease in probe signal in later cycles. This 'hook' effect is a result of increased specific product accumulation and competitive displacement of probes as the product strands anneal upon cooling (Fig. 1.5). This effect does not reduce the ability to accurately quantify initial target numbers. A hook is a good indicator that the amplification is preceding with a high efficiency.

One of the useful features of hybridization probes generally is the ability to carry out a multiplex qualitative analysis utilizing melting point analysis. For example, the same fluorescent dye on two different probes may be used to detect two amplification species in the same reaction providing their melting motifs can be resolved by the instrument and software. If the software can (a) determine more than one peak and (b) calculate the relative areas under these peaks using integral analysis, then a relative quantitative value for each species can be determined within a single reaction. Detection of internal controls using this approach is discussed in Lee *et al.* (2009).

Likewise, one probe may be used to detect strand variation by designing the probe to bind over a site where strand variation may occur. For example, allelic variation is readily detected using this approach. In the case of dual hybridization probes one probe is designed to bind to a conserved (non-variant) sequence that is adjacent to the sequence variation. This first probe is known as the 'anchor' probe as it is designed to hybridize to the target over a temperature range at which the TM of the second 'mutation' probe will change depending on the specific sequence. The mutation probe can be observed to melt at different temperatures depending on the presence of either variant. For example, the genotype for a given sample may be determined using this approach. A single peak would indicate a homozygous genotype, and if both peaks occurred then this would be a heterozygous genotype. This is a useful method

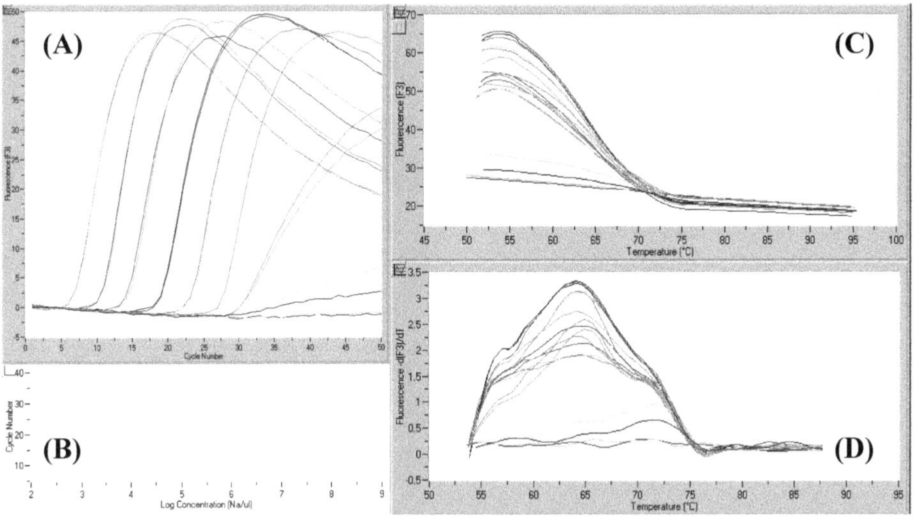

Figure 1.5 Amplification using dual hybridization probes on the LightCycler®: (A) the amplification plot shows 50 cycles of thermal cycling of replicates of eight dilutions of a 10-fold dilution series of a reference for the target, and no-template controls on the Roche LightCycler®. The amplification shows the data analysed using the background subtraction algorithm. The reaction utilised anti-Taq antibody hot-start and UNG carry-over protection. The reaction used the primers and probes for the Human β-globin gene using a 3′ labelled fluorescein donor probe and a 5′ Cy5.5 acceptor probe. The acceptor probe was blocked to prevent 3′ extension using an octendiol moiety. A 5′-3′ exonuclease Taq polymerase was used to illustrate that the 'hook' effect that is observed in linear hybridization probes as a result of probe displacement by product in later cycles. (B) The derived standard curve from the amplification plot. (C) The melt curve showing probe disassociation from product. (D) The first negative differential of fluorescence with respect to temperature showing the peak at which the maximum rate of probe dissociation occurs. A colour version of this figure is available in the plate section at the back of the book.

because a single probe can detect two variants, and this may be multiplexed by fluorescent emission to increase throughput if required. This approach can also be used to study the allele frequency in a mixed pool for a given population. For example, using the integral analysis of melting peaks described above allows a relative value for genetic variation, e.g. single nucleotide polymorphisms (SNPs), to be determined for a given population.

A less widely reported variation of the dual hybridization probe is one that uses two probes with the dyes in a *trans* configuration with respect to the probed strand (Bernard *et al.*, 1998). In this embodiment one 'free' probe is used in combination with a fluorescence-labelled amplimer. The energy from the dye on the amplimer may be transferred to the specific probe during the low temperature steps and the probe system may be similarly applied. Whilst this method is contemporary with the *cis* method it has been reported in fewer articles. The main drawback of the embodiment as reported in the original seminal paper was that it required internal labelling of the primer using an acceptor dye (Cy5). Internal labelling with dyes such as Cy5 is still not a general offering of many oligonucleotide synthesis houses. In addition the original method was reported for application on the LightCycler® system and the proprietary Roche dyes are also not generally offered in similar configurations. However, the authors have carried out a study of the labelling requirements and successfully produced

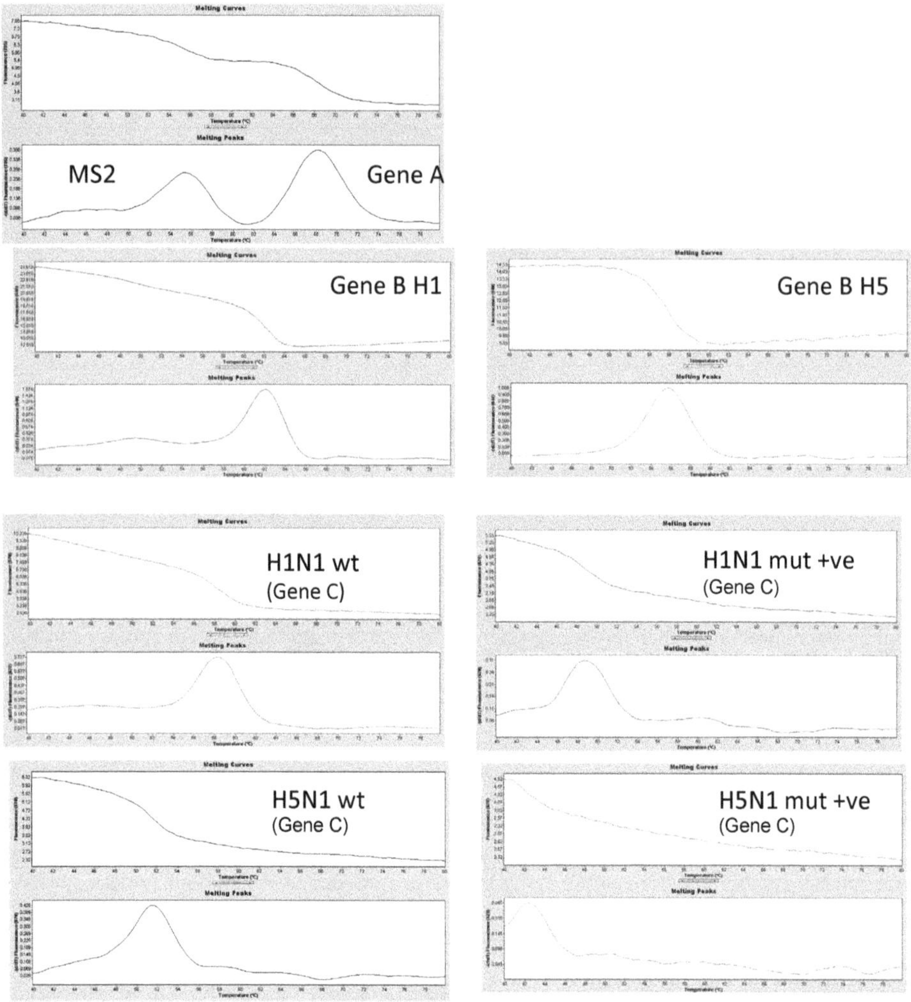

Figure 1.6 This shows experimental data for a multiplex assay using the *trans* dual Hybridization probe assay for characterization of pandemic Influenza A: Amplification was configured in the Universal Donor arrangement where a FAM label was present at the 3′-end of each specific probe. Reporter dyes (IDT dyes TEX615, TYE665 and TYE 705) were used at the 5′-end of specific primers. Signalling was achieved through energy transfer between the FAM label on the probe and the respective reporter on the primer. The amplification was carried out using the multiplex analysis of the LightCycler 2.0 which required the application of a colour correction process to the data generated using these fluorophores. The assay was designed to detect the following targets using the following reporters and melting point motifs: Pan Influenza A using TYE705 (~68°C). The assay was designed such that the primers and probes are directed against highly conserved regions of an Influenza A target (GENE A) (a). An MS2 process control (~56°C). MS2 assay component was designed such that the virus is introduced into the sample extraction process to control for all assay processes. Influenza HA subtyping component using Tex615 (b): this assay component was designed such that two assays (each two primers and one probe) are directed against highly conserved regions of a conserved target (Gene B) within Pandemic serotypes of Influenza A; HA1 (~62°C), HA5 (~56°C). Influenza NA subtyping and selection for oseltamivir resistance marker (polymorphism in H1N1-2009 and putative H5N1 mutation) using TYE665 in a conserved target GENE C. This component used a single assay (two primers and one probe) to amplify and type using highly conserved regions of

a multivalent assay for eight amplification species using only three reporter fluorophores which is described in Fig. 1.6. The system clearly has greater potential as this method does not use the full optical and analysis methods of the system utilized.

The impetus to construct an assay using this system was driven by the following factors. The method uses short probes based upon hybridization. Therefore, the kinetics of amplification and signalling are fast. The region of conserved sequence available to the probes may be very small. The primers and probes which are labelled only require a single donor fluorophore on the probe(s) and this may be on the 3′-end. This negates the additional requirement for protection from polymerase extension (steric blocking). This makes their implementation extremely low cost, in particular when compared to dual-labelled methods. Dyes may be shared between assays because the multiplex signals may be analysed by melting point. As the system uses FRET as the energy transfer system, rather than proximal quenching, colour compensation may be easily applied providing the universal donor is used on the probes (3′-fluoroscein) and the reporters are on the primers (5′-labels, for example). This allows for multiplexing by melting point analysis and fluorophore emission. The method is easily applied to parallel optic systems that use a single excitation source combined with multiple emission filters and an applied colour compensation algorithm. Such systems, e.g. the LightCycler® 2.0 and Enigma FL® and ML® platforms provide a rapid analysis since the emission energy at all wavelengths may be collected and analysed together. Channels that have only a single reporter may be analysed by their amplification signals and/or melt analysis. Channels that have two or more reporters may be analysed by end-point melt analysis. This combination of attributes is unique across the real-time methods. The main drawback of the method (in the described embodiment) is that primers contain a fluorophore label. Non-specific reaction artefacts may be generated that produce an amplification signal. However, the authors have found that careful design and empirical optimization can eliminate this occurrence. In addition, melt analysis may be used to discriminated them should this happen.

Yin–Yang probes are dual linear probes that work on the basis of competitive hybridization between themselves and the PCR product (Li *et al.*, 2002). This principle was first demonstrated for the end-point detection of PCR products in 1989 (Morrison *et al.*, 1989). Two labelled oligonucleotides are used. A fluorophore on the reporter probe is quenched by

---

the H1N1 and H5N1 genomes (c). The probe is designed to span this polymorphism, a missense mutation *TAT* to *CAT*; H1N1 (2009) wt (~58°C), H1N1-2009 mutation (~49°C), H5N1 wt (~51°C), H5N1 putative mutation (~42°C). All assay components (MS2, Gene A, B and C assays) were freeze dried into a environmentally stable cake which could be re-suspended with extracted viral material before amplification in a single reaction. Amplification is designed to be multivalent rather than a true multiplex where the relevant species amplify in the presence of a given viral isolate. Mixed samples may produce more complex amplification signals than these shown here. However, the assay illustrates how a small number of fluorophores (three) may be used to generate signal within a high level multiplex (eight possible targets). The data shown here are laboratory data from a LightCycler® 2.0 instrument but the final build is applied on the Enigma ML® point of use instrument. This instrument includes an automated data interpretation algorithm such that the operator receives clinically relevant information rather than experimental data. The assay may be applied to either screen, test, or confirm individuals for the pandemic strains (in 2009) of influenza A in combination with the oseltamivir resistance polymorphisms in H1N1-2009 and putative H5N1. This may be used directly by the clinician to administer the appropriate chemotherapy. The authors would like to thank Enigma Diagnostics Ltd (Dr Martin Lee, Dr Mark Laverick, Dr Paul Martin, Mr David Styles and Dr Samanth Spiers) and GSK (Dr Heather Madsen and Mr Tom White) for the kind permission to publish these data.

an acceptor on a second shorter complementary probe. Upon the increase in PCR product accumulation the acceptor probe is displaced by the hybridization of the longer target. The same principle is applied to the 'molecular zipper' probe, which is in fact a generic priming method utilized in rolling circle amplification (RCA). Displacement of the complement is facilitated by the amplification process which creates a stable hybrid complement that binds preferentially over the quenching complement (Yi *et al.*, 2006).

A limitation of these approaches is that two oligonucleotide probes are required. The application of dual hybridization probes may be difficult for some applications where there is little or no conserved region for one or either probe. The empirical optimization of two probes may increase the assay optimization time. ResonSense®, Hybeacons® and Light-up® probes are examples of single linear probe systems that have been described for a limited number of applications. The experimental considerations for the design and implementation of these probes will be similar to dual hybridization probes because of their similar mode of operation. These probes will be briefly described.

ResonSense® uses a single linear probe and a DNA-binding dye as a FRET pair. Signal is generated when the probe binds to the accumulating product and energy transfer occurs between the dye on the probe and the DNA binding agent in the DNA duplex (Cardullo *et al.*, 1988; Lee *et al.*, 1999; Taylor *et al.*, 2001). ResonSense® has been reported for use in real-time quantitative PCR (Lee *et al.*, 2002). The ResonSense® probe may be labelled with the fluorophore at either end or internally. A 3′ probe label will prevent primer extension. If the label is at the 5′-end or is internal then an additional 3′-blocker molecule will be required. The probe system may be used in either a universal donor arrangement (Lee *et al.*, 2002), or in a universal acceptor arrangement using a non-fluorescent quenching moiety. For example, the minor groove binding dye SYBR®Gold may be used to transfer energy to probes labelled with Cy3, Cy5, Cy5.5, to name a few. The broad emission spectra of SYBR®Gold allows for a variety of acceptor dyes to be utilized. SYBR®Gold and related dyes have two excitation peaks, one of which is in the UV. The use of a UV excitation source would allow for efficient excitation of the SYBR®Gold whilst minimizing the excitation of much longer wavelength dyes as described. Alternatively, the drop in donor dye emission using dyes such as fluorescein can be monitored by using a longer wavelength binding dye such as SYTO 63. This Universal Acceptor embodiment provides a method of applying ResonSense® on instruments other than the LightCycler®. The utility of ResonSense® for genotyping applications using LATE-PCR as a PCR end-point has also been reported (Sanchez *et al.*, 2006).

Hybeacons™ are linear probes that use a fluorescent moiety attached to an internal nucleotide that exhibits enhanced fluorescence with hybridization to a specific target (French *et al.*, 2001). The probe is 3′-protected to prevent primer extension during the course of the PCR. The increase in signal allows for direct quantification of initial target copy numbers and qualitative analysis of closely related sequences using melting point methodology. The simple structure of the probe and mechanism of signal generation make it ideal for multiplexing in strand variation analysis. Indeed, a six colour multiplex has been reported by applying a novel labelling strategy (Richardson *et al.*, 2010).

Light Up™ probes (Svanik *et al.*, 2000a,b; Isacsson *et al.*, 2000) are peptide nucleic acid (PNA) probes that use a DNA binding agent to report the hybridization of the PNA to specific DNA sequence. The application of Light Up™ probes for real-time PCR has been described (Wolffs *et al.*, 2001). In this probe method the probe binds to the target and brings the binding agent spatially close to the probe allowing it to bind with enhanced fluorescence. The dye reported in Light-Up™ probes is Thiazole orange (Lee *et al.*, 1986). PNA has been shown previously to be a useful molecule in some PCRs, being used to clamp amplification (Kyger *et al.*, 1998). The PNA has been shown not to clamp the PCR in this application due to the probe target sequence not being located near the amplimer regions.

The molecular beacon and other conformational probes

The use of a conformational change in fluorescent labelled probes to detect nucleic acid hybridization was first reported in 1987 (Heller *et al.*, 1987), and was later proposed for a homogeneous assay, in the context of nucleic acid amplification, in 1995 (Livak *et al.*, 1995; Young, 1996). The temperature dependent change in FRET occurring as a DNA probe, with a donor and acceptor label at either end, hybridizes to its specific complement is sufficient to provide a signal (of the event) in a homogeneous assay format. The Eclipse® probe system works on this principle but utilizes a minor-groove binder moiety to stabilize the structure (Afonia *et al.*, 2002). It is chemically similar to the MGB probes that may be used in the 5′-nuclease assay but the probe is typically labelled with the MGB and dark acceptor on the 5′-end to prevent the probe being hydrolysed. This type of probe can be used for a number applications and melting point analysis can be used to identify sequence variation as previously described.

The molecular beacon was described first in 1995 (Tyagi *et al.*, 1996, 1998; Salvatore *et al.*, 2006) and has been reported for a number of applications including direct 16SRNA detection (Schofield *et al.*, 1997), RNA detection in nucleic acid sequence-based amplification (NASBA) (Leone *et al.*, 1998), and recently the detection of DNA binding proteins (Heyduk *et al.*, 2002).

The chemical structure of the molecular beacon may be identical to that of the probe used in the 5′-nuclease assay in that it may be labelled at the 5′ and 3′-ends with a donor and an acceptor moiety respectively, and may also be protected from primer extension at the 3′-end. An alternative embodiment uses a single fluorophore label in the loop rather than two labels in the stem. The signalling is generated because the fluorescence is quenched when hybridized to target as opposed to fluorescing in the ssDNA loop (Hwang *et al.*, 2004).

The process of detection is different from TaqMan in that the molecular beacon is designed not to be consumed during the course of the reaction. The 5′ and 3′-sequences of the beacon are designed to be complementary, using a sequence not present in the target, such that the two fluorescent moieties are brought close enough to allow effective energy transfer in a 'hairpin' structure. The amplification is detected by the sequence dependent opening of a 'loop' sequence of the probe that is complementary to the target. Therefore these probes may be monitored for certain applications using melting point methodology (Robinson *et al.*, 200). Signalling may be poor in some molecular beacon assays since the stem-loop may preferentially fold, being an intramolecular process as opposed to the probe target duplex binding being a biomolecular process (Tsourkas *et al.*, 2003). Nucleic acid analogues have been used to improve the template probe hybridization (Ortiz *et al.*, 1998). One approach to increase the signal–noise ratio in a molecular beacon is the inclusion of non-nucleoside linkages below the stem structure which effectively reduce the chances of base mispairing (Häner *et al.*, 2010). These reduce unfavourable structures in the energy transfer pairing scheme that could produce a high background.

The difference in mode of operation of a hydrolysis probe and a probe such as a molecular beacon merits some discussion. Although the probe is chemically similar in structure, the process of detection is different. However, when the process is carried out using similar reaction conditions the signal in both assays will contain a component derived from both the conformational change in hybridization and a component derived from any strand specific hydrolysis by the 5′–3′ exonuclease activity of the enzyme contributing to the increase in the Förster distance of the fluorophores. The contribution of either component will depend on the assay design and the enzyme used, as some enzymes have different activities and may entirely lack 5′–3′-exonuclease activity, e.g. Stoffel fragment of *Taq* polymerase. The advantages and application of a molecular beacon to utilize both signalling processes has been reported (Kong. *et al.*, 2002) and the utility of either process for SNP scoring has been discussed (Täpp *et al.*, 2000).

Typically a universal acceptor arrangement is used since it can be carried out on all instruments that can analyse the 5′-nuclease assay. The original work used EDANS(5-((2-aminoethyl)amino)naphthalene-1-sulfonic acid) that is excited in the UV and emits in the blue region of the visible spectrum, and DABCYL (((4-(dimethylamino)phenyl)azo) benzoic acid) a dark acceptor. The use of a dark quencher for molecular beacons is still commonplace; however, the donor moiety is more commonly one that can be used on current instruments, although any FRET dye pair will work and some have been reported (Zhang *et al.*, 2001). The use of different fluorophores for quantitative multiplexing is possible (Vet *et al.*, 1999). A review of the use of such structures for a number of applications is given in Broude *et al.* (2002).

One application of the hairpin structure has been the development of a generic detection chemistry. This has been available in the past as Sunrise® and is now available as Amplifluor®. This chemistry uses a molecular beacon structure that is present as a tail on one amplimer (Nazarenko *et al.*, 1997). During PCR the *Taq* polymerase synthesizes the complement sequence so that upon hybridization the hairpin structure does not form, and the donor and acceptor are liberated as the basis of detection. A further simplified embodiment uses a unique sequence that is incorporated into the tail of one amplimer called a 'z' sequence. A generic Amplifluor® probe called a UniPrimer® can initiate primer extension since it has specificity for the z-sequence. When the complement is synthesized in the next cycle of PCR then the two dyes are liberated as previously described. The authors have seen many probe methods applied in a similar manner to the UniPrimer® including strand-specific approaches such as the 5′-nuclease assay (Whitcombe *et al.*, 1998; Zahng *et al.*, 2003).

Recent improvements include the development of the Lux® primer method (Nazarenko *et al.*, 2002). This uses the same principal as Amplifluor®, but only one fluorophore is required. The fluorophore must be two or three nucleotides from the 3′-end attached at the C5 position of thymidine. The primer has a hairpin structure such that the hairpin loop must terminate with CG or GC. This structure causes intrinsic *or* 'proximal' quenching of the dye. Upon extension of the complementary sequence a dsDNA structure promotes fluorescence of the dye. A generic system based upon energy transfer and the displacement of a complementary probe (rather than a hairpin) has been reported for application in rolling circle amplification (RCA) (Yi *et al.*, 2006). Whilst the 'molecular zipper' has been applied in RCA the principle could equally be applied to PCR.

Whilst Amplifluor® is a generic detection approach, the probe having no specificity for the internal sequence of the amplicon; it does offer some advantages over DNA binding approaches for detection. Firstly the signal–noise ratio of a FRET-based assays such as this is typically better than that obtained using a DNA binding dye, although non-specific priming may still generate signal. Secondly the probe may be a common probe sequence. A similar approach using a primer-incorporated sequence in a generic context utilizing common hydrolysis probes has also been reported for single tube genotyping (Whitcombe *et al.*, 1998). Finally, this approach allows for a generic approach to multiplexing using different fluorophore emissions.

The conformational principle has been applied to other 'me-too' approaches. Whilst each approach may offer slightly different utility, their main advantage is possibly with respect to freedom to practice the process. Some approaches, such as that reported by Xiao *et al.* (2009) have merit for applications where the PCR process itself is not essential.

Self-probing amplicons

A self-probing amplicon is an amplicon that has a probe chemically incorporated into the PCR product. The principle was first reported in 1999 and used a molecular beacon that was connected to the 5′-end of an amplimer via a linker molecule ((Newton *et al.*, 1993; Whitcombe *et al.*, 1999a,b). This probe is an oligonucleotide containing a sequence that is complementary to a region synthesized by the extension of the amplimer to which it is attached. The probe sequence also contains stem sequence elements at either end that hybridize and bring two fluorophores in a FRET pair spatially close together. The linker (PCR stopper) prevents the enzyme generating the complement of the probe during the extension of the reverse strand. Following denaturation of the duplex at the high temperature step of PCR, the probe binds to its complementary sequence that now forms part of the same molecule. The energetics favour this structure rather than that of the complementary stem structures within the probe sequence. The binding of the probe sequence to target increases the Förster distance allowing signal generation. The self-probing unimolecular mechanism has been proven to be the dominant mechanism of signal generation, rather than the bimolecular mechanism of the probe binding to other amplicons (Thelwell *et al.*, 2000). This is a key advantage over molecular beacons. It has also been shown that the cleavage of such probes does not occur due to the 5′–3′ exonuclease activity of the enzyme (Thelwell *et al.*, 2000).

In the original work the probe used fluorescein as the donor moiety and methyl red as a non-fluorescent acceptor. These probes are known commercially as Scorpion® probes.

One advantage of this method is that the probe is made in the same synthesis as the amplimer and therefore a separate probe is not required. The speed of signal generation has been reported to be much faster than that of the respective native molecular beacon which utilizes bimolecular probing (Whitcombe *et al.*, 1999a; Thelwell *et al.*, 2000), and actually more efficient when using short annealing times (Thelwell *et al.*, 2000). Since the probe becomes linked to the sequence it is probing, the event is in effect a zero order reaction (Saha *et al.*, 2001). The equal (1:1) stoichiometry of the probe to target means that the amplification signal is directly proportional to the amount of initial target species. This has also been illustrated utilizing a ResonSense® probe attached to an amplimer (Lee *et al.*, 2002). This type of probe has been termed Angler®. These types of assays will be of most use in improving throughput on faster thermal cyclers.

Other embodiments of the Scorpions® assay include the duplex Scorpion primer method (Solinos *et al.*, 2001). In this method the amplimer is labelled with a linear probe that has specificity for the strand the amplimer will generate. This probe contains one fluorophore of a FRET pair. A second probe that is the complement of the probe is free in solution. This probe is labelled with the second fluorophore in the FRET pair. Prior to the priming by the amplimer the probes interact and FRET occurs. When the amplimer extends the first probe is able to bind to the newly generated complement in an intramolecular fashion. The first probe is then spatially separated from the second and the change in FRET indicates amplification.

IntraTaq™ is the self-probing embodiment of the 5′-nuclease assay that has been shown to be more efficient in terms of the signal–noise ratio and the speed of the reaction (Solinas, *et al.*, 2002). In theory it should be possible to link any probe system to the amplimer in a similar fashion, and one would expect a commensurate improvement in signal with respect to speed for assays based on hybridization. One advantage of the self-probing amplicon that is not immediately apparent is that the probe is better controlled for, compared to a separate probe, since it is synthesized on the same molecule (Lee *et al.*, 2002). The main disadvantage of such probes is that the synthesis is more complex and therefore more costly, and the availability is limited to those synthesis companies that can reproduce these chemistries.

Intellectual property

In this section the authors do not provide definitive advice on what methodologies an individual may or may not have legal rights to use with respect to the available technologies. Regional variations in patent law mean that what intellectual property is protected varies greatly. In some territories patent protection may not have been sought so there will be no barriers to the use of the technology. Additionally, where patents are granted, they may not provide the same protection since the filing may have been examined and opposed on the basis of different prior art. At times patent status may also be the subject of appeals such that the protection of the intellectual property may change. To summarize, the patent landscape, and therefore the freedom to use the PCR process and various probe technologies, will vary depending where and when the process is to be used.

The history of the development of the PCR process has influenced the licensing programme structure of the core technologies. Cetus Corporation worked with Applied Biosystems on the development of early thermal cycling instrumentation and they protected the associated IP. The process became such a valuable tool to the life science researcher that when Roche acquired the process from Cetus the process was made available through an open licensing programme. Researchers that purchase licensed enzymes, reagents and thermal cycling instruments have the rights to carry out the process for their research activities. The authors have had contact with many groups who have misunderstood the licensing programme and have assumed that rights of use are automatically obtained for all areas under this programme. This has been observed mostly in academic, government, or non-profit making organizations that may have less commercial interest than pharmaceutical and diagnostics companies. Other activities such as service diagnostics require the end user to either obtain an end 'field' licence or purchase licensed diagnostic platforms and reagents. However, the key patents covering the core process are reaching the point where they will lapse and the requirement for licensing will diminish over the next few years. After this period licensing will only be a requirement if one uses technology covered by more recent filings such as those patents covering newer technologies such as new enzymes, some real-time processes and instruments. There is an 'extended licensing' programme from the key stakeholders covering their real-time processes and modified enzymes many of which are only now being given a granted patent status. This should in effect continue the research open licensing programme with additional provision for some real-time methods. By far the largest application area for real-time methods is in measurement of gene expression. However, many researchers will be developing a process that they would like to freely use. These researchers should be aware of the licensing programmes that make the technology available for use and be aware of third party IP and how that may be accessed. Given the cost and development times of validated assays, a naive consideration of IP could prove costly. Therefore these users should consider and further research the information provided here.

PCR technology has been a commercial success for many players. The patent landscape is complex and the exploitation of new technology has created a competitive litigious market for nucleic acid amplification technology. Some long court cases have proved costly. The requirement for a licence can be broken down into the key technology groups required. These are summarized below. The authors recommend that a detailed appraisal of the key patents status should be carried out by end users who plan to use a PCR process for application in fields other than purely research.

Enzymes

Taq polymerase and other enzymes that fall within the scope of the *Taq* patents will require the user to have a licence for use until the associated patents lapse. However, there is at least one commercial enzyme that is claimed to fall outside of this patent. Over the last 10 years the chemical hotstart methods have become the choice for many users. The processes to make and use these enzymes constitute new IP and therefore researchers may require licensing to use them.

Instruments

The automation patent covering the process of amplification in an automated process under computer control varies geographically. Therefore the requirement for an instrument to be licensed will depend upon the heating technology it uses, and specific embodiments covered by other key patents. Optical instruments are protected in some territories and therefore freedom to use these machines will vary such that the choice of instrument may change the requirement for licensing in some geographical regions. Some instruments purchased prior to the grant of the associated patents may not be licensed for some real-time methods. Indeed some licensed instruments with this capability were sold with a licence that did not cover their use for optical analysis because of the early status of these patent families.

Probe methods

Many probe methods have been developed. The rights to use probes for real-time PCR will depend upon which probes you choose since some probe technologies, albeit functioning by a different process, may well be covered by the main real-time patents. For example, in some patent territories probes with two or more fluorogenic labels and 3′ protection will fall within granted patents. Therefore the requirement to research your application and choose a suitable probe method may be invaluable in obtaining the rights of use.

Dye technology

The dye labels on purchased oligonucleotides may have been synthesized with a limited licence. Fortunately there are many dyes, some of which are generic for popular labels and may be more generally available for application.

Core technology

Many improvements to the PCR portfolio have come from third parties. These have been in the areas of enzyme 'hot start', carryover prevention and allele-specific priming methods to name a few.

Sequences

Many primer and probe sequences and genomes are the subject of granted patents. The ability to use these depends upon the local patent law. The right to those sequences for certain applications may require a licence.
To summarize, the choice of technologies that you choose to carry out you real-time method may reduce the cost of applying your process. There are some useful websites that can help researchers understand the patent landscape.

Summary and future improvements

There are many chemistries that may be used effectively to carry out the real-time detection of PCR amplification. Since the first edition of this text was published in 2004 there has been little improvement in the technology offering over that provided by existing fluorescent methods. Chemistries may differ in their utility for both quantitative and qualitative applications. For example, the 5′-nuclease assay has achieved broad appeal across all technical and market sectors using fluorogenic PCR. This commercial success is largely the result of early adoption, dominant marketing and support from many instrument suppliers. Few methodologies, with perhaps with the exception of the dual hybridization probe approach, offer a sensible solution across most quantitative and qualitative applications. However, the costs associated with the intellectual property rights to practise these established methodologies, in particular within the diagnostics sector, have been a driving force in the development of the many 'me-too' approaches presented here. Such approaches often have little technical merit over the established core real-time methodologies. As these methods enter their third decade, patent protection will cease, and the commercial value of these newer methods will diminish. The commercial freedom for any user to exploit their work beyond the research sector is now ensured by careful planning and commercial research.

However, the expansion of other enabling technology such as novel bases, new fluorophores and better instrumentation has improved the possibilities for those who use both the established and new methodologies. The user should look beyond the specific probe embodiment of these augmented technologies when considering their own requirements.

The authors would direct new users to the use of established processes to ensure a high rate of experimental success. In particular those methods that use linear probes are simple to design using known principles and generic molecular software tools. Therefore they have a have a high rate of success. Methods that do not utilize two labels within a single oligonucleotide will generally be of a lower synthesis cost for an equivalent yield of probe, even if the method requires more than one oligonucleotide.

The suitability of all assays will depend upon the application and the instrument platform used, the suitability of the software for correct data analysis, and the target sequence. The specifications of new instruments are improving, generally becoming faster and more flexible in operation. New PCR real-time detection methods such as fluorescence polarization, time resolved fluorescence and application of quantum dot technology are possible but are yet to be fully realized within mainstream laboratories. The authors would expect that advantages of bioluminescent methods, with respect to the relative simplicity of luminometry compared with fluorimetry, will be realized in the near future with emerging commercial systems. One can appreciate the possibilities by looking at the development of the parallelism achieved in bioluminescent sequencing.

Such methods could improve the speed, robustness, and costs associated with real-time PCR.

Acknowledgements

The authors would like to thank the editors for the invitation to submit a revised version for the second edition.

References

Afonia, I.A., Reed, M.W., Lusby, E., Shishkina, I.G., and Belousov, Y.S. (2002). Minor groove binder-conjugated DNA probes for quantitative DNA detection by hybridisation-triggered fluorescence. BioTechniques *32*, 940–949.

Bassam, B.J., Allen, T., Flood, S., Stevens, J., Wyatt, P., and Livak, K.J. (1996). Nucleic acid sequence detection systems: Revolutionary automation for monitoring and reporting PCR products. Australasian Biotechnology *6*, 285–294.

Becker A., Reith, A., Napiwotzki, J., and Kadenbach, B. (1996). A quantitative method of determining initial amounts of DNA by polymerase chain reaction cycle titration using digital imaging and a novel DNA stain. Anal. Biochem. *237*, 204–207.

Bernard, P.S., Lay, M.J., and Wittwer, C.T. (1998). Integrated amplification and detection of the C677T point mutation in the methylenetetrahydrofolate reductase gene by Fluoresence Resonance Energy Transfer and probe melting curves. Anal. Biochem. *255*, 101–107.

Bustin, S.A. (2000). Absolute quantification of mRNA using real-time reverse transcriptase polymerase chain reaction assays. J. Mol. Endo. *25*, 169–193.

Broude, N.E. (2002). Stem-loop oligonucleotides: a robust tool for molecular biology and biotechnology. Trends Biotechnol. *20*, 249–256.

Cardullo, R.A., Agrawal, S., Flores, C., Zamecnik, P.C., and Wolf, D.E. (1988). Detection of nucleic acid hybridisation by non-radiative fluorescence resonance energy transfer. Proc. Natl. Acad. Sci. U.S.A. *85*, 8790–8794.

Clegg, R.M. (1992). Fluorescence resonance energy transfer and nucleic acids. In Methods in Enzmology, Lilley, D.M.J., and Dahlber, J.E., eds. (Academic Press, New York) *211*, 353–388.

Dubertret, B., Calame, M., and Libchaber, A. (2001). Single-mismatch detection using gold-quenched fluorescent oligonucleotides. Nat. Bio. *19*, 365–370.

French, D.J., Archard, C.L., Brown, T., and McDowell, D.G. (2001). Hybeacon™ probes: a new tool for DNA sequence detection and allele discrimination. Mol. Cell. Probes *15*, 363–374.

Gelfand, D.H. (1989). Taq D.N.A polymerase. In P.C.R Theory, Principles, and Application to D.N.A Amplification. (1988). In Erlich, H.A. (ed.) PCR Technology (Stockton Press, NY). p. 17–22.

Gelmini, S., Orlando, C., Sestini, R., Vona, G., Pinzani, P., Ruocco, L., and Pazzagli, M. (1997). Quantitative polymerase chain reaction-based homogeneous assay with fluorogenic probes to measure c-erB-2 oncogene amplification. Clin. Chem. *43*, 752–758.

Häner, R., Biner, S.M., Langeneger, S.M., Meng, T., and Malinovskii, V.L. (2010). A highly sensitive, excimer-controlled molecular beacon. Angew. Chem. *49*, 1–5.

Heller, M.J., and Jablonski, E.J. (1987). Fluorescent stokes shift probes for polynucleotide hybridisation assays. European patent application 86116652.8.

Heyduk, T., and Heyduk, E. (2002). Molecular beacons for detecting DNA binding proteins. Nat Biotech. *20*, 171–176.

Higuchi, R., Dollinger, G., Walsh, P.S., and Griffith, R. (1992). Simultaneous amplification and detection of specific DNA sequences. Biotechnology *10*, 413–417.

Higuchi, R., Fockler, C., Dollinger, G., and Watson, R. (1993). Kinetic PCR analysis: Real-time monitoring of DNA amplification reactions. Biotechnonology *11*, 1026–1030.

Holland, P.M., Abramson, R.D., Watson, R., and Gelfand, D.H. (1991). Detection of specific polymerase chain reaction product by utilising the 5′–3′ exonuclease activity of *Thermus aquaticus* DNA polymerase. Proc. Natl. Acad. Sci. U.S.A. *88*, 7276–7280.

Hwang, G.T., Seo, Y.J., and Kim, B.H. (2004). A highly discriminating quencher-free molecular beacon for probing DNA. J. Am. Chem. Soc. *126*, 6528–6529.

Isacsson, J., Cao, H., Ohlsson, L., Nordgren, S., Svanik, N., Westman, G., Kubista, M., Sjöback, R., and Sehlstedt, U. (2000). Rapid and specific detection of PCR products using light-up probes. Mol. Cell. Probes *14*, 321–328.

Ju, J., Ruan, C., Fuller, C.W.,Glazer, A.A., and Mathies, R.A. (1995). Fluorescence energy transfer dye-labeled primers from DNA sequencing and analysis. Proc. Natl. Acad. Sci. U.S.A. *92*, 4347–4351.

Kandimalla, E.R., and Agrawal, S. (2000). 'Cyclicons' as hybridisation-based primer-probes: synthesis, properties and application in real-time PCR. Bioorg. Med. Chem. *8*, 1911–1916.

Kong, D.M., Gu, L., Shen, H.X., and Mi, H.F. (2002). A modified molecular beacon combining the properties of Taqman probe. Chem. Commun. *8*, 854–855.

Koshkin, A.A., Rajwanshi, V.K., and Wengel, J. (1998). Novel convenient syntheses of LNA [2.2.1] bicyclo nucleosides. Tetrahedron Lett. *39*, 4381–4384.

Kutyavin, I.V., Afonina, I.A., Mills, A., Gorn, V.V., Lukhtanov, E.A., Belousov, E.S., Singer, M.J., Walburger, D.K., Lokhov, S.G., Gall, A.A., *et al.* (2000). 3′ minor groove binder-DNA probes increase sequence specificity at PCR extension temperatures. Nucleic Acids Res. *28*, 655–661.

Kyger, E.M., Krevolin, M.D., and Powell, M.J. (1998). Detection of the hereditary hemochromatosis gene mutation by real-time fluorescence polymerase chain reaction and peptide nucleic acid clamping. Anal. Biochem. *260*, 142–148.

Lee, L.G., Chen, C.H., and Chiu, L.A. (1986). Thiazole orange: a new dye for reticulocyte analysis. Cytometry. *7*, 508–517.

Lee, L.G., Connell, C.R., and Bloch, W. (1993). Allelic discrimination by nick-translation PCR with fluorogenic probes. Nucleic Acids Res. *21*, 3761–3766.

Lee, L.G., Livak, K.J., Mullah, B., Graham, R.J., Vinayak, R.S., and Woudenberg, T.M. (1999). Seven-colour, homogeneous detection of six PCR products. Biotechniques 27, 342–349.

Lee, M.A., Brightwell, G., Leslie, D., Bird, H., and Hamilton, A. (1999). Fluorescent detection techniques for real-time multiplex strand specific detection of *Bacillus anthracis* using rapid PCR. J. Applied. Micro. *87*, 218–223.

Lee, M.A., Siddle, S.L., and Hunter, R.P. (2002). ResonSense®: Simple linear probes for quantitative homogeneous rapid polymerase chain reaction. Anal. Chimic. Acta. *457*, 61–70.

Lee, M.A., Leslie, D., and Squirrell, D. J. (2009). Internal and external controls for reagent validation. In Real-time PCR: Current Technology and Applications, Logan, J., Edwards, K., and Saunders, N., eds. (Caister Academic Press, Norfolk, UK), pp. 95–104.

Leone, G., Schijndel, H., Gemen, B., Kramer, F.R., and Schoen, C D. (1998). Molecular beacon probes combined with amplification by NASBA enable homogenous, real-time detection of RNA. Nucleic Acids Res. *26*, 2150–2155.

Letertre, C., Perelle, S., Dilasser, F., Arar, K., and Fach, P. (2003). Evaluation of the performance of LNA and MGB probes in 5′-nuclease PCR assays. Mol. Cell. Probes *17*, 307–3111.

Li, Q., Luan, G., Guo, Q., and Liang, J. (2002). A new class of homogeneous nucleic acid probes based on specific displacement hybridisation. Nucleic Acids Res. *30*, e5.

Livak, K.J., Flood, S.A.J., Marmaro, J., Giusti, W., and Deetz, K. (1995). Oligonucleotides with fluorescent dyes at opposite ends provide a quenched probe system useful for detecting PCR product and nucleic acid hybridisation. PCR. Meth. Appl. *4*, 357–362.

Logan, M.J., and Edwards, K. (2009a). An overview of PCR platforms. In Real-time PCR: Current Technology and Applications, Logan, J., Edwards, K., and Saunders, N., eds. (Caister Academic Press, Norfolk, UK), pp. 7–23.

Logan, M.J., and Edwards, K. (2009b) Performing real-time PCR. In Real-time PCR: Current Technology and Applications, Logan, J., Edwards, K., and Saunders, N., eds. (Caister Academic Press, Norfolk, UK), pp. 85–94.

Matayoshi, E.D., Wang, G.T., Krafft, G.A., and Erickson, J. (1990). Novel fluorogenic substrates for assaying retroviral proteases by resonance energy transfer. Science *247*, 954–958.

Morrison, L.E., Halder, T.C., and Stols, L.M. (1989). Solution-phase detection of polynucleotides using interacting fluorescent labels and competitive hybridisation. Anal Biochem. *183*, 231–244.

Morrison, T.B., Weis, J.J., and Wittwer, C.T. (1998). Quantification of low-copy transcripts by continuous SYBR®Green 1 monitoring during amplification. Biotechniques *24*, 954–962.

Mouritsen, P., Noerholm, M., Nielson, P.S., Jacobsen, N., Lornholt, C., Pfundheller, H.M., and Tolstrup, N. (2005). ProbeLibrary: A new method for faster design and execution of quantitative real-time PCR. Nat. Methods *2*, 313–316.

Mullis, K.B., and Faloona, F.A. (1987). Specific synthesis of DNA *in vitro* via a polymerase catalyzed chain reaction. Meth. Enzymol. *155*, 335–350.

Nazarenko, I.A., Bhatnagar, S.K., and Hohman, R.J. (1997). A closed tube format for amplification and detection ofDNA based on energy transfer. Nucleic Acids Res. *25*, 2516–2512.

Newton, C.R., Graham, A., Heptinstall, L.E., Powell, S.J., Summers, C., Kalsheker, N., Smith, J.C., and Markham, A.F. (1989). Analysis of any point mutation in DNA. The amplification refractory mutation system (ARMS). Nucleic Acids Res. *17*, 2503–2516.

Nazarenko, I., Pires, R., Lowe, B., Obaidy, M., and Rashtchian. (2002). Effect of primary and secondary structure of oligodeoxyribonucleotides on the fluorescent properties of conjugated dyes. Nucleic Acids Res. *30*, 2089–2195.

Newton, C.R., Holland, D., Heptinstall, L.E., Hodgson, I., Edge, M.D., Markham, A.F., and Mclean, M.J. (1993). The production of PCR products with 5′ single-stranded tails using primers that incorporate novel phosphoramidite intermediates. Nucleic Acids Res. *21*, 1155–1162.

Nurmi, J., Wikman, M., Karp, M., and Lövgren, T. (2002). High-performance real-time quantitative RT-PCR using lanthanide probes and a dual temperature hybridization assay. Anal Chem. *74*, 3525–32.

Ogura, M., and Mitsuhahi, M. (1994). Screening method for a large quantity of polymerase chain reaction products by measuring YOYO-1 fluorescence on 96-well polypropylene plates. Anal Biochem. *218*, 458–459.

Ortiz, E., Estrada, G., and Lizardi, P.M. (1998). PNA molecular beacons for rapid detection of PCR amplicons. Mol. Cell. Probes *12*, 219–226.

Richard, J.A., Gerowaska, M., Shelbourne, M., Franch, D., and Brown, T. (2010). Six-colour HyBeacon probes for multiplex genetic analysis. Chembiochem *11*, 2530–2533.

Ririe, K.M., Rasmussen, P.R., and Wittwer, C.T. (1997). Product differentiation by analysis of DNA melting curves during the polymerase chain reaction. Anal. Biochem. *245*, 154–160.

Robinson, J.K., Mueller, R., and Filippone, L. (2000). New molecular beacon technology. Am. Lab. *32*, 30–24.

Saha, B.K., Tian, B., and Bucy, R.P. (2001). Quantitation of HIV-1 by real-time PCR with a unique fluorogenic probe. J. Virol. Meth. *93*, 33–42.

Saiki, R.K., Scharf, S., Faloona, F.A., Mullis, K.B., Horn, G.T., Erlich, H.A., and Arnheim, N. (1985). Enzymatic amplification of beta-globin genomic sequences and restriction site analysis for diagnosis of sickle cell anemia. Science *230*, 1350–1354.

Salmon, M.A., Vendrame, M., Kummert, J., and Lepoivre, P. (2002). Detection of apple chlorotic leaf spot virus using a 5′ nuclease assay with a fluorescent 3′ minor groove-DNA binding probe. J. Virol. Meth. *104*, 99–106.

Salvatore, A.E.M., Tyagi, S., and Kramer, F.R. (2006). Real-time assays with molecular beacons and other fluorescent nucleic acid probes. Clinica Chimica Acta *363*, 48–60.

Sanchez, J.A., Abramowitz, J.D., Salk, J.J., Reis, A.H., Rice, J.E., Pierce, K.E., and Wangh, L.J. (2006). Two-temperature LATE-PCR endpoint genotyping. BMC Biotech. *6*, 44.

Schofield, P., Pell, A.N., and Krause, D.O. (1997). Molecular Beacons: Trial of a fluorescence-based solution hybrization technique for ecological studies with ruminal bacteria. App. Env. Micro. *63*, 1143–1147.

Selvin, P.R. (1995). Fluorescence resonance energy transfer. Meth. Enzymol. *246*, 300–334.

Solinas, A., Brown, L.J., McKeen, C., Mellor, J.M., Nicol, J.T.G., Thelwell, N., and Brown, T. (2001). Duplex scorpion primers in SNP analysis and FRET applications. Nucleic Acids Res. *29*, E96.

Solinas, A., Thelwell, N., and Brown, T. (2002). Intramolecular TaqMan probes for genetic analysis. Chem. Commun. 2272–2273.

Svanik, N., Westman, G., Wang, D., and Kubista, M. (2000a). Light-Up probes: Thiazole orange-conjugated peptide nucleic acid for detection of target nucleic acid in homogeneous solution. Anal. Biochem. *281*, 26–35.

Svanik, N., Nygren, J., Westman, G., and Kubista, M. (2000b). Free-probe fluorescence of light-up probes. J. Am. Chem. Soc. *123*, 803–809.

Taylor, M.J., Hughes, M.S., Skuce, R.A., and Neill, S.D. (2001). Detection of Mycobacterium bovis in bovine clinical specimens using real-time fluorescence and fluorescence energy transfer probe rapid-cycle PCR. J. Clinic. Micro. *39*, 1272–1278.

Täpp, I., Malmberg, L., Rennel, E., Wik, M., and Syvänen, A.-C. (2000). Homogeneous scoring of single nucleotide polymorphisms: comparison of the 5′-nuclease TaqMan® assay and molecular beacon probes. Biotechniques *28*, 732–738.

Thelwell, N., Millington, S., Solinas, A., Booth, J and Brown, T. (2000). Mode of action and application of scorpion primers to mutation detection. Nucleic Acids Res. *28*, 3752–3761.

Tsourkas, A., Behlke, M.A., Rose, D., and Gang, B. (2003). Hybrisidisation kinetics and thermodynics of molecular beacons. Nucleic Acids Res. *31*, 1319–1330.

Tyagi, S., and Kramer, F.R. (1996). Molecular beacons: probes that fluoresce upon hybridisation. Nat. Biotech. *14*, 303–308.

Tyagi, S., Bratu, D.P., and Kramer, F.R. (1997). Multicolour molecular beacons for allele discrimination. Nat. Biotech. *16*, 49–53.

Vet, J.A.M., Majithia, A.R., Marras, S.A.E., Tyagi, S.., Dube, S., Poiesz, B.J., and Kramer, F.R. (1999). Multiplex detection of four pathogenic retroviruses using molecular beacons. Proc. Natl. Acad. Sci. U.S.A. *96*, 6394–6399.

Whitcombe, D., Brownie, J., Gillard, H.L., Mckechnie, D., Theaker, J., Newton, C.R., and Little, S. (1998). A homogeneous fluorescence assay for real-time PCR amplicons: its application to real-time, single tube genotyping. Clin. Chem. *44*, 918–923.

Whitcombe, D., Theaker, J., Guy, S.P., Brown, T., and Little, S. (1999a). Detection of PCR products using self-probing amplicons and fluorescence. Nat. Biotech. *17*, 804–807.

Whitcombe, D., Kelly, S., Mann, J., Theaker, J., Jones, C., and Little, S. (1999b). Scorpions™ primers – a novel method for use in single tube genotyping. Am. J. Hum. Genet. *65*, 2333.

Wittwer, C.T., Ririe, K.M., Andrew, R.V., David, D.A., Grundy, R.A., and Balis, U.J. (1997). The LightCycler™: a microvolume multisample fluorimeter with rapid temperature control. Biotechniques *22*, 176–181.

Wolffs, P., Knutsson, R., Sjöback, R., and Rådström, P. (2001). PNA-based light-up probes for real-time detection of sequence-specific PCR products. Biotechniques *31*, 766–771.

Xiao, Y., Plakos, K.J.I., Lou, X., White, R.J., Jiangrong, Q., Lacxco, P., K.W., and Soh, H.T. (2009). Fluorescence detection of single-nuleotide polymorphisms with a single self-complementary, triple-stem DNA probe. Angnew. Chem. Int. Ed. Engl. *48*, 4354–4358.

Yi, J., Zhang, W., and Zahng, D.Y. (2006). Molecular zipper: a fluorescent probe for real-time isothermal DNA amplification. Nucleic Acids Res. *34*, e81.

Young, D. (1996). Viral load quantitation: An integral part of future health management. Australasian Biotechnology *6*, 295.

Zahng, P., Beck, T., and Tan, W. (2001). Design of a molecular beacon DNA probe with two fluorophores. Angew. Chem. Int. Ed. Engl. *40*, 402–405.

Zahng, Y., Zahng, D., Wenquan, L., Chen, J., Peng, Y., and Cao, W. (2003). A novel real-time quantitative PCR method using attached universal template probe. Nucleic Acid Res. *31*, 20.

Internal and Other Controls for Real-time PCR Validation

2

Martin A. Lee, Dario L. Leslie and David J. Squirrell

Abstract

A range of factors can cause false-negative results in real-time PCR through effects on one or more of the reaction components. Consequently applications requiring a high level of confidence need to be designed to control for the occurrence of false-negatives. Whilst an external, or batch, control is often used, the ideal control is an internal one included in the reaction cocktail in a multiplex assay. Here we discuss the application and development of molecular mimics as controls in real-time PCR and explain concepts and experimental considerations to aid in the optimization of controlled multiplexed assays.

Introduction

In real-time PCR assays there is inherent scope for variability in the outcome because the reagent mixes are relatively complex, with a dozen or more components, and they are often made up in very small volumes, limiting the accuracy of their composition. In addition, the PCR process is exponential, so variations in starting conditions tend to be amplified, affecting the course followed by individual reactions. The course may include assay failure if problems with the sample, the reagents, or the instrumentation have occurred.

In many applications, including infectious disease testing in clinical and veterinary diagnostics, food quality control and environmental analysis, the majority of PCR results are genuinely negative, but the consequences of false-negatives may be serious. There is therefore a pressing need to control for the occurrence of false-negatives.

For most molecular tests the use of reference material in a batch test is the only control that can be implemented. However, PCR and other nucleic acid amplification techniques provide not only extremely sensitive detection, but also have the advantage that they can be multiplexed to include an internal control (IC). An IC is a second target molecule that can be amplified with, but distinguished from, other products in the same tube. Ideally this should be able to control for almost all of the reagents in a reaction cocktail and for variations in instrument operation. The exception the IC cannot control for is a specific probe for another target. The IC can be produced as a synthetic molecule that gets co-amplified with the target species, using the same set of amplimers, but comprising an internal sequence differing from the target to allow it to be bound by a distinguishable probe. In this, the preferred embodiment, the IC acts as a 'molecular mimic'.

Before real-time PCR, IC approaches used amplicons with different molecular masses and post-amplification analysis by a separation method. For example, the molecular mimic could contain both priming sites, but an internal sequence changed by insertions or deletions (Ursi *et al.*, 1992). The products of amplification could be analysed by the use of agarose gel electrophoresis with ethidium bromide staining. There are a large number of such examples reported in the literature. The main drawbacks of the approach were the time-consuming nature of the analysis method, the tendency of smaller amplicons to be amplified more efficiently, and unpredictable generation of specific and control product hetero-duplexes confusing the data interpretation (Henley *et al.*, 1996).

Because of its sensitivity, PCR is routinely used in applications where there may be only a few molecules of the target species present in the sample. Particularly where the test is for an RNA species, loss of the target during sample preparation and PCR set-up can lead to inaccurately low or false-negative outcomes. To cover this eventuality, the IC may additionally be used as a 'process control' and introduced with the sample to indicate loss of target molecules during sample storage or sample processing. This can ensure that methods to overcome false-negatives caused by degradation of the target have succeeded. These methods include: the addition of carrier RNA or DNA; the use of an RNAse inhibitor; the use of reagents treated and tested for removal of RNase and DNase activity; and always wearing disposable gloves when handling any components so as to avoid transfer of RNAse or DNAse activity from the skin.

False-positives are also a problem in nucleic acid amplification based tests, but are not addressable through the use of an IC. Measures for controlling these are covered in the endnote to this chapter.

Real-time PCR approaches with fluorescence reporting are particularly useful for high-confidence applications because the assays can be carried out in the same tube, in multiplex, with measurements *in situ* so the tube never has to be reopened. This closed tube format allows an IC to control for negatives whilst significantly reducing opportunities for cross-contamination by amplification products and thereby reducing the risk of false-positives. Most commercial fluorimeter instruments can co-amplify and detect at least two targets in a quantitative manner. However, increasing the level of multiplexing adds significantly to the complexity of designing and optimizing the assay. Multiplexing of real-time assays is discussed in detail in Lee *et al.* (2009) and reference should be made to that chapter for optimizing multiplex assays.

In this chapter we present various strategies for producing internal controls for different fluorescent reporting arrangements and discuss some experimental considerations for their optimization. The experimental approaches described here are based largely on our own experiences and we make some suggestions for the development and implementation of controlled assay systems. Whilst users will find that for many applications the use of a homologous or competitive control is not necessary, the information will be of equal use to those developing heterologous controls and synthetic standards alike. Competitive homologous internal controls are the focus of this chapter. In the literature, the term internal control is often used incorrectly so we begin with some definitions.

Nomenclature

The use of a second amplicon can serve to control, normalize or standardize the result. This chapter only covers the first application, the use of molecular mimics to check reaction outcomes, but the strategies described are equally valid in the development of internal and external mimics for other applications, such as their use as external or internal standards. The nomenclature used to describe the types of second amplicon is summarized in Table 2.1. It is easy to see how an internal control using homologous amplimers will add more

Table 2.1 Definitions and types of standard, control and reference nucleic acids.

Type	Second amplicon type			
	Standard	Control	Reference	
	A standard is a nucleic acid preparation that has a set or known concentration. Unknown samples may be compared to one or more standards and an absolute value for the unknown can be determined by numerical interpolation/extrapolation from experimentally determined values	A control is an amplicon added to the PCR and allowed to amplify to verify the integrity of one or more reagent(s) in the cocktail	A nucleic acid target that is used to compare the sample for either a qualitative or relative quantitative analysis.	
External: amplified in a different well	The standards are amplified separately from the target species in different wells on the same instrument*	"Batch control" This involves the use of a control in a separate well on the instrument. This approach cannot control for false negatives in different samples. It controls only for successful thermal cycling and the integrity of a common core reaction mix*	A related nucleic acid that is used for a qualitative comparison of amplification.	
Internal: amplified in the same well	Homologous		Exogenous	This is an internal reference that is added to the sample and is co-amplified with the target nucleic acid using a second set of amplimers
	A competitive standard that has the same amplimers as the specific target	A competitive control that is included in the same well of the instrument that has the same amplimers as the specific assay and controls for the core reagents and primers, and the thermal cycling		
	Heterologous		Endogenous	A 'housekeeping gene'. This is usually a gene that is naturally present in the sample at constant levels such that the up or down regulation of the target species can be relatively quantified using a second set of amplimers.
	A non-competitive standard with different amplimers to those of the specific target	A control that is included in the same well of the instrument but has different amplimers to that of the specific assay that controls for both the thermal cycling and the core reagents		

*External standards and controls may also be homologous, or heterologous, to avoid the risk of cross-contamination.

confidence to a result by confirming that conditions in the PCR are adequate for allowing amplification to occur. However, this controls only for the amplification and, where a nucleic acid reporter probe is also involved, this has to be controlled for as a separate factor.

Strategies for development of internal controls and their analysis

The strategy for detecting an internal control depends upon the instrument's capabilities and the fluorescent reporting chemistry. The simplest approach is to use a sequence-specific probe for the IC with a reporting dye of different emission wavelength to that of the probe for the specific target. This requires that the emission wavelengths of the dyes are either spectrally separated enough to not overlap, or that the emissions can be separated by the instrument through the means of a dye de-convolution algorithm. For example on the ABI 7700, a 5′ nuclease assay may use FAM as the specific reporter probe and a dye such as VIC, JOE or HEX as the reporter for the control probe. The cycle-by-cycle increase in the emissions of the dyes reports amplification of the respective species. This strategy may be applied to virtually all fluorescent chemistries and instruments.

The signal from the IC does not necessarily need to be quantitative since it serves only to report the integrity of the PCR reagents and the amplification conditions. The ability to detect the species at the end of thermal cycling should suffice for most applications. For assays based on hybridization run on an instrument that can determine melting point (T_m), discrimination between target and control species can be obtained by designing the control amplicon and/or probe to have a T_m different to that of the specific nucleic acid target. This may be achieved using DNA-binding dyes for amplicons with different melting points (Lee *et al.*, 1999), or labelled probes for amplicons with different internal sequences. When using specific probes melting point can be used in tandem with the use of different emission wavelengths. The merits of sequence-specific probes over DNA binding dyes have been discussed in Chapter 1. Whilst the use of DNA binding dyes as probes has some disadvantages in terms of specificity relative to sequence-specific probes, the correlation between amplification of the IC and the specific target is enhanced as they are detected through by the same signalling reagent. Equally, the target sequence with a polymorphism introduced into it can used in the design of the IC so that a single hybridization probe can be used for both IC and target amplicons with T_m used to differentiate the binding partner. This represents one of the few approaches where a sequence-specific reporter probe can be controlled for as well as the amplification. In this embodiment the function of the probe function can be better controlled than in approaches with independent probes for target and IC. Using dual hybridization probes in either quantification and/or melting point analysis it is possible to use one common probe, either the donor or the acceptor, to provide a better approach for control of sequence-specific probes. The methods for generating these mimics are now briefly described.

Synthesizing and optimizing molecular mimics for use as internal controls

There are several recombinant DNA techniques that can be used to generate a homologous internal control mimic. In each case the objective is to create a synthetic target that can co-amplify with the specific amplicon and that has a 'motif' which may be easily distinguished from that of the specific amplicon. This distinction can be achieved by the use of dyes with different emission wavelengths and/or a shift in the melting point as described above. Several methods have been reported. These include: insertion of the specific target into a vector and insertion/deletion of a sequence between the amplimer regions (Ursi *et al.*, 1992; Zimmerman *et al.*, 1996; Brightwell *et al.*, 1998), the amplification of a generic target from a related sequence (deWit *et al.*, 1993), the use of overlapping and tailed PCRs (Müller *et al.*, 1998; Sachadyn *et al.*, 1998), site-directed mutagenesis (Nash *et al.*, 1995), random misprimed products (Mergenthaler *et al.*, 2004) and polymerase extension of oligonucleotides and subsequent PCR (Rosenstraus *et al.*, 1998). In addition a complete synthesis of the gene is possible for smaller amplicons (Zimmerman *et al.*, 2000) and some commercial suppliers will now make the gene and supply it in a vector ready for use. Since the ability to discriminate amplicons by size is not a requirement, the amplicon may be similar, or the same in size, and thus achieve a similar amplification efficiency. Gene synthesis is now available commercially and so the entire sequence may be requested in a choice of vectors. The only constraints are cost and time. However, two main techniques that we use to generate mimics are:

- Direct insertion of amplimer and probe sequences into cloning vectors using recombinant techniques. This can be done by amplifying amplimer/probe sites, or the use of oligonucleotide linkers. In either case the use of restriction overhangs will assist in the correct directional insertion of the fragments.
- Generation of targets using a PCR approach of amplimers tailed with specific amplimer sequences to amplify alternative probe regions. The products contain the specific amplimer sequences incorporated into the PCR product. This product may be subsequently cloned into the desired vector (Fig. 2.1). The use of the additional overhangs with unique restriction sites will assist in directional cloning that may be required for some applications (e.g. sense-specific RT-PCR).

For reverse transcriptase PCR (RT-PCR) a mimic may be made using the methods described and subsequent insertion of the product into a vector. *In-vitro* transcription using the T7/T3 promoters allows the RNA molecule to be then synthesized in quantity. The use of modified nucleotides that make the transcripts resistant to RNase(s) could form one strategy that would allow for a more stable RNA control (Capodici *et al.*, 2002). Alternatively, the sequence could be cloned into an RNA virus such as the bacteriophage MS2. Armoured RNA™ is a product available from Ambion INC (Pasloske *et al.*, 1998). This is a method for packaging recombinant RNA into pseudoviral particles using a plasmid system whereby the MS2 bacteriophage coat protein is located downstream of the promoter and the target gene of interest. RNA encoding the coat protein and the target sequence is transcribed under the control of the inducer. Coat protein is translated and subsequently binds to recombinant RNA to form the product. The packaged molecule is resistant to RNase digestion yet easily extracted for amplification. The process has been shown to stabilize the RNA compared to *in vitro* transcripts made from plasmid and is an ideal method for RNA IC production. The authors would expect that any recombinant RNA virus containing the control sequence would improve the stability of the nucleic acid over naked RNA and an alternative protected RNA control system has been developed based on cowpea mosaic virus (King *et al.*, 2007). It should be noted that for providing a robust and inexpensive exogenous IC for RT-PCR extraction or amplification control native MS2 sequences have proved to be useful (Dreier *et al.*, 2005).

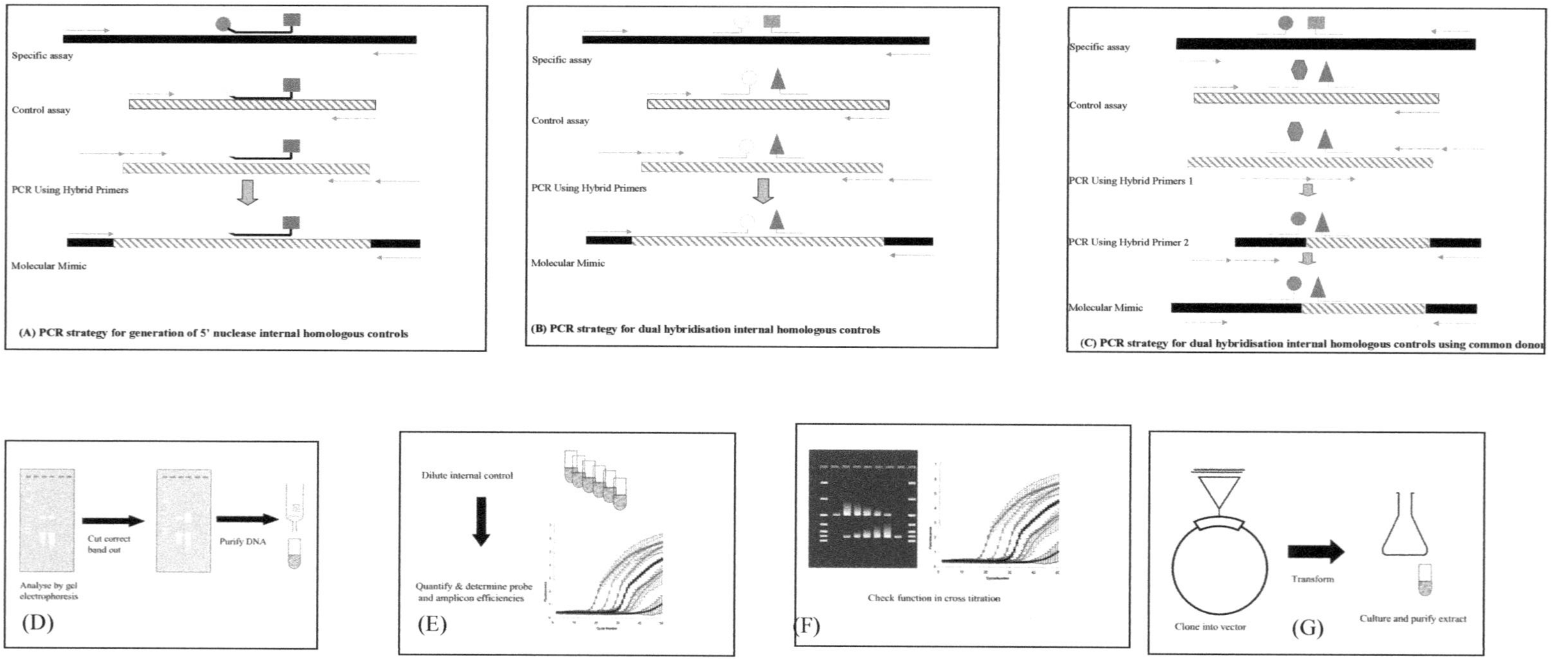

Figure 2.1 Schematic showing strategies for rapidly generating and evaluating competitive internal homologous controls by PCR. (A) For assays such as the 5′ nuclease where there is only one probe. (B) For dual hybridization assays using two additional probes. (C) Dual hybridization probes using one common probe for both the control and specific assay. (D) Amplify the IC using PCR and extract the correct band from the gel. (E) Serially dilute the product and determine the amplification and probe efficiencies (Fig. 2.2). (F) Cross-titre the IC with specific target to assess the efficacy of the IC in multiplex PCR before cloning into a suitable vector. (G) Transformation, culture, extraction and use. A colour version of this figure is available in the plate section at the back of the book.

Experimental considerations

In designing an experiment the sequence of the specific target and the assay are important considerations. (Reference should be made to Chapter 1 covering the utility, design and sequence dependency of different reporting chemistries.) The design of the assay is best facilitated by software tools that can automatically select amplimers and probes and predict (score) amplimer–amplimer and amplimer–probe interactions for the multiplexed assay. Here we highlight some of the considerations that we consider important in achieving a successful IC implementation.

Control sequence

The choice of control sequence is the most important factor since the other considerations are mostly dependent on this. With nucleic acid probes one suggested approach is to utilize a sequence that has been previously reported/optimized, or that is commercially available for the same assay type. For example, the ABI human β-actin sequence for a 5′ nuclease assay is useful because the probe is efficient and available as a kit. Ideally, the probe sequence should not be of the same origin (or endogenous to the sample) as the target for the specific test. For example, it would not be appropriate to use the β-actin sequence for a human test. Providing the probe sequence has been shown to work, any other sequence would be suitable, but preferably the GC content for the sequence should be similar to that of the specific target.

Amplicon efficiency

Whilst the differences in amplification efficiencies of the target and IC may be compensated for by changing the concentration of the initial target, they should ideally be similar and can be determined empirically. This is most important if the mimic is to be used for competitive quantification since the amplification efficiency can vary significantly between amplicons (Zimmerman *et al.*, 1996). It is useful to include native sequence either side of the amplimer when designing the mimic to ensure that the effects of neighbouring sequence is maintained. The amplification efficiency is best determined using a DNA binding dye such as SYBR®Green-1 combined with an efficient hot-start to amplify a broad range of concentrations across the dynamic range of the assay. Whilst using a probe directly to do this is satisfactory, checking the amplification efficiency without the probe allows it to be measured directly and is useful in understanding the efficiency of the probe when it is later evaluated. It will avoid dismissing a satisfactory reaction for the sake of iterating the probe design and/or as a result of a failure in probe synthesis. The efficiency (E) can be calculated by using the slope value for the derived standard curve using (Fig. 2.2);

$$E = 10^{-1/\text{slope}}$$

The efficiency of different amplicons for the same (or similar) dilution series can then be directly determined.

Probe efficiency

The efficiency of a sequence-specific probe is important if the signal from the IC is to be commensurate with that from the target of the assay. Probe efficiency can be affected by a number of factors that are discussed in Chapter 1. The probe sequence to be chosen

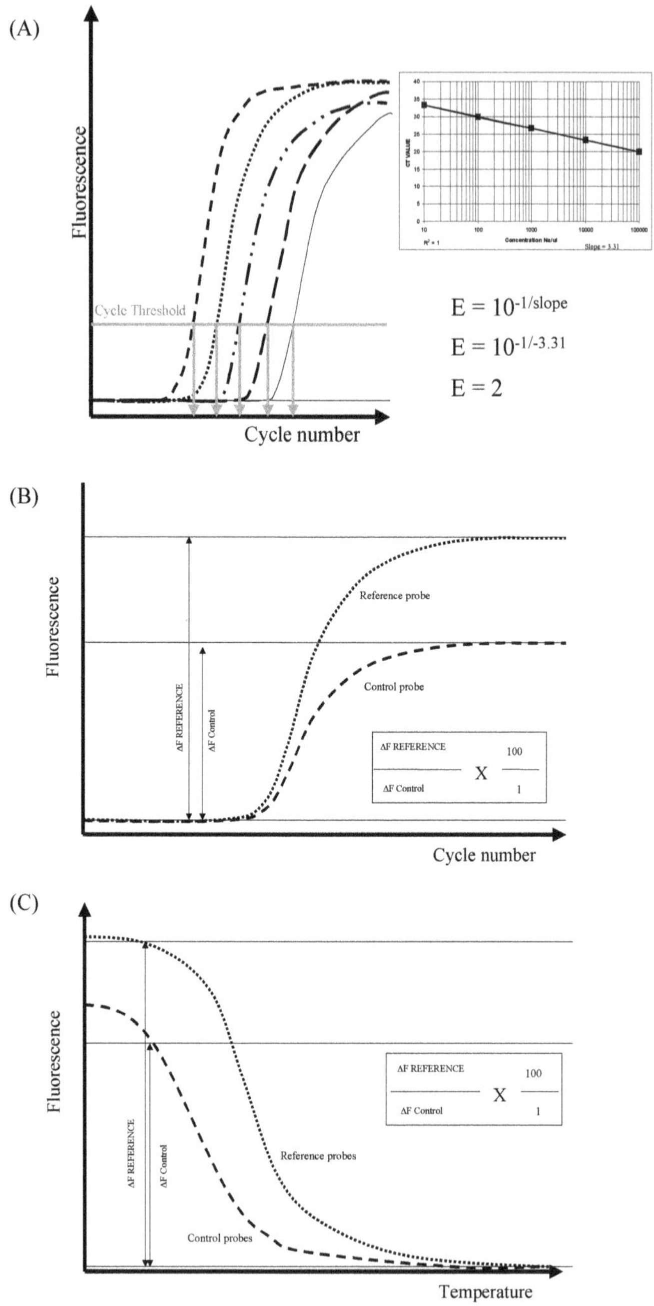

Figure 2.2 Determining and comparing amplification and probe efficiencies of different amplicons. (A) Determining amplification efficiency (E). Ideally, using a generic method such as a DNA binding dye, e.g. SYBR® Green I, the cycle threshold (CT) values are determined for a dilution series of target material over a range of concentrations and the plot of fluorescence intensity against cycle number generates the amplification curves. The derived CT values can be used to generate a standard curve. The slope of the curve can then be used to calculate E for the range tested. This value can be used to compare E with other amplicons using

should be checked for complementarities to other probes, amplimers and amplicons in the reaction. This is most important when the probe melting point approach is used since the signal from a control probe that is too large may swamp a low signal obtained in positives. When assessing potential probes for the mimic sequence it is useful to score the probe efficiency relative to a reference assay. For example, the percentage of signal–noise ratio for both 5′ nuclease assay specific and control probes can be compared to that of another assay such as the ABI Human β-actin probe. For this to be done the reaction for either probe must be taken to completion (plateau). These data may be expressed as:

$$\frac{\text{Fluorescence signal value specific assay/control assay at (cycle 50} - \text{cycle 1)}}{\text{Fluorescence signal value } \beta\text{-actin (cycle 50} - \text{cycle 1)}} \times \frac{100}{1}$$

This provides a comparative means for matching candidate control sequences to given assays. Likewise, the same may be done for comparing dual hybridization probes using:

$$\frac{\text{Fluorescence signal value specific assay/control assay at } (95°\text{C}-60°\text{C}^1)}{\text{Fluorescence signal value } \beta\text{-globin } (95°\text{C}-60°\text{C}^1)} \times \frac{100}{1}$$

This approach may be carried out using any reference probe and is equally applicable to other assay systems to estimate the relative probe efficiency (see Fig. 2.2). The quotient used for comparison should be based on the values derived from the maximum signal (when the target is in excess of the probe) subtracted from the minimum signal (the background) for each probe. The minimum signal is usually at the start of the amplification or when the probe is dissociated from the target at elevated temperatures.

Control vector

The choice of control vector is important since the vector may affect the stability and amplification efficiency. Plasmid constructs are commonly used and the efficiency of amplification may drop dramatically at lower target numbers (< 100 copies per reaction). Linearizing the plasmid by digestion at a unique restriction site in a vector can improve amplification efficiency. The efficiency of closed plasmids is often seen to drop off dramatically at ~ 100 to 1000 copies. The use of other vectors such as bacteriophage λ vectors and M13 (ssDNA rescue by co-infection with helper phage) may also be useful approaches for generating large amounts of linear IC molecules. PCR product itself is fine for use as a control although a vector represents the best method of safe storage of the desired sequence. We recommend the initial amplification of control DNA by PCR since the products may be used to evaluate potential mimics prior to the cloning of the molecule

---

the same or similar dilution series. (B) Determining probe efficiency. Using the 5′ nuclease assay and other assays by comparing the total change in fluorescence (as a function of a reference probe/assay) when the reaction is allowed to go to completion. (C) Using probes based on hybridization by comparing (as a function of a reference probe/assay) the change in fluorescence when the probe is hybridized (at a specific temperature) to when it is unhybridized (at high temperatures).

into a vector. However, the use of an *in-vivo* amplification method for the final synthesis of DNA to be used for controls is superior in that individual clones may be readily sequenced to check their correct identity and this process avoids the spurious artefacts that can be generated in PCR. These artefacts can be minimized during the evaluation stages by cutting the predicted PCR band out of an agarose gel and using any one of a number of methods for cleaning up the product. These include column purification and/or dialysis, before dilution and evaluation as suitable mimics.

Number of molecules

When using a competitive control the number of molecules to be added into the reaction must be determined empirically so that the required sensitivity of the specific assay is not compromised. The control must be amplified effectively in the absence of specific target. With increasing numbers of specific target molecules in the reaction the control amplification may be outcompeted. We have observed that it is possible to stop this if the T_m of the specific amplicon is higher than that of the control. Reducing the denaturing temperature of the thermal cycle (which is possible on some thermal cyclers) to below that of the specific target, but higher than that of the control, suppresses specific amplification thus allowing the IC to amplify. However, this is not necessary since the amplification of the control only serves to validate the reagents in the absence of the specific target. A cross-titration of specific target to control should be carried out to determine the correct amount of control molecules to include in a reaction. In practice, this amount is routinely determined to be between 10 and 100 copies per reaction if the specific assay is to maintain a sensitivity of ten or more copies of specific target. Accurate determination of the IC mimic concentration is therefore important. The use of fluorescent dyes and standards is inherently more accurate than ultra-violet optical density quantification at 260 nm. An alternative method for statistically determining the correct concentration of a stock of internal controls using PCR is provided by Rosenstraus *et al.* (1998). This method describes using a dilution series of the control preparation and performing multiple amplifications. At a given dilution the amplification will become stochastic owing to the presence of low or no copies of the template in the reaction in accordance with Poisson's law. The relationship between the average number of internal control copies in a dilution (C) and the probability (P_n) that no molecules exist in a sample of this dilution is given by:

$$C = -\ln(P_n)$$

where P_n is determined by counting the number of negative replicates used to calculate C.

Target nucleic acid

The control must be RNA if it is to be used for controlling the RT step. If a specific primer initiated RT step is used then utilizing the same primers in the mimics will control for both the RT and the PCR and negates the requirement for a second mimic for the PCR step in combined RT-PCR. Depending on the application, controls at both the RNA and DNA levels may be required in a two-step RT-PCR. The RT efficiency of the mimic and the specific target must be similar if the mimic is to be used as a standard in any application. When the RT-PCR is designed to be sense specific, directional cloning and the choice of vector orientation is important.

Dye/channel assignment

The most efficient dye and/or detector should be used for the specific target assay. This is of importance where a single excitation source is used in a universal acceptor arrangement (see Chapter 1), in which case it is usually the shorter wavelength dye that will be most efficiently excited. When a universal donor arrangement is used, it will be the acceptor to which energy is most efficiently transferred from the donor. In either multiplex arrangement, the optimum channel will depend on the instrument's specification and the dyes used.

Synthesis

If the IC is to be a synthetic oligonucleotide then the authors would recommend using a different supplier to the company that will be making your amplimers. The risk of cross-contamination by the IC of an amplimer is such as to have been reported numerous times in PCR discussion forums..

Summary and future improvements

The further development of instruments and methods of detection will increase the ability to multiplex PCR to higher levels in a quantitative manner. This introduces new possibilities for controlling for and/or normalizing sample extraction, and/or RT-PCR steps. This will be achieved through a combination of competitive (homologous) and non-competitive (heterologous) internal controls for each process. It is not always possible to control for these processes with one control since each may be subjected to different inhibitory factors. The ability to do this will improve the usefulness of the PCR for both research and diagnostic applications. The ability to control for all such processes will be of most use in emerging applications where remote testing in the non-laboratory environment necessitates the use of internal validation.

End note: causes of false-positives and measures to reduce their occurrence

False-positives are a significant problem in real-time PCR and in other analysis methods that depend upon nucleic acid amplification. They are often caused by cross-contamination from positive samples or reaction products, and sometimes from unwitting contamination of PCR reagents.

Preventative methods against cross-contamination are essential for eliminating the occurrence of false-positives. These are most effective when used in combination and include: good laboratory practice (see Espy *et al.*, 2006) in terms of segregating clean and contaminated materials and work areas, especially separation of PCR preparation and analysis areas; the use of closed tube assay formats to contain reaction products; using cabinets with UV treatment for PCR set-up; UV laboratory air scrubbers for removing aerosolized nucleic acids; and using uracil N-glycosylase carry over prevention chemistry (Longo *et al.*, 1990).

As mentioned above concerning synthesis of the IC, synthesis of assay target nucleic acids in the same facility as used to produce the amplimers should be strictly avoided. Contamination from workers in production plants making sampling kits, reagents and consumables has been known to cause false-positives in PCR testing of human samples. PCR users should also be aware that reaction components such as polymerase enzymes and hotstart antibodies will have been derived from natural sources that may contain, for example, *E. coli* or mouse DNA as contaminants.

In isothermal methods that are not dependent upon thermal cycling, amplification can commence as soon as reaction mixtures are brought together. This means that, for example, the pipette tips used to load the reaction vessels may well contain residues in which amplified products and by-products can be generated and thereby pose a contamination risk.

In batch assays, therefore, a no template control should routinely be incorporated in every run to control for false-positives.

References

Brightwell, G., Pearce, M., and Leslie, D. (1998). Development of internal controls for PCR detection of *Bacillus anthracis*. Mol. Cell. Probes *12*, 367–377.

Capodici, J., Kariko, K., and Weissman, D. (2002). Inhibition of HIV-1 infection by small interfering RNA-mediated RNA interference. J. Immunol. *169*, 5196–5201.

Dreier, J., Störmer, M., and Kleesiek, K. (2005). Use of bacteriophage MS2 as an internal control in viral reverse transcription-PCR assays. J. Clin. Micro. *43*, 4551–4557.

Espy, M.J., Uhl, J.R., Sloan, L.M., Buckwalter, S.P., Jones, M.F., Vetter, E.A., Yao, J.D.C., Wengenack, N.L., Rosenblatt, J.E., Cockerill, I.I.I., F.R., *et al.* (2006). Real-time PCR in clinical microbiology: applications for routine laboratory testing. Clin. Microbiol. Rev. *19*, 165–256.

Henley, W.N., Schuebel, K.E., and Nielsen, D.A. (1996). Limitations imposed by heteroduplex formation on quantitative RT-PCR. Biochem. Biophys. Res. Comm. *226*, 113–117.

King, D.P., Montague, N., Ebert, K., Reid, S.M., Dukes, J.P., Schädlich, L., Belsham, G.J., and Lomonossoff, P. (2007). Development of a novel recombinant encapsidated RNA particle: evaluation as an internal control for diagnostic RT-PCR. J. Virol. Methods *146*, 218–225.

Lee, M.A., Brightwell, G., Bird, H., Leslie, D., and Hamilton, A. (1999). Fluorescent detection techniques for real-time multiplex strand specific detection of *Bacillus anthracis* using rapid PCR. J. Applied. Microbiol. *87*, 218–223.

Lee, M.A., Leslie, D., and Squirrell, D.J. (2009). Internal and external controls for reagent validation. In Real-time PCR: Current Technology and Applications, Logan, J., Edwards, K., and Saunders, N., eds. (Caister Academic Press, Norfolk, UK), pp. 95–104.

Longo, M.L., Berninger, M.S., and Harley, J.L. (1990). The use of uracil glycosylase to control carry-over contamination in polymerase chain reactions. Gene *93*, 125–128.

Mergenthaler, E., Fodor, M., and Sűle, S. (2004). A new method to develop internal controls for the determination of phytoplasma concentration. Acta. Phyto. Ent. Hung. *39*, 1–7.

Müller, F.-M.C., Schnitzler, N., Coot, O., Kockelkorn, P., Haase, G., and Li, Z. (1998). The rationale and method for constructing internal control DNA used in Pertussis polymerase chain reaction. Diag. Microbiol. Infect. Dis. *31*, 517–523.

Mullis, K.B., and Faloona, F.A. (1987). Specific synthesis of DNA *in vitro* via a polymerase catalyzed chain reaction. Meth. Enzymol. *155*, 335–351.

Nash, K.A., Klein, J.S., and Inderlied, C.B. (1995). Internal controls as performance monitors and quantitative standards in the detection by polymerase chain reaction of herpes simplex virus and cytomegalovirus in clinical specimens. Mol. Cell. Probes *9*, 347–356.

Pasloske, B.L., Walkerpeach, C.R., Obermoeller, R.D., Winkler, M., and DuBois, D.B. (1998). Armoured RNA technology for production of ribonuclease-resistant viral RNA controls and standards. J. Clin. Micro. *36*, 3590–3594.

Rosenstraus, M., Wang, Z., ChangS.-Y., DeBonville, D., and Spadoro, J.P. (1998). An internal control for routine diagnostic PCR: design properties, and effect on clinical performance. J. Clin. Microbiol. *36*, 191–197.

Rossen, L., Norskov, P., Holmstrom, K., and Rasmussen, O.F. (1992). Inhibition of PCR by components of food samples, microbial diagnostic assays and DNA extraction solutions. International J. Food. Microbiol. *17*, 37–45.

Sachadyn, P., and Kur, J. (1998). The construction and use of a PCR internal control. Mol. Cell. Probes *12*, 259–262.

Saiki, R.K., Scharf, S., Faloona, F.A., Mullis, K.B., Horn, G.T., Erlich, H.A., and Arnheim, N. (1985). Enzymatic amplification of beta-globin genomic sequences and restriction site analysis for diagnosis of sickle cell anemia. Science *230*, 1350–1354.

Ursi, J.-P., Ursi, D., Ieven, M., and Pattyn, S.R. (1992). Utility of an internal control for the polymerase chain reaction. Application to detection of Mycoplasma pneumoniae in clinical specimens. APMIS *100*, 635–639.

Wilson, I.G. (1997). Inhibition and facilitation of nucleic acid amplification. Appl.Environ. Microbiol. *63*, 3741–51.

deWitt, D., Wooton, M., Allan, B., and Steyn, L. (1993). Simple method for production of internal control DNA for Mycobacterium tuberculosis polymerase chain reaction assays. J. Clin. Micro. *31*, 2204–2207.

Zimmerman, K., and Mannhalter, J.W. (1996). Technical aspects of quantitative competitive PCR. Biotechniques *21*, 268–279.

Zimmerman, K., Rieger, M., Groß, P., Turecek, P.L., and Schwarz, H.P. (2000). Sensitive single-stage PCR using custom synthesised internal controls. Biotechniques *28*, 694–702.

Analysis of mRNA Expression by Real-time PCR

3

Stephen A. Bustin and Tania Nolan

Abstract

The last few years have witnessed the transformation of the real-time, fluorescence-based reverse transcription polymerase chain reaction (RT-qPCR) from an experimental technology into a mainstream scientific tool for the detection and quantification of RNA with an enormous range of uses in basic research, molecular medicine and biotechnology. The continuous improvement of reagents and instruments, combined with the trend towards high throughput and miniaturization, is likely to reinforce that pre-eminence and continue to open up new application areas. Nonetheless, although in principle undoubtedly a straightforward technology, the reliability of RT-qPCR assays depends on a series of sequential steps that include careful experimental design, optimization and validation, which must be implemented pragmatically to obtain meaningful, biologically relevant data.

Introduction

The real-time quantitative reverse transcription polymerase chain reaction (RT-qPCR) (Gibson *et al.*, 1996) offers a unique combination of conceptual simplicity, cost-efficiency and accuracy, making it relevant for every aspect of biological investigation (Kubista *et al.*, 2006). Its homogeneous (closed tube) format, use of fluorescent reporter molecules and lack of requirement for post-PCR handling makes RT-qPCR assays convenient, easily automated and much less prone to contamination (Bustin, 2000; Ginzinger, 2002). This convenience is enhanced by its accuracy, speed and wide dynamic range, making it today's method of choice for any application that aims to detect RNA, assess its abundance or distinguish small sequence changes (Bustin *et al.*, 2005). Consequently RT-qPCR has become the method of choice for acquiring quantitative data in basic research, detecting genetically manipulated organisms (Buh Gasparic *et al.*, 2010) and is increasingly used for routine clinical diagnostic assays (Bustin and Mueller, 2005) and as a standard exploratory tool in pharmaceutical microbiology and drug development (Brazeau, 2004)

Recent developments

RT-qPCR technology continues to evolve in response to the need for ever-more specific, convenient and cost-effective applications, with several recent developments improving assay performance and accuracy as well as simplifying data analysis:

- The introduction of more reliable and labour-saving, yet less expensive reagents, coupled with the launch of affordable, but more robust, qPCR thermal cyclers simplifies reaction assembly and improves flexibility, performance and consistency of qPCR assays. GMP-compliant manufacturing and formulation facilities permit the production of oligonucleotides with rigorous quality control and the documentation required for clinical trials and diagnostic expression analysis.

- The extension of qPCR into digital PCR has unlocked another vast range of applications for the technology. In digital PCR a sample is terminally diluted across a large number of separate PCR reactions and results are recorded at the reaction endpoint. Most will contain no template molecules and give a negative amplification result. A successful amplification is counted as one molecule; hence recording the number of positive PCR reactions allows accurate quantification of the individual template molecules present in the original sample
- Another recent development has been termed qPCR tomography (Sindelka *et al.*, 2008) which combines the use of laser microdissection with RT-qPCR analysis to obtain expression profiles from within single cells.
- Automated workflows (Grisold and Kessler, 2006; Marshall *et al.*, 2007), more reliable multiplexing assays (Westh *et al.*, 2009; Beck *et al.*, 2010), the launch of instruments using 384- and 1536-well formats and of miniaturized qPCR systems (Morrison *et al.*, 2006), including microfluidic digital PCR (Zhang and Xing, 2007), has resulted in the expansion of qPCR technology into high-throughput applications. Miniaturisation offers several potential advantages such as short assay time, low reagent usage and rapid heating/cooling rates, as well as integration of multiple processing modules to further reduce size and power consumption. (Marcus *et al.*, 2006; Neuzil *et al.*, 2006). High-density array-base formats and support for faster assay conditions enable shorter assay times, allowing the parallel screening of hundreds of targets at vastly reduced reagent cost where RT-qPCR can provide an alternative to cDNA microarrays for accurate expression analysis with high sensitivity (Dixon *et al.*, 2009; Wagner and Arany, 2009) or even be used for routine diagnostic services (Gunson *et al.*, 2008). They also allow powerful multiway profiling, where quantification of reporter genes permits the characterization of samples as functions of multiple factors such as time, genetic make-up or drug load (Stahlberg *et al.*, 2008).
- The increased emphasis on high-throughput profiling ties in with the increased understanding for the need for constant and consistent quality qPCR data control. This has resulted in the development of software programs that aim to minimize subjective human input and maximize objective multilevel analysis. The publication of numerous advanced algorithms (Burns *et al.*, 2005; Larionov *et al.*, 2005; Jin *et al.*, 2006; Nordgard *et al.*, 2006; Smith *et al.*, 2007; Batsch *et al.*, 2008; O'Connor and Runquist, 2008; Platts *et al.*, 2008; Rutledge and Stewart, 2008a; Spiess *et al.*, 2008; Verderio *et al.*, 2008a) and their emergent inclusion in the development of more intuitive analysis software helps with assay set-up (Gallup and Ackermann, 2008), project management and data analysis (Hellemans *et al.*, 2007) to allow robust, accurate and biologically relevant quantification.

Quality considerations

Reliable and reproducible RT-qPCR assays require meticulous adherence to standard operating procedures for sample acquisition and handling, as well as the preparation, quality assessment and storage of RNA (Bustin, 2010; Bustin *et al.*, 2009, 2011b; Huggett and Bustin, 2011).

The consistency and reliability of RT-qPCR assays depends on the proper execution of a number of steps, in particular those relating to sample selection, template quality, assay design and data analysis (Nolan *et al.*, 2006a) as well as the correct application of statistical models and methods for data analysis (Fu *et al.*, 2006). If correct procedures are adhered to, results are generally robust, reproducible and accurate. Specifically, the reliability of the RT-qPCR assay is determined by careful consideration of the quality of the RNA template (Fleige and Pfaffl, 2006; Fleige *et al.*, 2006; Nolan *et al.*, 2006b; Vermeulen *et al.*, 2011), the choice of cDNA priming strategy (Bustin and Nolan, 2004) and reverse transcriptase (Stahlberg *et al.*, 2004a,b), the characteristics of the primers (Pattyn *et al.*, 2003, 2006) and the validity of the normalization method (Pfaffl *et al.*, 2002, 2004; Szabo *et al.*, 2002; Vandesompele *et al.*, 2002a; Akilesh *et al.*, 2003; Andersen *et al.*, 2004; Haller *et al.*, 2004; Abruzzo *et al.*, 2005; Kanno *et al.*, 2006). For diagnostic laboratories, external quality assurance programmes have been shown to be useful (Marubini *et al.*, 2004; Raggi *et al.*, 2005; Ramsden *et al.*, 2006; Verderio *et al.*, 2008b; Raymaekers *et al.*, 2009). Not surprisingly, the pervasive penetration of this technology has led to the development of numerous, distinct and frequently divergent experimental protocols that often generate discordant results (Bustin, 2002, 2005; Bustin *et al.*, 2005). Fortunately, the resulting uncertainty has increased the awareness of a need for common guidelines, in particular those relating to quality assessment of every component of the RT-qPCR assay and appropriate data analysis (Nolan *et al.*, 2006a). This clear requirement for improved consistency of gene expression measurements is particularly relevant in relation to human clinical diagnostic assays (Bustin and Mueller, 2005, 2006). Certainly, it is apparent that while qPCR is driving PCR-based innovations, effective guidelines regulating its use are essential for the future of molecular diagnostics within biomedical sciences (Valasek and Repa, 2005; Picard *et al.*, 2006). The MIQE guidelines, discussed in Chapter 11, are a list of recommendations that cover the critical steps that make up a qPCR assay and provide standards that will help make RT-qPCR a more robust and reliable technique.

Sample selection

Optimal sample quality is a prerequisite for generating valid quantitative data. Hence sample selection and collection as well as RNA quality control, discussed below, are critical parameters in test performance and must be optimized (Muller *et al.*, 2004; Benoy *et al.*, 2006). In general, extraction of RNA from tissue culture, blood and serum is relatively straightforward, while there are significant problems associated with the extraction of RNA from solid tissue, faeces, semen, plants and soil samples. A critical consideration concerns the quantification of RNA from complex tissue samples, as these usually contain several different cell types that may express the target RNA(s) at different levels of abundance. This inevitably results in the averaging of the expression of different cell types and the expression profile of a specific cell type may be masked, lost or ascribed to and dismissed as illegitimate transcription. This is particularly relevant when comparing gene expression profiles between normal and cancer tissue since normal cells adjacent to a tumour may be phenotypically normal but genotypically abnormal or exhibit altered gene expression profiles due to their proximity to the tumour. This problem can be addressed by microdissection, in particular laser capture microdissection (Emmert-Buck *et al.*, 1996), and significant differences have been detected in the gene expression profiles of microdissected and bulk tissue samples (Fink *et al.*, 2002; Oji *et al.*, 2003). Incidentally, when dealing with very small amounts of sample, e.g. single cells, or minute amounts of laser capture microdissected fresh/frozen sample, it is best to carry out the RT-qPCR directly from lysed tissue without going through an extraction process (Hartshorn *et al.*, 2005). A further refinement combines RT-qPCR with in situ hybridization on adjacent tissue sections and so allows simultaneous spatiotemporal monitoring and quantification of gene expression changes (Haupt *et al.*, 2006). However, it is essential to realize that quantification from very few cells also poses problems, as different cells of the same type may not express the same set of mRNAs (Peixoto *et al.*, 2004). In addition, it could be expected that identical cells samples would contain identical copy numbers of a target mRNA or more likely a normal distribution however, it has been demonstrated that mRNA distribution is often in a lognormal fashion (Bengtsson *et al.*, 2008). Microdissection can be carried out on either fresh/frozen or archival samples.

Choice of chemistry

The appropriate detection chemistry depends on the application: The use of a double stranded DNA binding dye in the PCR mix, e.g. SYBR Green I dye, is the most cost-effective chemistry and is widely used for initial investigations, large scale screening studies and for primer optimization steps. While it may be possible to use legacy assays, it is advisable to adopt a new design that is specifically designed for RT-qPCR. These assays are very simple to design (Bustin *et al.*, 2011a) and there are several web-based options that are freely available, including from commercial companies such as OligoArchitect™ (sigma.com/designmyprobe). Although the primers alone determine the specificity of product detection, information regarding product size and amplicon population is determined from a post amplification melting curve analysis. The introduction of high-resolution melting (HRM) can be used to identify very small differences between targets and is sensitive even to single base differences (Wittwer *et al.*, 2003). Other non-fluorescent probe-based chemistries include Plexor (www.promega.com), LUX primer (www.invitrogen.com) and Snapback primers (Zhou *et al.*, 2008).

At very low (< 1000 copies) target concentration there is a greater probability of non-specific amplification and problems with primer dimer products becoming more pronounced. In this case the use of a probe to detect amplicons may be preferable. The most widely used probe system is the dual-labelled fluorescent hydrolysis probe, which can be synthesized with added locked nucleic acid (LNA; Sigma Custom Products), minor groove binder (MGB; Exiqon) or ZNA modifications (Sigma Custom Products). LNA is a nucleic acid analogue that contains a 2′-O–4′-C methylene bridge, which restricts the flexibility of the ribofuranose ring and locks the structure into a rigid bicyclic formation, conferring enhanced hybridization performance. MGBs are dsDNA-binding agents that are attached to the 3′ end of hydrolysis probes and also stabilize the hybridization reaction. In practice, both modifications allow for the design of shorter probes, which increases design flexibility and specificity. This is particularly useful when detecting single nucleotide polymorphisms or when targeting AT-rich sequences and when targets are highly structured or repetitive. Zipped Nucleic Acids (Paris *et al.*, 2010) consist of a stretch of DNA with the addition of spermine units to the 5′ or 3′ end. The development of intercalating nucleic acids

(Christensen and Pedersen, 2002; Christensen *et al.*, 2004; Filichev *et al.*, 2004) provides additional alternatives for the improvement of assay design.

Amplicons and primers

There is no consensus opinion concerning the best amplicon location to target on a mRNA molecule. Where possible, assays should be designed to span intron/exon boundaries, selecting the longest possible intron, as this avoids signal generation from amplified genomic DNA (gDNA). This is essential when targeting specific splice variants of a mRNA. It is most economical to place the downstream primer across the splice junction, as this allows the use of a single (expensive) probe for the detection of possible splice variants in separate assays and, at the same time, minimizes multiple target amplification in any one assay. However, we have noticed that, very often, it is not possible to design an optimal assay around splice junctions and if the choice is between a poor (i.e. inefficient and/or insensitive) assay designed across a splice junction and an efficient one contained within a single exon, it is better to use the latter. In this case it is advisable to treat the sample with DNase I and essential to add a 'no RT' control to determine the contribution of any gDNA contamination to the mRNA quantification. For some specialist applications, e.g. the quantification of mRNA in a single fertilized egg, or cases where the target is a highly expressed gene from a single copy gDNA sequence, the relative difference in the abundance of gDNA and mRNA is such that any gDNA will not interfere significantly with accurate quantification of mRNA. Since there are so many variables to consider when designing an assay it is critical to ensure that as much information is provided as possible.

The amplicon length should be minimized because this tends to increase amplification efficiency. In addition, small amplicons are more tolerant of reaction conditions because they are more likely to be denatured during the 92–95°C step of the PCR, allowing the probes and primers to compete more effectively for binding to their complementary targets. They also require shorter polymerization times for amplicon replication, making amplification of gDNA contaminants containing the intronic sequences less likely. From a practical perspective it is also easier and cheaper to synthesize a shorter oligonucleotide, when a synthetic sense-strand oligonucleotide is used to generate amplicon-specific standard curves and positive controls. When it is impossible to find an acceptable short amplicon the design parameters relating to amplicon size should be changed little by little, until the smallest acceptable amplicon is found. Whenever possible, the amplicon should be around 100 bp, no more than 150 bp in length when using a probe and up to 250 bp when using SYBR Green I binding dye detection. Amplicons should have a G/C content of 40–60% to ensure complete denaturation during thermal cycling, leading to an efficient reaction. Furthermore, highly GC-rich sequences are susceptible to non-specific interactions that result in non-specific signal in assays utilizing SYBR Green I dye.

Primer selection, especially of the reverse primer used in the RT step, is critical since it affects the sensitivity of the RT-PCR assay (Raengsakulrach *et al.*, 2002). It has also been shown that tissue-specific factors such as polysaccharides or proteins can influence amplification kinetics in a sequence-specific manner and that this affect can be mitigated, in part, by appropriate primer selection (Tichopad *et al.*, 2004). The structure of the RNA target at the primer-binding site must be taken into account, as this affects the accessibility of the target to the primers. Even so-called single-stranded RNA is typically extensively folded and selection of a primer binding site at a position that is normally double stranded can result in a very inefficient assay (Fig. 3.1). For RNA viruses there is the additional problem of different viral serotypes resulting in sequence variability and it may be necessary to use a nested RT-qPCR assay with universal primers that bind to target sequences which are shared by all the serotypes followed by a serotype-specific primer pair (Aradaib, 2009; Zaayman *et al.*, 2009).

The choice of 3′-terminal position is crucial for the control of mispriming during the PCR. Primers with one or more G or C residues at that position will have increased binding efficiency due to the stronger hydrogen bonding of G/C residues. It also helps to improve the efficiency of priming by minimizing any 'breathing' that might occur. However, clamping the 3′ of the primer should be avoided because ambiguous binding of oligonucleotides to the target site may result in misprimed elongation ('slippage effect'). A reasonable compromise is to aim for two or three G/Cs within the five 3′ bases of the primer. Similarly, it is best to avoid long runs of a single base (i.e. more than three or four, especially G or C) since homopolymeric runs can also cause the 'slippage effect'.

Direct repeats in the target sequence should be avoided since stable hybridization to secondary binding sites in repetitive regions results in non-productive binding of primers to non-specific regions of the sequence so that efficiency of DNA amplification and detection decreases.

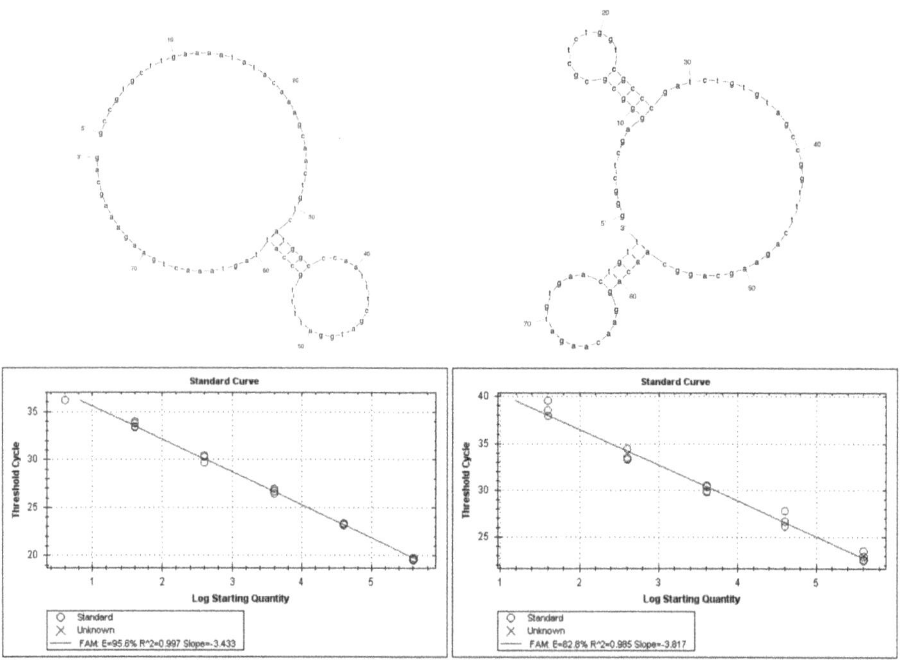

Figure 3.1 Importance of secondary structure at primer binding site. The amplicon on the left has a stem/loop structure within the amplicon, with both upstream and downstream (RT) primers able to hybridize to an open structure. A dilution curve results in an efficient qPCR assay (96%) and tight triplicate replicates (R^2=0.997). The amplicon on the right has stem loop structures at the primer binding sites, resulting in a significantly poorer amplification efficiency (82.8%) and more variation amongst the replicates (R^2=0.985).

Primers must have no intraprimer homology to avoid the formation of secondary structures that can interfere with annealing to the template. They are generally between 15 and 25 bases long when composed entirely of DNA, to maximize specificity, with a G/C content of around 50%. This design contributes to a higher yield of full-length product after synthesis and purification. For primers binding to very AT-rich sequences it is advantageous to substitute one or more of the bases with a locked nucleic acid LNA analogue to reduce the overall primer length while maintaining an appropriate T_m.

Primers should also not have any secondary structure (i.e. inverted repeats) or have sequence complementarity at the 3′ ends that could form dimers. Owing to competition between intermolecular (primer-template, probe-target) hybridization, and intramolecular hybridization, inverse repeats can cause inefficient priming and probing of the target sequence. PCR and/or probe binding can completely fail because of formation of stable hairpins at the binding region, or inside the amplicon in general. Primer dimers have a negative ΔG value, so primers should be chosen with a ΔG value no less negative than −10 kcal/mol.

Quantitative PCR assays using hydrolysis probes are usually carried out as two-step reactions, with a denaturation step followed by a combined annealing/polymerization step, during which fluorescence is measured. Most detection systems rely upon binding of the probe to the template before hybridization of the primers. It is for this reason that the probe T_m should be around 10°C higher than the T_m of the primers, usually in the 68–70°C range. The exception to this is the design of Scorpion primers (Whitcombe *et al.*, 1999) where the probe region hybridizes to the newly formed template and so has a T_m that is the same or 1°C lower than the primers. Probes should be designed with the 5′ end as close to the 3′ end of one of the primers as is possible without them overlapping. This ensures rapid cleavage by the DNA polymerase.

The probe should be no more than 30 nucleotides in length to maximize quenching, with a G/C content of around 50%. If the target sequence is AT-rich, incorporate analogues such as LNA or minor groove binders. Guanine (G) can quench fluorescence and so is avoided as the 5′ base. In this position the G would continue to quench the reporter even after cleavage, resulting in reduced fluorescence values (ΔR), which could result in reduced sensitivity. Any runs of four or more identical nucleotides, especially G, should be avoided because these can cause secondary structure of the probe and reduce hybridization efficiency. Where an alternative sequence cannot be selected, disruption of a series of Gs by the substitution of an inosine can significantly improve probe performance.

Oligonucleotide suppliers are best chosen based on consistent oligo quality, service and support. If using amplicon-specific oligonucleotides to generate standard curves, it is essential to order those oligonucleotides on a different date or to request that these are produced at a different location, so as not to permit any chance of contamination of the primers with that oligonucleotide. It is also advisable to order a number of potential primers for a single target, optimize the assay and then order the probe. This helps to ensure that the best assay is selected. Different batches of primer and probe can generate different results, hence it is essential to verify batch concentrations and compare results obtained with any new batch of oligonucleotide.

Artificially synthesized amplicon target oligonucleotides should be PAGE purified, probes should be HPLC purified, whereas primers can be supplied in a standard desalted or reverse phase cartridge purified format. It is essential to aliquot all oligonucleotides, as this prevents contamination of a whole batch and reduces the need to repeatedly freeze/thaw them. Forward and reverse primers can be stored at 10 μM or mixed together after optimization at the required working stock concentrations. Probes should be stored frozen at −70°C both when lyophilized or in solution (at 5 μM). The probe must be protected from light, but long-term stability is variable, with some probes stable for several years, and others hydrolysed after 6 months.

RNA assessment

The quality of the RNA template is an essential determinant of the reproducibility, reliability and biological relevance of any subsequent RT-qPCR assay. RNA extraction is easiest from fresh, or frozen material, with the main concern relating to the maintenance of RNA integrity. RNA is easily degraded and it is easy to co-purify inhibitors of the RT, or PCR steps, which will generate inaccurate results (Huggett *et al.*, 2008). The diversity of the different sources of mRNA poses a real practical problem: they include not just relatively simple *in vitro* samples such as tissue culture cells, but *in vivo* biopsies obtained during endoscopy, post-surgery, post-mortem and from archival materials. Naked RNA is extremely susceptible to degradation by endogenous ribonucleases (RNases) that are present in all living cells. Therefore, the key to the successful isolation of high-quality RNA and to the reliable and meaningful comparison of qRT-PCR data is to ensure that neither endogenous nor exogenous RNases are introduced during the extraction procedure. Two related problems are that different mRNAs display differential stability and that mRNA expression profiles can change rapidly after cells or tissue samples have been collected, but before they have been frozen. This implies that quantification of a single transcript or determination of relative transcript levels could be seriously affected by the sampling procedure adopted and makes it essential that careful consideration is given to how best to conserve the original expression profile during sample extraction and subsequent processing. This is a particular problem with cells extracted from whole blood (Keilholz, 1998) and mRNA profiles can change over several orders of magnitude even in the short time it can take from collecting the blood to processing it in the laboratory. The main causes of the observed differences are RNA degradation and the induction of certain genes, e.g. Cox-2, as well as the method of RNA preparation (de Vries *et al.*, 2000). Therefore, appropriate sample acquisition has a major influence on the quality of the RNA and subsequently on any result of qRT-PCR assays (Vlems *et al.*, 2002a,b) and once a biological sample has been obtained, immediate stabilization of RNA is the most important consideration (Grotzer *et al.*, 2000; Madejon *et al.*, 2000; Rainen *et al.*, 2002; Muller *et al.*, 2004).

The analysis of RNA extracted from archival material constitutes an exception from the highest quality rule. RNA extracted from formalin-fixed, paraffin-embedded material is a little more difficult to analyse, as extensive degradation can occur before (Mizuno *et al.*, 1998) or during (Klimecki *et al.*, 1994) the formalin fixation process. Furthermore, formalin fixation generates cross-links between nucleic acids and proteins and covalently modifies RNA, making subsequent RNA extraction, reverse transcription and quantification analysis problematic (Masuda *et al.*, 1999). Not surprisingly, fixatives are important (Goldsworthy *et al.*, 1999) and different tissue preparation methods will invariably lead to different results from different laboratories. There are numerous reports that describe the successful extraction and quantification of RNA and miRNA (Doleshal *et al.*, 2008) from formalin-fixed, paraffin-embedded (FFPE) archival materials (Coombs *et al.*, 1999; Godfrey *et al.*, 2000; Specht *et al.*, 2001; Macabeo-Ong *et al.*, 2002; von Smolinski *et al.*, 2005; Yang *et al.*, 2006; Li *et al.*, 2008). Formalin-fixed tissues are ideal for retrospective clinical studies of disease mechanisms and as the use of PCR technology has become more common in molecular testing, it has enhanced the clinical utility of these tissues. This is important, since such studies have the potential to enable the correlation of molecular findings with the response of the patient to treatment and eventual clinical outcome. RT-qPCR assays, with amplicon lengths of below 100 bp, are ideally placed to amplify the usually degraded RNA from these archival samples whose average size is 200–250 nucleotides. However, care must be taken when interpreting the results obtained from archival material, as gene expression profiles from FFPE samples do not exactly correlate with the profiles generated from the corresponding frozen samples; for example, in a study demonstrating that 64 genes were differentially expressed in matching fresh-frozen normal colon and cancer samples, only 38 genes were identified with differential expression in the corresponding FFPE samples. Furthermore, only 28 of these genes were in common (Bibikova *et al.*, 2004a). Thus, any results obtained using FFPE samples require independent experimental determination (Bibikova *et al.*, 2004b), but also may underestimate or report misleading changes in gene expression patterns. On the other hand, variation appears to be constant for each assay, allowing a standardized assessment of RT-qPCR assay efficiencies and the calculation of correction factors. These have the potential to make FFPE-derived expression data comparable to those obtained from fresh material as well as among each other (Koch *et al.*, 2006).

Just as real-time chemistries have revolutionized the PCR reaction, so the introduction of the Bioanalyser 2100 (Agilent), Experion (Biorad) and Lab901 (Agilent) has revolutionized the ability to analyse the quality of RNA samples. These instruments are rapidly becoming the tools of choice for analysing RNA preparations destined to be used in RT-qPCR assays. All three systems use small bench top instruments that integrate sample detection, quantification and quality assessment. They do this by using a combination of microfluidics, capillary electrophoresis and fluorescent dyes. Prior to RNA quality analysis on the BioRad or Agilent systems, a chip containing an interconnected network of microchannels fabricated in glass is filled with a sieving polymer and fluorescence dye. Following chip preparation, either 12 (NanoChip kit) or 11 (PicoChip kit), sample wells together with a marker ladder, are filled and the chip becomes an integrated electrical circuit. The chip is placed into its appropriate instrument and electrodes fit into the wells of the chip. A voltage gradient is applied and, similar to slab gel electrophoresis, RNA molecules are separated by size. During the chip run, a fluorescent dye intercalates directly with the RNA and all bands pass a detector and are detected by laser-induced fluorescence. RNA ladders containing six RNA fragments are used as a reference for data analysis and the unknown samples are compared to the ladder fragments to determine the concentration of the unknown samples and to identify the rRNA peaks. The main differences with the Lab901 ScreenTape System are that it consists of 16 individual separation channels that have been preloaded with reagents and that one RNA sample can be analysed at a time. There is broad correlation between the different formats, although the Lab901 system gave better indications of performance classification for some of the criteria measured (Wilkes *et al.*, 2010). These systems have a wide dynamic range and can quantify as little as 5 ng/µl, but are most accurate when working at concentrations above 50 ng/µl. An electropherogram of fluorescence versus time and a virtual gel image are generated and the software assesses RNA quality by using the areas under the 28S and 18S rRNA peaks to calculate their ratios. However, it is worth noting that, despite sophisticated algorithms, these profiles provide a measure of rRNA integrity, which may or may not reflect the quality of the (largely unmeasured) mRNA. An Agilent Bioanalyzer analysis of 24 colorectal cancer samples is shown in Fig. 3.2.

Accurate delineation of mRNA quality is likely to demand additional quality assessments similar to the 3′:5′ ratios used during microarray experiments. A method using the relative expression of the GAPDH transcript has been described (Nolan *et al.*, 2006a) and modified for the successful, large-scale screening of primary neuroblastoma tumour samples (Vermeulen *et al.*, 2011). The data obtained are independent of ribosomal RNA (rRNA) integrity and provide a quantifiable measure of the degradation of the transcripts of interest.

Although the microfluidic instruments can be used for quantification, more reproducible measurements are achieved by using either the Nanodrop (www.nanodrop.com) system or, for higher throughput applications, the Ribogreen RNA (www.invitrogen.com) staining dye that binds to RNA and produces a fluorescent signal. The concentration of test samples is determined by comparison of the fluorescent signal to that of serial dilutions of controls forming a standard curve series. This can be used in conjunction with a fluorescent plate reader or some qPCR instruments. Note that it is essential that a single method of RNA quantification is selected and that this is used exclusively throughout a study. It has been clearly demonstrated that different methods of quantification give widely differing results (Bustin, 2005).

RNA integrity

RNA quality measurements are defined by both purity (absence of protein and DNA contamination, absence of inhibitors) and integrity. Traditionally RNA quality has been determined by analysis of the A260/A280 ratio and/or analysis of the rRNA bands on agarose gels, an approach updated with the introduction of the Agilent Bioanalyser/Lab901/BioRad Experion microfluidic capillary electrophoresis systems. In a report that claims that it is possible to calculate a more objective measure of RNA quality by measuring several characteristics of the electropherogram generated by the Bioanalyser 2100, the fraction of the area in the region of 18S and 28S RNA, the relative height of the 28S peak, the area containing molecules which are defined as having a 'fast' mobility and marker height, are used to calculate and assign an RNA integrity number (RIN) to each RNA sample (Schroeder *et al.*, 2006). However, an exhaustive analysis of the influence of RNA integrity on RT-qPCR assay performance reaches a different conclusion (Fleige and Pfaffl, 2006; Fleige *et al.*, 2006). The authors extracted RNA samples extracted from numerous tissue types, subjected them to controlled degradation and analysed them using the Agilent Bioanalyser. The samples had RINs between 10 (apparently intact material) and 4.0 (almost no evidence of rRNA bands). The quantity of individual transcripts in each of these samples was then determined using RT-qPCR assays. In some tissues the quantity of the measured transcripts detected was unaffected by the RIN whereas in others there was a linear relationship and in others a threshold response. Critically, the relationship between transcript quantity and RIN was different for different tissues and there was not a predictable relationship between these factors. The authors conclude that moderately degraded RNA samples can be reliably analysed and quantified, as long as the amplicons are kept short (< 250 bp) and expression

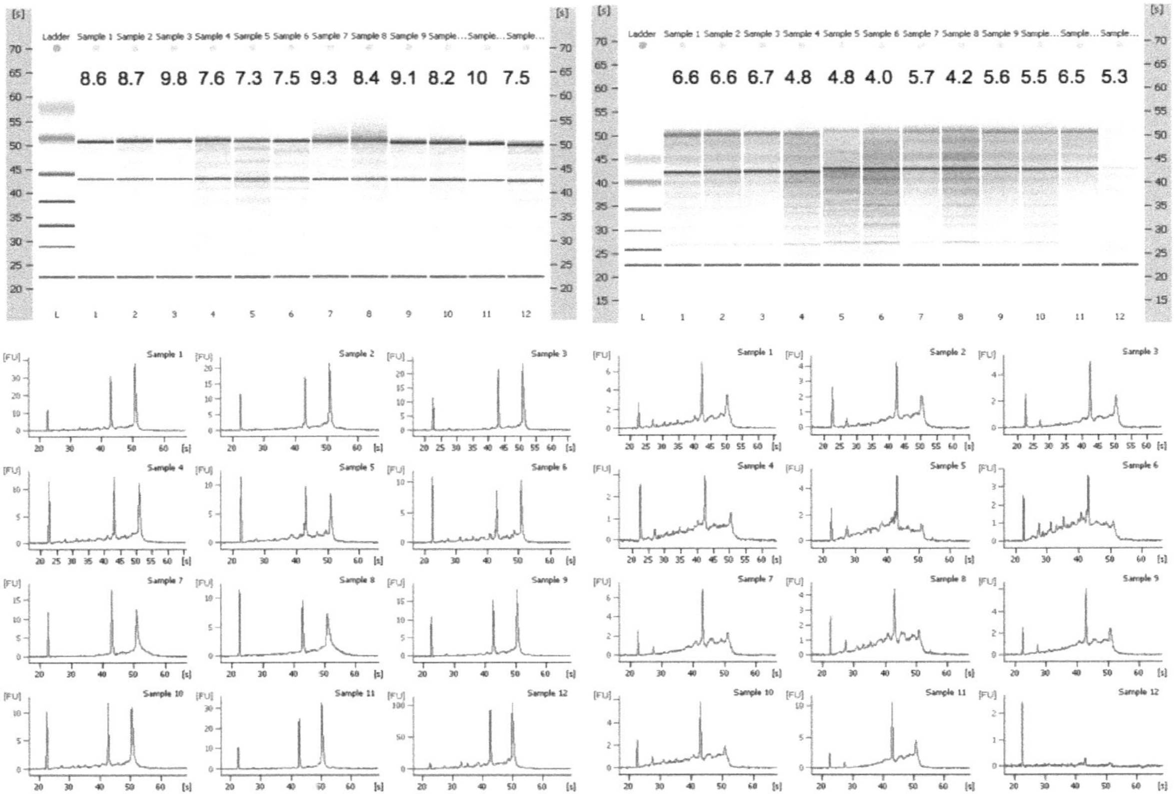

Figure 3.2 Analysis of RNA extracted from colorectal cancer samples using the Agilent 2100 Bioanalyzer. The gel view shows a representation of the RNA, with the 18S and 28S peaks prominent. The samples on the right have additional bands and the 28S band is less distinct, suggesting poorer integrity of the rRNA. The electropherogram uses the distinct differences between the samples to calculate the RIN numbers shown. The amount of small RNA (first peak), which can include 5.8S and 5S ribosomal peaks and transfer RNA, is highly dependent on the preparation protocol.

is normalized against an internal reference. The discrepancy between the two reports could be due to the relatively poor correlation coefficient (0.52) between RIN and expression values of the reference genes reported by the authors advocating the use of RINs. Similarly, a very recent study shows that the microfluidic analysis by the Experion instrument was a less reliable measure of sample quality than measurement of the 3′/5′ ratio of HPRT (Vermeulen *et al.*, 2011).

Inhibitors

Inhibitory components can result in a significant reduction in the sensitivity and kinetics of qPCR (Cone *et al.*, 1992; Liss, 2002; Guy *et al.*, 2003; Perch-Nielsen *et al.*, 2003; Lefevre *et al.*, 2004; Radstrom *et al.*, 2004; Sunen *et al.*, 2004; Jiang *et al.*, 2005; Huggett *et al.*, 2008). Inhibitors may be present in reagents used during nucleic acid extraction or co-purified components from the biological sample, for example bile salts, urea, heparin or IgG. Even the RT enzyme that remains in the cDNA sample and bound to cDNA also has an inhibitory effect on PCR (Chandler *et al.*, 1998). Blood is well-known for containing numerous inhibitors of the PCR reaction (Akane *et al.*, 1994; Fredricks and Relman, 1998; Al Soud and Radstrom, 2001) and we have observed that haem can affect qPCR reading by differentially enhancing FAM signals while reducing HEX fluorescence. The presence of inhibitors can produce inaccurate quantitative results and, at worst, a high degree of inhibition will create false-negative results. The most common procedure used to account for any differences in PCR efficiencies between samples is to amplify a reference gene in parallel to the reporter gene and relate their expression levels. However, this approach assumes that the two assays are inhibited to the same degree and it is clear that this is not the case (Huggett *et al.*, 2008). The problem is even more pronounced in 'absolute' quantification, where an external calibration curve is used to calculate the number of transcripts in the test samples, an approach that is commonly adopted for quantification of pathogens. Some, or all, of the biological samples may contain inhibitors that are not present in the nucleic acid samples used to construct the calibration curve, leading to an underestimation of the mRNA levels in the test samples (Stahlberg *et al.*, 2005). The increasing interest in extracting nucleic acids from formalin-fixed paraffin-embedded (FFPE) archival material will undoubtedly lead to an exacerbation of this problem. Obviously, such inhibitors are likely to distort any comparative quantitative data. However, a survey of practices adopted by qPCR researchers has revealed that only 6% of researchers test their nucleic acid samples for the presence of inhibitors (Bustin, 2005).

Various methods can be used to assess the presence of inhibitors within biological samples. The PCR efficiency in a test sample can be assessed by serial dilution of the sample (Stahlberg *et al.*, 2003), although this is impossible when using very small amounts of RNA extracted, for example, from single cells or from laser capture microdissected sections. When a large number of samples are to be analysed this approach becomes impractical. Furthermore, there are mathematical algorithms that provide a measure of PCR efficiency from analysis of the amplification response curves (Liu and Saint, 2002a,b; Ramakers *et al.*, 2003; Tichopad *et al.*, 2003; Rutledge, 2004; Rutledge and Stewart, 2008a,b) although these are generally suited to SYBR Green I dye based assays and are generally less effected when using probe based systems. We have observed that standard curve measurements are a simple and effective means for identification of samples that contain inhibitors, since both changes in efficiency (gradient) and shifts to higher cycles (efficiency is maintained but delays in initial amplification causes reduced sensitivity) are identified. Internal amplification controls (IAC) that co-purify and co-amplify with the target nucleic acid can be used to detect inhibitors as well as indicate template loss during processing (Pasloske *et al.*, 1998). Another approach utilizes a whole bacterial genome to detect inhibition from clinical samples (Cloud *et al.*, 2003). As an alternative, we have developed a method, which we have called the 'SPUD' assay, that is suitable for the detection of inhibitors in nucleic acids extracted from all tissue with the exception of potato (Nolan *et al.*, 2006b). This is a simple qPCR based assay that is amenable to high-throughput screening consisting of a predetermined, universal template sequence (SPUD template, based on a potato sequence), primers and the SPUD detecting probe. The sequences have been designed such that they are compatible with studies on all organisms (with the exception of potato). In the absence of RNA (or cDNA) sample template amplification of the SPUD target results in a quantification cycle, *Cq* (Bustin *et al.*, 2009) that reflects the optimal reaction conditions for the chosen SPUD template concentration. When RNA (or cDNA) sample is added, any observed deviation from this control *Cq* value indicates that the presence of some component within the sample has inhibited the qPCR reaction. The method can also be used with an artificial RNA template to monitor both the RT and qPCR steps.

Different polymerases react differently to inhibitors and it is worthwhile checking each template preparation for inhibition by testing several polymerases for their efficiency at amplifying the template. Sequence-dependent PCR amplification bias has been observed and results in incorrect or ambiguous quantification (Shanmugam *et al.*, 1993) and can even result in the preferential amplification of one allele over another (Weissensteiner and Lanchbury, 1996). AmpliTaq Gold and the *Taq* polymerases are totally inhibited in the presence of 0.004% (v/v) blood, whereas HotTub (*Tfl*) from *Thermus flavus, Pwo,* r*Tth,* and *Tfl* DNA polymerases are able to amplify DNA in the presence of 20% (v/v) blood without reduced amplification sensitivity; the DNA polymerase from *Thermotoga maritima* (Ultma) appears to be the most susceptible to a wide range of PCR inhibitors. HotTub and *Tth* are the most resistant to the inhibitory effect of K^+ and Na^+ (Al Soud and Radstrom, 1998) and biological samples (Poddar *et al.*, 1998). Thus, the PCR-inhibiting effect of various components in biological samples can, to some extent, be eliminated by the use of the appropriate thermostable DNA polymerase. One thing to bear in mind is that some enzymes are in fact enzyme mixes that combine a highly processive enzyme with a proof-reading enzyme to balance fidelity and yield (Barnes, 1994).

Reverse transcription

The most appropriate method for cDNA priming remains a contentious issue. Direct comparison of the different methods available reveals that there is no one universally best method, and that results are target and enzyme-dependent (Bustin and Nolan, 2004; Stahlberg *et al.*, 2004a,b; Tichopad *et al.*, 2009). Random or oligo-dT priming both allow a representative pool of cDNA to be produced during a single reaction. However, it has been shown that RT priming using random hexamer primers does not necessarily result in equal efficiencies of reverse transcription for all targets in the sample and that there is not a linear correlation between input target amount and cDNA yield when specific targets are measured (Bustin and Nolan, 2004; Lacey *et al.*, 2005). The linearity of cDNA synthesis may be restored by addition of PEG or linearized acrylamide (Gaillard *et al.*, 1999; Stahlberg *et al.*, 2004a) although we have observed that even after such additions, not all genes are detected at the appropriate dilution from the same cDNA series. A comparison of the efficiency of RT priming by random primers of varying lengths showed that 15-nucleotide-long random oligonucleotides consistently yielded at least twice the amount of cDNA as random hexamers (Stangegaard *et al.*, 2006). The 15-mers were more efficient at priming, resulting in reverse transcription of > 80%, while random hexamers induced reverse transcription of only 40% of the template. Not surprisingly, this resulted in the detection of one order of magnitude more genes in whole transcriptome DNA microarray experiments. Since the authors did not address the question of linearity it remains to be seen whether this change overcomes that particular limitation of the random hexamer approach. Unfortunately, we have not observed the same beneficial effect and found that specific priming, followed by random nonomers resulted in the most sensitive detection of targets from colon cancer RNA samples, with 15-mers the least efficient primers (Fig. 3.3).

The choice of oligo dT primers requires the use of non-degraded RNA. Even when intact mRNA is used, the cDNA molecules may be truncated, since the RT enzyme cannot proceed efficiently through highly structured regions; accordingly assays should be targeted towards the 3′ end of the transcript. This is an unsuitable choice for experiments that require examination of splice variants, sequences with long 3′ UTR regions or those without polyA sequences. Furthermore, oligo-dT priming is not recommended when using RNA extracted

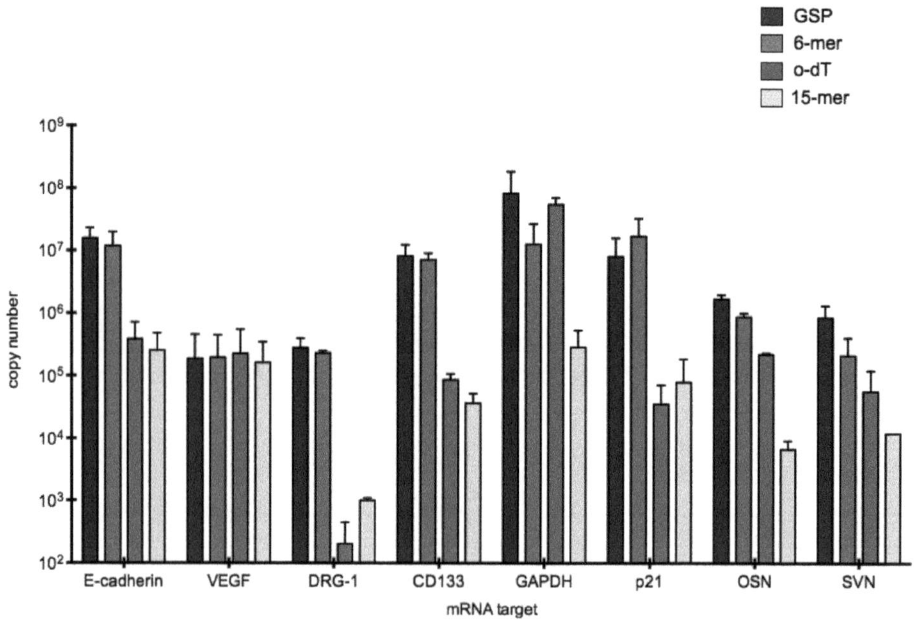

Figure 3.3 Choice of priming strategy affects RT-qPCR results. Four priming strategies were compared for the reverse transcription of eight target mRNAs. The cDNA from each was analysed using qPCR. In each case, gene-specific priming resulted in the most efficient reverse transcription, although the use of random hexamers generated very similar results. Reverse transcription from random 15-mers was consistently the least efficient priming method. Oligo-dT priming was less consistent, resulting in greater variability of quantification and relative quantity between targets. A colour version of this figure is available in the plate section at the back of the book.

from paraffin tissue sections, since formalin fixation results in the loss of the polyA tails on mRNA (Lewis and Maughan, 2004).

Target specific primers are the most unambiguous method for converting mRNA into cDNA (Lekanne Deprez *et al.*, 2002) and are the recommended choice when RNA quantity is not a limiting factor. In a comparison of priming methods, well-designed specific primers were also the most sensitive. Interestingly, it has been reported that specific primers can be applied in tandem for the reliable and specific amplification of 72 genes from limiting amounts of RNA using a multiplexed tandem PCR approach (Stanley and Szewczuk, 2005). Nevertheless, as with random priming there may be differences in the efficiencies at which individual RT reactions occur. These variations must be controlled for by reference of unknown samples to a calibrator sample (when using $\Delta\Delta Cq$ analysis) or to standard curves. Specific priming of RNA dilutions results in a linear response of target cDNA yield (Bustin and Nolan, 2004) and so a further advantage of using this priming method is that the efficiency of the combined RT-qPCR reaction can be confirmed by analysis of the gradient of the standard curve. The inclusion of a calibrator sample or standard curve on every plate is important since these serve as controls for measuring inter-assay variability that may occur when multiple samples are run on different plates.

Fortunately, whereas the RT step is highly variable, the qPCR segment of the assay is remarkably reproducible when run under optimal conditions (Kubista *et al.*, 2006), although unsurprisingly, greater variability is observed between replicates when the assay is run under suboptimal conditions (Hilscher *et al.*, 2005).

Choice of enzymes

The RT-qPCR assay can be performed either as a single RT and PCR enzyme procedure in a single tube or as a separate RT and PCR enzyme procedure in a single or a two tube format.

Single RT and PCR enzyme

A single enzyme such as *Tth* polymerase is able to function both as an RNA- and DNA-dependent DNA polymerase and can be used in a single tube without secondary additions to the reaction mix. The main advantages of the single enzyme approach are the reduced hands-on time and potential for contamination. There are several disadvantages:

1. Since all reagents are added to the reaction tube at the beginning of the reaction, separate optimization of the two reactions is not feasible.
2. The assay can only be carried out using target-specific primers.
3. The assay may be less sensitive, due possibly to the less efficient RT activity of *Tth* polymerase when compared with other RT enzymes (Cusi *et al.*, 1994; Easton *et al.*, 1994).
4. The most thorough study comparing the two procedures found that the one-step reaction was characterized by extensive accumulation of primer dimers, which could obscure the true results in quantitative assays (Vandesompele *et al.*, 2002b).

Two enzyme procedures: separate RT and PCR enzymes

Reactions can be either 'uncoupled' or 'coupled'. In the 'uncoupled' alternative the reverse transcriptase is used to synthesize cDNA in a first tube, under optimal conditions, using random, oligo-dT or sequence-specific primers. An aliquot of the RT reaction is then transferred to another tube (containing the thermostable DNA polymerase, DNA polymerase buffer and PCR primers) and PCR is then carried out under optimal conditions for the DNA polymerase. In the 'continuous or coupled' alternative the reverse transcriptase is used to synthesize cDNA in the presence of high concentrations of dNTPs and either target-specific or oligo-dT primers. Following the RT reaction, PCR buffer (without Mg^{2+}), a thermostable DNA polymerase and target-specific primers are added and the PCR is performed in the same tube. Interassay variation of two enzyme protocols can be very small when carried out properly, with correlation coefficients ranging between 0.974 and 0.988 (Vandesompele *et al.*, 2002b). The disadvantages of this approach are:

1. In two-enzyme/one-tube assays, a template switching activity of viral RTs can generate artefacts during transcription (Mader *et al.*, 2001).
2. Two-enzyme/two-tube reactions present additional opportunities for contamination.
3. The RT can inhibit the PCR assay, resulting in reduced amplification efficiency and an underestimation of target quantity (Suslov and Steindler, 2005).

On balance, the increased flexibility and potential for optimization makes the use of a two enzyme protocol preferable to the single enzyme procedure.

PCR

When selecting a PCR enzyme, there are two important considerations; first, an absolute requirement for a thermostable polymerase with a 5′–3′ nuclease activity when carrying out 5′-nuclease assays using hydrolysis probes (TaqMan) and second, selection of a hot start *Taq* polymerase, either complexed with an antibody or chemically modified to inhibit polymerase activity at low temperatures. This is particularly important when targeting low copy number targets using SYBR Green I dye chemistry. We have used enzymes and kits from all the major suppliers and have found no major, consistent differences in quality but note that antibody inactivated enzymes require a shorter period of activation. Again, it is important to use the buffer supplied with the enzyme, since the manufacturer has optimized it for that enzyme. Some manufacturers include PCR enhancer solutions in their kits. It is important to read the instructions carefully, since they can affect the reaction conditions. For example, if designed for amplification of G/C-rich regions, they may lower the DNA melting temperature, thus lowering the primer annealing temperature. Combined reagent mastermixes are increasingly popular. The advantage of using these is that there is a reduction in potential errors that could be introduced when combining components prior to reactions. In many cases these buffers are stabilized with components that will also disrupt secondary structures in templates and so can give improved results. In the case of difficult templates, a sequence specific effect with specific reagent blends will be observed and so trying a different blend may alleviate problems. We have also found that the 'fast' reagents offered by all suppliers are better at amplifying difficult templates or from samples showing inhibition. Blended buffers are provided in a convenient format and are suitable for routine or high-throughput assays. The disadvantage is that there is less flexibility for optimization of challenging simplex or multiplex assays. It is important to check the mixture components with particular attention to matching the $MgCl_2$ concentration to the detection chemistry used and to the presence and concentration of optical normalization reference dyes, such as ROX.

Protocols

PCR optimization

The thermodynamic stability (ΔG) of a duplexed primer/target structure differs for different primers and varies with primer concentration. Therefore, it is important to use primers at concentrations that result in optimal hybridization and priming. Although optimization of PCR reactions used to be an essential part of assay development, recent trends towards high throughput and rapid data reporting have resulted in the elimination of this step in many laboratories. Indeed, some manufacturers claim it is not necessary to optimize primer concentrations when using their particular mastermixes. However, it is clear that the rationale underlying the original recommendations remain valid and that optimization of primer concentration can significantly improve assay detection sensitivity (Fig. 3.4).

The protocol below utilizes SYBR Green I dye to detect amplified products and to visualize melting of the amplicons after the reaction. Post-amplification melting is performed by gradually increasing the temperature of the reaction and assessing the fluorescence due to amplicon-bound dye at each temperature. As the temperature is increased the fluorescence decreases due to separation of the double strands of the DNA. The temperature at which this occurs depends on amplicon size and sequence. In this way the melting characteristic of primer dimer and specific products can be distinguished, since primer dimers are generally shorter and therefore melt at lower temperatures than longer, target amplicon products (Fig. 3.5).

1. Selection of target sample to use during optimization. When using an artificial amplicon or linearized plasmid as the template to determine the PCR efficiency prepare a stock at around 10^4–10^5 copies/µl. While working with the concentrated, artificial template is it absolutely critical to find a working area that is well away from the qPCR set up area and to take all precautions necessary to ensure that reaction contamination does not occur. Test the amplification of the stock template using 200 nM of each primer to establish that amplification of the sample results in *Cq* values of close to 20. If the concentration is outside of this range assume that a 10-fold difference in concentration corresponds to 3.323 cycles and adjust the sample concentrations accordingly. Alternatively, if cDNA synthesized from universal RNA is to be used, prepare 6–8 fivefold serial dilutions. Test the amplification of the dilution series of template using 200 nM primers and select the diluted sample that results in *Cq* values closest to 20.
2. Prepare individual working dilutions of forward and reverse primers for the optimization matrix. Prepare 20 µl of stocks of 0.5 µM, 1 µM, 3 µM, 6 µM and 9 µM each primer.
3. Prepare a matrix of primer concentrations by adding 2.5 µl of the stock concentration stated in the appropriate column of primer to two qPCR reaction tubes, such that each reaction is prepared in duplicate. Hint: Refer to Table 3.1 for the final concentrations of primers. Concentrations are stated as forward primer/reverse primer in nM in the final qPCR reaction.

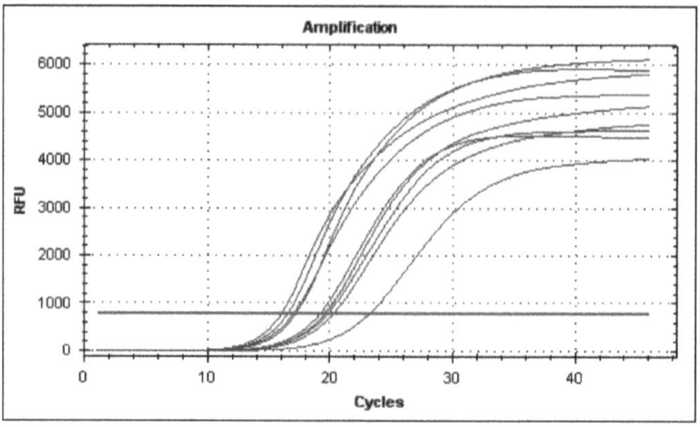

Figure 3.4 The difference in primer binding efficiency affects PCR kinetics. The amplification plots were generated by amplifying the same target using different primer concentrations between 100nM and 300nM as described in Table 3.1. In this example, the difference between the most and least efficient primer concentration combinations is 7 *Cqs*.

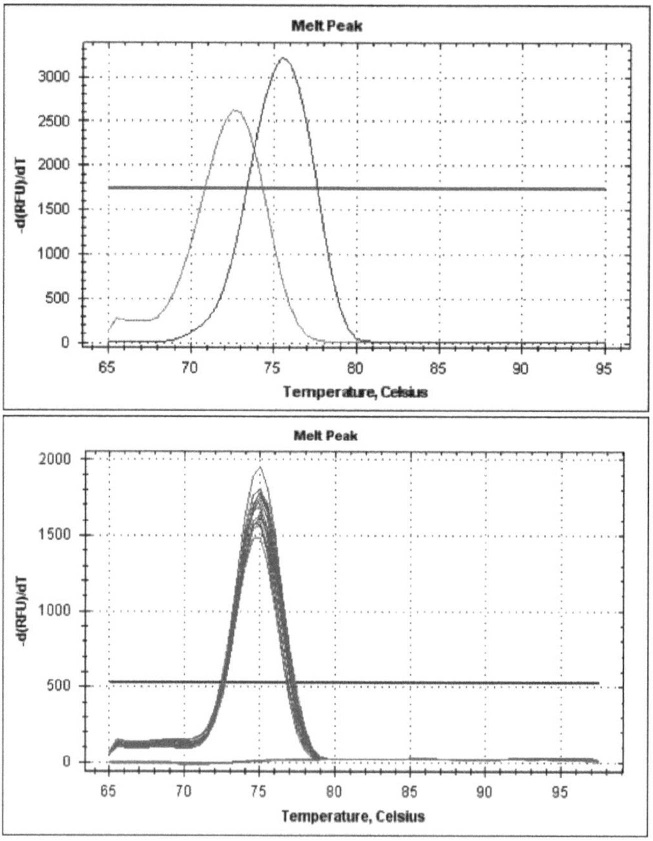

Figure 3.5 (A) SYBR Green I dye melt curve demonstrating the appearance of primer dimers in a poorly designed assay. The target amplicon has a T_m of around 76°C (right), whereas the primer dimers melt at 72°C (left). (B) SYBR Green I dye melt curve demonstrating the absence of primer dimers in a well-designed assay. The amplicon T_m is 75°C.

Table 3.1 Forward and reverse primer concentrations for optimization assay

Reverse primer (μM)	Forward primer (μM)				
	0.5	1	2	3	6
0.5	50/50	100/50	200/50	300/50	600/50
1	50/100	100/100	200/100	300/100	600/100
2	50/200	100/200	200/200	300/200	600/200
3	50/300	100/300	200/300	300/300	600/300
6	50/600	100/600	200/600	300/600	600/600

4 Prepare a mastermix for 55 reactions according to Table 3.2A.

Table 3.2A qPCR mastermix for primer optimization assay

	Stock concentration	Volume to add for single 25μl reaction
Water		2.125μl
SYBR Green I Dye Mastermix (qPCR)	2×	12.5μl
Reference dye (ROX)[a]	0.03μM (optional depending on instrument)	0.375μl
Sample	10^4 to 10^5 copies in 5μl diluted cDNA synthesis reaction or alternative template from step 1	5μl

[a]Note that the given protocol includes the use of the ROX reference dye; if this is not required, adjust water volume accordingly.

5 Add 20 μl mastermix to each primer pair in qPCR tubes. Mix, avoiding bubbles and briefly spin reaction tubes to ensure components are in the bottom of the tubes.

6 Repeat the set up but use a mastermix that does not contain template to create no template controls for each primer concentration.

7 Run in qPCR instrument using three step protocol as shown in Table 3.2B.

Table 3.2B qPCR three-step protocol

1 cycle	Activation	95°C, 2–5min, depending on mastermix
40 cycles	Denaturation	95°C, 30s
	Annealing/	60°C, 30s (collect data during each cycle)
	Extension	72°C, 30s
1 cycle	Melt curve	Between 55°C and 95 C (collect data at each temperature)

8 Refer to the data analysis section for general guidelines. Select the primer combination that fulfils the following criteria in the given order of priority:

9 Absence of primer dimers; use the melt curve profile.

10 Lowest *Cq*: This determines the primer combination resulting in the most sensitive (and efficient) reaction.

11 Highest end point fluorescence (the highest dRn). This is likely to signify the highest number of amplicon products being formed.
12 Absence of signal in the NTC or high $Cq > 35$.

Determination of assay sensitivity and reproducibility

1 Prepare a 10-fold serial dilution of the template described in step 1 of the PCR optimization above, using a minimum of five samples between 10^7 and 10^3 copies in 5 µl. It is possible to quantify a dynamic range of 10–12 logs on most systems (i.e. from 1 to between 10^{10} and 10^{12} target copies). Ensure that the range spans the *Cq* values anticipated for the experimental analysis, repeating the dilution series using different concentrations if required.
2 Add 5 µl of each dilution point to qPCR reaction tubes, including each sample in duplicate.
3 Prepare a qPCR mastermix according to Table 3.3A using the primer concentrations determined from the primer concentration optimization process.

Table 3.3A qPCR mastermix for determination of assay efficiency and sensitivity

	Stock concentration	Volume (µl) to add for single 25µl reaction
Water		2.125µl
SYBR Green I Dye Mastermix (qPCR)	2×	12.5µl
Reference dye (ROX)[a]	0.03µM (optional depending on instrument)	0.375µl
Primers	As determined during primer optimization process	5µl

[a]Note that the given protocol includes the use of the ROX reference dye, determine whether this is required for the instrument and omit, replacing with water, if desired.

4 Add 20 µl mastermix to each sample in qPCR tubes. Mix, avoiding bubbles and spin places briefly to gather mix in the bottom of the tubes.
5 Run in qPCR instrument using three-step protocol as shown in Table 3.3B.

Table 3.3B qPCR three-step protocol

1 cycle	Activation	95°C 10min (or 2min verify enzyme hot start requirement)
40 cycles	Denaturation	95°C 30s
	Annealing/	60°C 30s (collect data during each cycle)
	Extension	72°C 30s
1 cycle	Melt curve	Between 55°C and 95°C (collect data at each temperature)

6 Follow the steps outlined in the data analysis section.

7 Use the instrument software to plot *Cq* against log target concentration and determine the slope and R^2 values. Confirm that the slope lays between −3.2 and −3.5 and that the R^2 value is above 0.98, preferably above 0.99 as described for standard curve analysis.

Perform the qPCR as soon as possible after preparing the reaction mixes. Some assays are stable for 12 hours if stored at 4°C overnight; however others are more sensitive to storage. Reactions containing glycerol and DMSO or made from commercial master mixes can be stored at −20°C.

RT-qPCR assay

Reverse transcription

See flow chart for a summary of the steps (Fig. 3.6).

Variation between RT experiments

When using random primers, the efficiency of the RT reaction may vary between reactions. It also appears that the RT efficiency for different targets within the same reaction is variable (Lacey *et al.*, 2005) and it is important to control for these factors by following the steps below:

- When possible perform RT reactions on all samples at the same time.
- Use the same amount of RNA in all reactions.

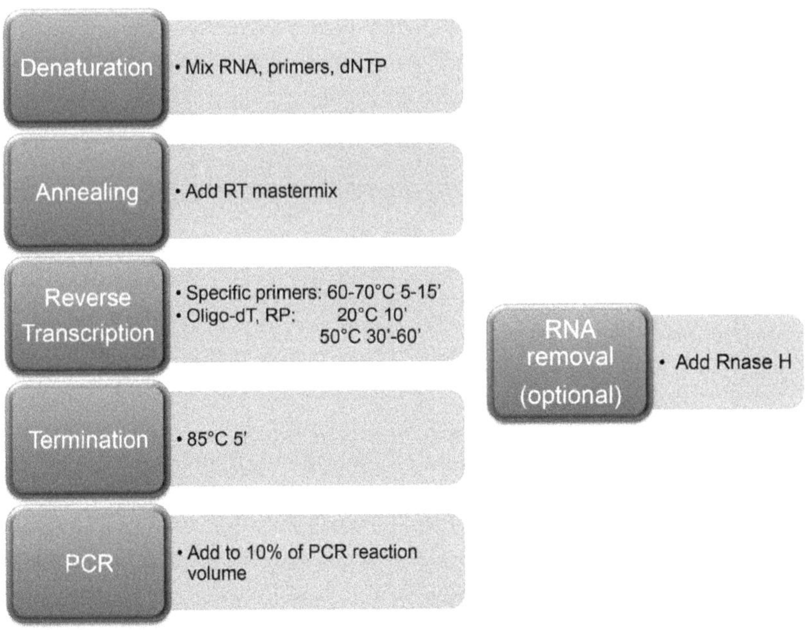

Figure 3.6 Outline of reverse transcription protocol.

- When multiple samples are processed in different batches, include a common positive reference sample alongside all batches.
- Quantify target genes relative to a standard curve including the target copy number in the positive control samples from each batch.
- Express quantity of target in samples relative to the quantity measured in the control sample included in the same RT batch (Lacey *et al.*, 2005).

1 Keep reagents and microfuge tubes on ice. Mix and briefly spin each component and prepare a mastermix in a microfuge tube by combining the reagents in the order listed in Table 3.4A. Note that this protocol is adapted for use with specific primers, random primers or oligo-dT priming. In some cases a combination of oligo dT and random primers may be appropriate.

Critical: Prepare no-RT controls in duplicate: To prepare each no-RT control, add 18 µl mastermix to 7 µl water prior to adding RT enzyme to the mastermix.

Table 3.4A RNA/primer mix components

Component	Volume per reaction
Total RNA 1–200ng (max. 500ng)	× (max 7µl)
3′ (antisense) primer (2µM) or	1µl
Random 15-mer (50ng/µl) *or*	1µl
Oligo-dT (500ng/µl)	1µl
Water	to make a total volume 10µl

2 Incubate at 65°C for 10 minutes, then snap cool on ice for 5 minutes.

3 Prepare an RT mastermix for the required number of reactions by combining the reagents as listed in Table 3.4B.

Table 3.4B RT mastermix components

Component	Volume per reaction
10×RT buffer	2.5µl
$MgCl_2$ (25µM)	5µl
DTT (0.1M)	2.5µl
RT enzyme (200U/µl)	1µl
Water	to 15µl

4 Add 15 µl of the RT mastermix to the RNA/primer mix. Gently mix the tube contents and briefly spin.

4a *Specific priming*: incubate at 60°C for 15 minutes.

4b Oligo dT priming or random priming: incubate at 20°C for 10 minutes followed by 50°C for 60 minutes.

5 Terminate the reaction at 85°C for 5 minutes, and then place on ice. Collect by brief centrifugation.

6 Optional step: add 1 µl (2 U) of *E. coli* RNase H and incubate at 37°C for 15 minutes.
7 Store at −20°C or proceed to qPCR.

PCR

Option A: qPCR using SYBR Green I dye detection

1a Make a master mix by adding the reagents in the order shown in Table 3.5A, mix gently by repeatedly pipetting up and down making sure that there are no bubbles, (add DNA polymerase after mastermix production if separate components are being used and mix gently again). Buffer, MgCl2, dNTPs, SYBR Green I dye and DNA polymerase are combined in commercial 2 × mastermix preparations.

Table 3.5A PCR mastermix components (SYBR Green I dye)

Master mix	Final concentration	Volume per reaction (µl)
dH_2O		6.625[a]
2×buffer	1×	12.5
Stock Primers (or 10µM for a final concentration of 100nM)	Refer to optimization results	(0.25µl each primer) 0.5
ROX diluted (optional)	Refer to instrument manufacturer's recommendations	0.375[a]

[a]Note that this protocol includes the use of the reference dye ROX; if this is not required, adjust water volume accordingly.

2a Add 5 µl template to the reaction tubes or wells.
3a Add 20 µl volume of reaction mix to each tube or well containing template. Cap carefully, spin briefly to ensure there are no bubbles, and place the tubes, strips, rotor or 96/384-well plate into the real time thermal cycler.
4a Perform a three-step PCR reaction according to the thermal profile in Table 3.5B.

Table 3.5B qPCR three-step protocol

1 cycle	Activation	95°C	10min (or 2min verify enzyme hot start requirement)	
40 cycles	Denaturation	95°C	30s	
	Annealing	60°C[b]	30s	Collect data
	Extension	72°C	30s	Collect data
Followed by a melting profile (as below or according to the instrument manufacturer's instructions)				
1 cycle	95°C		1min	Collect data
41 cycles	55°C		30s	
	repeat and increase temperature by 1°C per cycle			Collect data

[b]Annealing temperature is primer design dependent. A two step protocol with combined annealing and extension steps into a single 62°C incubation for 60 seconds can be used when primers are being evaluated to be used with a TaqMan probe.

Option B: qPCR using dual-labelled probe detection

1b. Make a master mix by adding the reagents in the order shown in Table 3.6A, mix gently by repeatedly pipetting up and down, making sure that there are no bubbles. (Add DNA polymerase after preparing the rest of the mastermix, if separate components are being used and mix gently again.)

Table 3.6A PCR mastermix components (TaqMan)

Master mix	Final concentration	Volume per reaction (µl)
dH_2O		5.625[a]
2×buffer	1×	12.5
Stock Primers (or 10µM for 100nM)	Refer to optimization results	(0.25 each primer) 0.5
ROX diluted	Refer to instrument manufacturer's recommendations	0.375[a]
5µM Probe	200nM	1

[a]Note that the given protocol includes the use of the reference dye ROX. For alternative volumes or to exclude, adjust water volume accordingly.

2b Add 5 µl template to the reaction tubes or wells.

3b Add 20 µl of reaction mix to each tube or well containing template. Cap carefully, spin briefly to ensure there are no bubbles, and place the tubes, strips, rotor or 96/384-well plate into the real time thermal cycler.

4b Perform a two step PCR reaction according to the thermal profile outlined in Table 3.6B.

Table 3.6B qPCR two step protocol

1 cycle	Activation	95°C	10min (or 2min verify enzyme hot start requirement)	
40 cycles	Denaturation	95°C	30s	
	Annealing/extension	60°C	1min	Collect data

Note: reactions containing hydrolysis probes are usually performed using a two-step PCR profile. During this reaction the double-stranded template is melted at 95°C and the annealing and extension steps both occur during a single incubation at 60°C. Although this is suboptimal for amplification by *Taq* polymerase it is believed to encourage more efficient cleavage of the internal probe, resulting in maximum fluorescent signal per cycle. Experiments including other probe based chemistries, e.g. Molecular Beacons, Scorpions, can be run under these conditions but a three step protocol is optimal (Table 3.6C).

Table 3.6C qPCR three-step protocol

1 cycle	Activation	95°C	10min (or 2min verify enzyme hot start requirement)	
40 cycles	Denaturation	95°C	30s	
	Annealing	60°C or as appropriate for primers	30s	Collect data
	Extension	72°C	30s	Collect data

Data analysis

The analysis of qPCR data can be highly subjective, since *Cq* values can be altered by changes to baselines and threshold levels. Therefore it is important to understand that *Cq* values, in themselves are not sufficient information. Although software default options can assuage some subjectivity, it is also true that these settings are not always appropriate. Clearly, it is important to understand the analysis procedure in order to be able to generate reliable data. The following steps constitute a basic data analysis workflow:

1 Examine amplification plots of standard or reference samples and identify abnormal plots. A normal amplification plot is defined as consisting of a horizontal baseline region, a log phase of amplification, then a linear amplification phase, followed by a plateau. Note samples that record a *Cq*, but where the amplification plot clearly does not show the appropriate characteristics (Fig. 3.7), these should be subjected to further investigation.

2 The threshold settings are subjective. Select a setting that is within the logarithmic region of amplification. This is best viewed as the linear region of the plot of log normalized fluorescence versus cycle (log dRn vs. cycle). The precise setting should be within a range where all the amplification plots to be analysed are parallel, hence the setting represents cycles during which the amplification efficiencies are equivalent for each reaction to be compared. Select the threshold setting where the replicates are closest to each other.

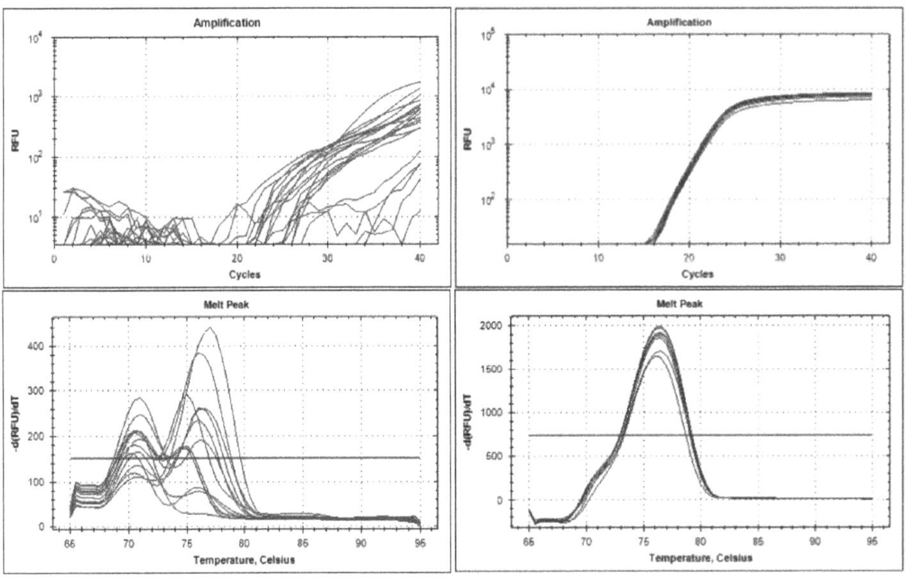

Figure 3.7 Example of problematic amplification reaction resulting in inappropriate *Cq* values. (A) Clearly, there is a problem with the amplification of the unknown samples, as the plots are of an unusual shape. (B) The melt curves are also indistinct, with more than one peak per sample, and different T_m values for the same amplicon. This is often due to the presence of inhibitors in the assay, and a 10-fold dilution of the RNA results in efficient amplification plots (C) and good melt curves (D) on the right.

3 Check slope is between –3.2 and –3.5 and reproducibility of replicates $R^2 > 0.98$.
4 Check controls for expected results.
5 Check melt curve profiles for SYBR Green I dye detection and ensure that products are specific.
6 Where possible, ensure that the *Cq* of the positive control (or dilutions of standard curve) is within one cycle of that recorded during a previous experiment and that the initial and final fluorescence levels are similar.
7 Ensure that all data points to be recorded are within the dynamic range defined by the standard curve.
8 Examine replicates. All replicates should be within 0.5 *Cq* of each other. At low *Cq* the tolerance should be lower than at high *Cq*. Above cycle 35 the variability will be greater and quantification may be unreliable.
9 Ensure a minimum *Cq* difference of 5 between any NTC signal and a sample data point (Bustin and Nolan, 2004) (Fig. 3.8). When using SYBR Green I dye take care with *Cq* values when primer dimer products are evident in the sample after melt curve analysis.

In general, the higher the fluorescence released per cycle, the more sensitive and accurate the measurement.

Normalization

All RT-qPCR assay results are subject to variability caused by technical, as well as biological, variation. It is essential that technical variability is kept at a minimum so as to optimize the chances of identifying biologically relevant changes in mRNA levels. Consequently, data normalization is an essential part of a meaningful RT-qPCR assay and requires standardization. Unfortunately, normalization is a rather problematic area (Vandesompele *et al.*, 2002a; Smith *et al.*, 2003; de Kok *et al.*, 2005; Rubie *et al.*, 2005; Argyropoulos *et al.*, 2006; Kanno *et al.*, 2006; Port *et al.*, 2007; Guenin *et al.*, 2009; Marullo *et al.*, 2010) and, as yet, there is no universally accepted method for data normalization that accounts for all variables encountered during the course of a RT-qPCR experiment. The use of RT-qPCR to quantify miRNAs adds to the concern as to which method is the most appropriate to ensure reliable normalization of non-coding RNAs (Peltier and Latham, 2008; Mestdagh *et al.*, 2009; Chang *et al.*, 2010). One report clearly demonstrates that the use of non-miRNA reference genes has a pronounced effect on the differential expression of tested miRNAs, an effect especially pronounced when the overall abundance of miRNAs varies, e.g. in experiments affecting the miRNA processing machinery, or in comparisons involving multiple tissues (Git *et al.*, 2010). All proposed methods represent a compromise and the selection depends upon the study aims and acceptable tolerance (Huggett *et al.*, 2005).

Normalization against input RNA amount is feasible and, if validated, perfectly acceptable (Bustin, 2002). However, there are problems when comparing tissues that are proliferating at different rates, since they will contain different amounts of rRNA and the mRNA/rRNA ratios are likely to be different. Furthermore, differences in cDNA synthesis are not taken into account and in some situations (e.g. embryology) may generate irrelevant results.

Normalizing to an endogenous reference gene (it is best not to use the term 'housekeeping gene') or genes is the most common method for internally controlling for error in RT-qPCR and is currently the preferred option, as long as it is carried out correctly (Goossens *et al.*, 2005) The inevitable consequence of inappropriate normalization is the acquisition of

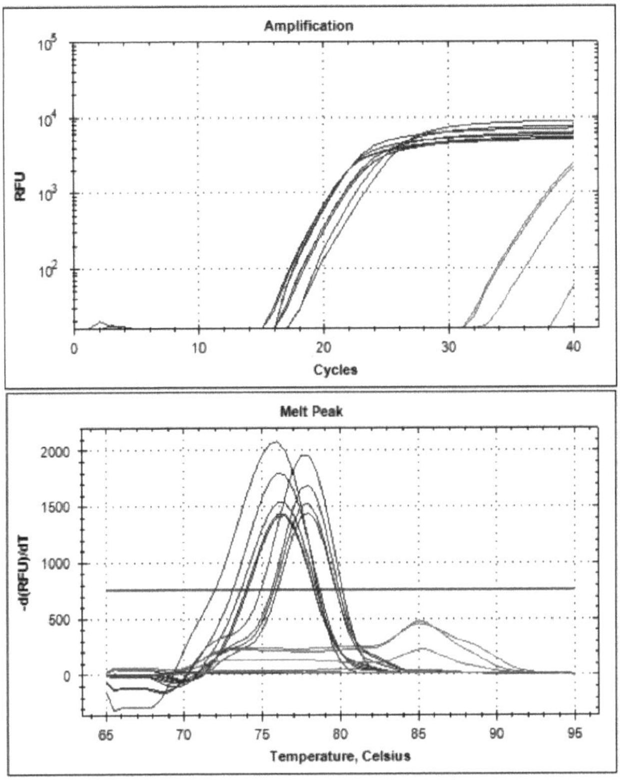

Figure 3.8 Analysis of a SYBR Green I dye reaction with positive NTCs. In this example, the replicate NTCs are recording *Cq* values of >33, whereas the lowest *Cq* generated by the unknowns is at 23, i.e. the d*Cq* is >5. The melt curves also clearly distinguishes between the two amplicons and the NTCs. Hence it is acceptable to use these data for quantification. The positive *Cq*s are caused by primer dimers, and if the aim of the experiment was to quantitate very low abundance targets, it would probably be necessary to redesign the primers.

biologically irrelevant data (Tricarico *et al.*, 2002; Dheda *et al.*, 2004, 2005). When using the endogenous reference strategy target gene levels are expressed relative to those of internal reference genes and so all of the steps required for the final PCR measurement are controlled. The procedure is simplified as both the gene of interest and the reference gene are measured using real-time RT-PCR. However, it is essential that reference gene expression in the target tissue is carefully analysed (Perez-Novo *et al.*, 2005) and the minimum variability is determined and reported (Vandesompele *et al.*, 2002a). The currently accepted method combines the evaluation of a panel of several reference genes together with a method for selecting reference genes with the most stable expression, e.g. GeNorm medgen.ugent.be/~jvdesomp/genorm/or Bestkeeper (Pfaffl *et al.*, 2004). A recent development of this strategy is the suggestion that transcript levels can be normalized to the expression of small interspersed nuclear elements (SINEs) such as expressed Alu repeats in primates (Marullo *et al.*, 2010) or B-elements in mouse models (Huggett, 2011, personal communication). In this system a single assay is designed to target the repetitive element but detection is representative of several genes. Fluctuations in mRNA levels of different genes should result in the overall level of SINEs remaining constant. This appears to be the closest system to a universal normalization method though care must be taken that the model does not cause expression of repetitive elements, such as heat shock responses.

Normalization against gDNA is another option (Kanno *et al.*, 2006), particularly when performing RT-qPCR analyses on RNA obtained using kits such as the Invitrogen CellsDirect kit. The main problems here are (1) differential stability of DNA and RNA may distort quantification; (2) sample cannot be DNase treated; (3), whilst being an internal control, there is no equivalent RT step; and (4) high relative concentrations of gDNA may inhibit the RT-PCR. Finally, it is also possible to normalize against area dissected when using laser capture microdissection, or cell number when extracting RNA from nucleated blood cells.

Standard curve

Analysis of the data from the standard curve can provide a substantial amount of information about the assay. For this reason, initially, assays should be validated on a serial dilution of high quality template, even in situations where data collection will not require reference to a standard curve. The template material used to generate a standard curve should accurately reflect the sample complexity and so this means that it is preferable to use a total RNA preparation, either from the tissue sample under investigation or from a commercially supplied reference RNA, or spiking a known amount of cDNA, plasmid DNA or oligonucleotides into a total RNA solution. Each concentration should be amplified in triplicate to allow a determination of reproducibility. The standard curve is constructed from a measure of *Cq* against log template quantity. It is realistic to expect a linear dynamic range of at least four logs with highly reproducible quantification, with five to nine logs typical (Fig. 3.9). However, it is important to emphasize that if all the unknowns are clustered between *Cqs* of 20 and 30, there is little point in subsequently using a standard curve that is linear all the way from 11 to 35. As with everything else, common sense must be applied to the use of this tool. This defines the working dynamic range for the assay. One measure of assay efficiency is made by comparison of the relative *Cq* values for subsequent dilutions of sample. The efficiency of the reaction can be calculated by the equation: $E = 10^{(-1/\text{slope})} - 1$. The efficiency of the PCR should be as close to 100% as possible, corresponding to a doubling of the target amplicon at each cycle. Using this measure, an assay of 100% efficiency will result in a standard curve with a gradient of −3.323 (also see http://www.gene-quantification.de/efficiency.html). An optimized assay will result in a standard curve with a slope between −3.2 and −3.5. Reproducibility of the replicate reactions also reflects assay stability, with R^2 values of 0.98 or above and preferably closer to 0.99 being indicative of a stable and reliable assay. The intercept on the *Cq* axis indicates the *Cq* at which a single unit of template concentration would be detected and is therefore an indication of the theoretical sensitivity of the assay.

Fig. 3.9 illustrates a typical result obtained for amplicons detected using SYBR Green I dye. Note that the slopes of all amplification plots are identical, indicating that the amplification efficiencies of every sample are the same. The plateau of one of the plots shows a significantly reduced ΔR_n value, but clearly this does not interfere with the ability to quantify accurately from that sample. The NTCs in the assay on the left are negative, whereas the NTC for the assay on the right shows a positive *Cq* value of 42. This phenomenon is fairly common with SYBR Green I dye based assays and is due to the formation of primer dimers in the absence of template. We would suggest that the presence of the primer dimers is reported, together with the ΔCq between the highest *Cq* recorded by an unknown and

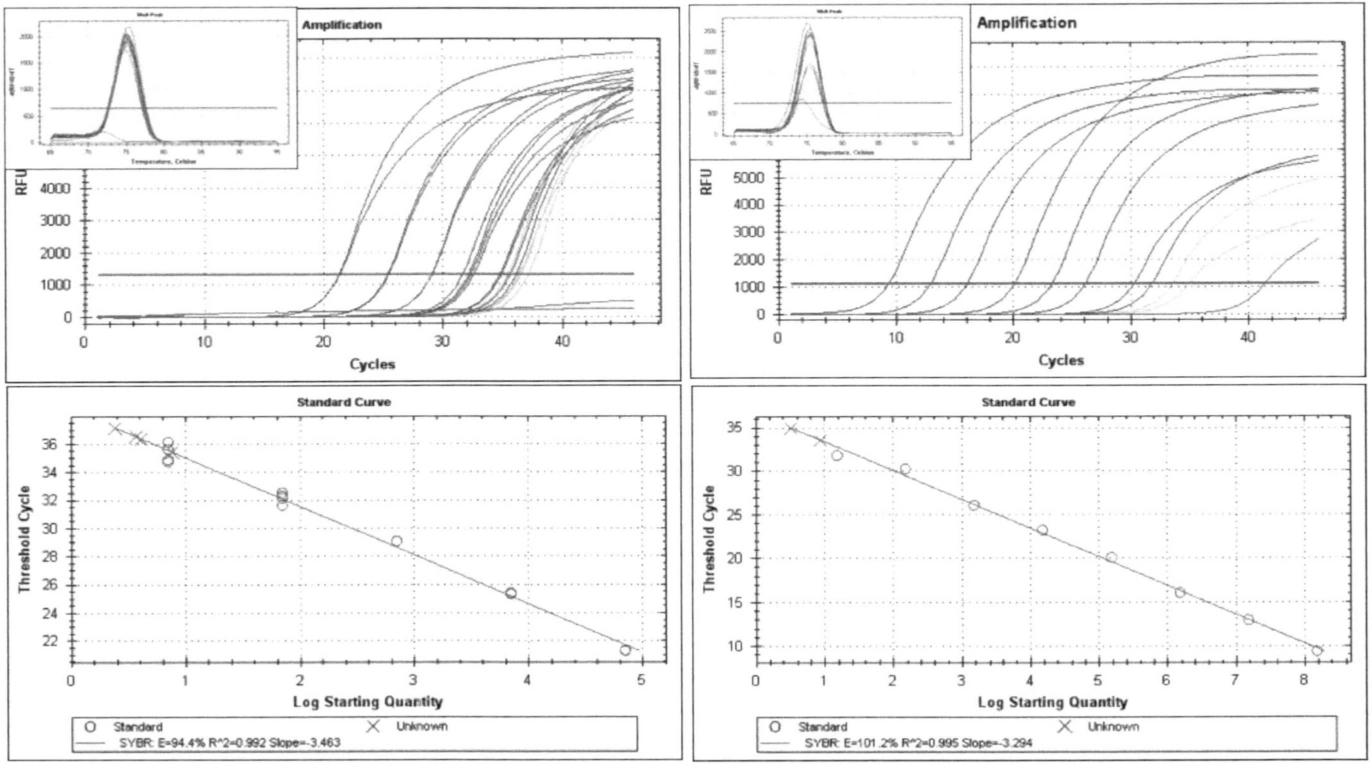

Figure 3.9 Standard curves are generated from a plot of *Cq* against the log copy number of a 10-fold serial dilution of template. They should be linear, with a slope of –3.323 and an R^2 of greater than 0.98 and preferably 0.99. In the example shown on the left, the slope of the SYBR Green I dye assay is –3.463 with an R^2 of 0.992, with a dynamic range of five logs. The NTC is negative even after 45 cycles. The example on the right is another SYBR Green assay that is also within the above specifications. In this case the NTC is positive, but the ΔCq between the lowest copy number standard (eight copies) and the NTC is >5.

the NTC. For probe-based assays, NTCs should always be negative, since the probe will not detect dimers and should not bind to non-specific amplicons.

Conclusions

qPCR is a remarkable technology that has enabled many of the advances made in our understanding of basic biological and disease processes and is increasingly applied to human and veterinary diagnostics, forensics and the detection of genetically modified organisms. Its applications are constantly expanding, posing new opportunities, but also new challenges. Analysis of expression profiles from single cells is one such challenge (Diercks *et al.*, 2009; Stahlberg and Bengtsson, 2010), nanolitre high-throughput RT-qPCR is another (Dixon *et al.*, 2009), proximity ligation assay (PLA) technology is expanding the range of qPCR applications to include the direct detection of proteins through the amplification of a surrogate DNA template after antibody binding (Swartzman *et al.*, 2010), even in single cells (Renfrow *et al.*, 2011) and qPCR is increasingly used to validate digital gene expression tag profiling data obtained using next generation sequencing technology (Wang *et al.*, 2010).

Unfortunately, the combination of ease of use and lack of rigorous standards of practice has resulted in widespread misinterpretation of RT-qPCR data and consequent publication of erroneous conclusions. Hence the need to make PCR-based assays more reliable, a Herculean task that requires both an appreciation and an understanding of numerous attributes that include biological concepts, statistics, mathematical modelling, technical know-how and a willingness to share that information. MIQE, discussed elsewhere (Bustin *et al.*, 2009), addresses a number of these issues. In addition, these problems are being addressed by a next generation of assays that aim to overcome the limitations that are inherent to any RT-PCR reaction. Developments in microfluidics, (Liu *et al.*, 2002), single cell analysis (Pierce and Wangh, 2011), digital PCR (McCaughan and Dear, 2010) and direct analysis of single molecules (Pushkarev *et al.*, 2009) are rapidly approaching a threshold that will make them practical and accessible to the wider scientific community. These advances this will eventually allow accurate and precise quantification without any of the drawbacks inherent in the need to convert RNA to DNA enzymatically, pre-amplify small amounts of target and utilize complex normalization procedures to obtain relative quantification data. That is the future. For now, qRT-PCR assays, when carried out appropriately, continue to be the method of choice for RNA detection and quantification.

References

Abruzzo, L.V., Wang, J., Kapoor, M., Jeffrey Medeiros, L., Keating, M.J., Edward Highsmith, W., Barron, L.L., Cromwell, C.C., and Coombes, K.R. (2005). Biological validation of differentially expressed genes in chronic lymphocytic leukemia identified by applying multiple statistical methods to oligonucleotide microarrays. J. Mol. Diagn. *7*, 337–345.

Abu Al-Soud, W., and Radstrom, P. (1998). Capacity of nine thermostable DNA polymerases To mediate DNA amplification in the presence of PCR-inhibiting samples. Appl. Environ. Microbiol. *64*, 3748–3753.

Akane, A., Matsubara, K., Nakamura, H., Takahashi, S., and Kimura, K. (1994). Identification of the heme compound copurified with deoxyribonucleic acid (DNA) from bloodstains, a major inhibitor of polymerase chain reaction (PCR) amplification. J. Forensic Sci. *39*, 362–372.

Akilesh, S., Shaffer, D.J., and Roopenian, D. (2003). Customized molecular phenotyping by quantitative gene expression and pattern recognition analysis. Genome Res. *13*, 1719–1727.

Al-Soud, W.A., and Radstrom, P. (2001). Purification and characterization of PCR-inhibitory components in blood cells. J. Clin. Microbiol. *39*, 485–493.

Andersen, C.L., Jensen, J.L., and Orntoft, T.F. (2004). Normalization of real-time quantitative reverse transcription-PCR data: a model-based variance estimation approach to identify genes suited for normalization, applied to bladder and colon cancer data sets. Cancer Res. *64*, 5245–5250.

Aradaib, I.E. (2009). PCR detection of African horse sickness virus serogroup based on genome segment three sequence analysis. J. Virol. Methods *159*, 1–5.

Argyropoulos, D., Psallida, C., and Spyropoulos, C.G. (2006). Generic normalization method for real-time PCR. Application for the analysis of the mannanase gene expressed in germinating tomato seed. FEBS J. *273*, 770–777.

Barnes, W.M. (1994). PCR amplification of up to 35-kb DNA with high fidelity and high yield from lambda bacteriophage templates. Proc. Natl. Acad. Sci. U.S.A. *91*, 2216–2220.

Batsch, A., Noetel, A., Fork, C., Urban, A., Lazic, D., Lucas, T., Pietsch, J., Lazar, A., Schomig, E., and Grundemann, D. (2008). Simultaneous fitting of real-time PCR data with efficiency of amplification modeled as Gaussian function of target fluorescence. BMC Bioinformatics *9*, 95.

Beck, E.T., Jurgens, L.A., Kehl, S.C., Bose, M.E., Patitucci, T., LaGue, E., Darga, P., Wilkinson, K., Witt, L.M., Fan, J., *et al.* (2010). Development of a rapid automated influenza A, influenza B, and respiratory syncytial virus A/B multiplex real-time RT-PCR assay and its use during the 2009 H1N1 swine-origin influenza virus epidemic in Milwaukee, Wisconsin. J. Mol. Diagn. *12*, 74–81.

Bengtsson, M., Hemberg, M., Rorsman, P., and Stahlberg, A. (2008). Quantification of mRNA in single cells and modelling of RT-qPCR induced noise. BMC Mol. Biol. *9*, 63.

Benoy, I.H., Elst, H., Van Dam, P., Scharpe, S., Van Marck, E., Vermeulen, P.B., and Dirix, L.Y. (2006). Detection of circulating tumour cells in blood by quantitative real-time RT-PCR: effect of pre-analytical time. Clin. Chem. Lab. Med. *44*, 1082–1087.

Brazeau, D.A. (2004). Combining genome-wide and targeted gene expression profiling in drug discovery: microarrays and real-time PCR. Drug Discov. Today *9*, 838–845.

Buh Gasparic, M., Tengs, T., La Paz, J.L., Holst-Jensen, A., Pla, M., Esteve, T., Zel, J., and Gruden, K. (2010). Comparison of nine different real-time PCR chemistries for qualitative and quantitative applications in GMO detection. Anal. Bioanal. Chem. *396*, 2023–2029.

Burns, M.J., Nixon, G.J., Foy, C.A., and Harris, N. (2005). Standardisation of data from real-time quantitative PCR methods – evaluation of outliers and comparison of calibration curves. BMC Biotechnol. *5*, 31.

Bustin, S.A. (2000). Absolute quantification of mRNA using real-time reverse transcription polymerase chain reaction assays. J. Mol. Endocrinol. *25*, 169–193.

Bustin, S.A. (2002). Quantification of mRNA using real-time reverse transcription PCR (RT-PCR): trends and problems. J. Mol. Endocrinol. *29*, 23–39.

Bustin, S.A. (2005). Real-time, fluorescence-based quantitative PCR: a snapshot of current procedures and preferences. Expert Rev. Mol. Diagn. *5*, 493–498.

Bustin, S.A. (2010). Why the need for qPCR publication guidelines? The case for MIQE. Methods *50*, 217–226.

Bustin, S.A., and Mueller, R. (2005). Real-time reverse transcription PCR (qRT-PCR) and its potential use in clinical diagnosis. Clin. Sci. (Lond.) *109*, 365–379.

Bustin, S.A., and Mueller, R. (2006). Real-time reverse transcription PCR and the detection of occult disease in colorectal cancer. Mol. Aspects Med. *27*, 192–223.

Bustin, S.A., and Nolan, T. (2004). Pitfalls of quantitative real-time reverse-transcription polymerase chain reaction. J. Biomol. Tech. *15*, 155–166.

Bustin, S.A., Benes, V., Nolan, T., and Pfaffl, M.W. (2005). Quantitative real-time RT-PCR – a perspective. J. Mol. Endocrinol. *34*, 597–601.

Bustin, S.A., Benes, V., Garson, J.A., Hellemans, J., Huggett, J., Kubista, M., Mueller, R., Nolan, T., Pfaffl, M.W., Shipley, G.L., *et al.* (2009). The MIQE guidelines: minimum information for publication of quantitative real-time PCR experiments. Clin. Chem. *55*, 611–622.

Bustin, S., Bergkvist, A., and Nolan, T. (2011a). In Silico Tools for qPCR assay design and data analysis. Methods Mol. Biol. *760*, 283–306.

Bustin, S.A., Benes, V., Garson, J.A., Hellemans, J., Huggett, J., Kubista, M., Mueller, R., Nolan, T., Pfaffl, M.W., Shipley, G.L., *et al.* (2011b). Primer sequence disclosure: a clarification of the MIQE guidelines. Clin. Chem. *57*, 919–921.

Chandler, D.P., Wagnon, C.A., and Bolton, H.J. (1998). Reverse transcriptase (RT) inhibition of PCR at low concentrations of template and its implications for quantitative RT-PCR. Appl. Environ. Microbiol. *64*, 669–677.

Chang, K.H., Mestdagh, P., Vandesompele, J., Kerin, M.J., and Miller, N. (2010). MicroRNA expression profiling to identify and validate reference genes for relative quantification in colorectal cancer. BMC Cancer *10*, 173.

Cloud, J.L., Hymas, W.C., Turlak, A., Croft, A., Reischl, U., Daly, J.A., and Carroll, K.C. (2003). Description of a multiplex Bordetella pertussis and Bordetella parapertussis LightCycler PCR assay with inhibition control. Diagn. Microbiol. Infect. Dis. *46*, 189–195.

Cone, R.W., Hobson, A.C., and Huang, M.L. (1992). Coamplified positive control detects inhibition of polymerase chain reactions. J. Clin. Microbiol. *30*, 3185–3189.

Coombs, N.J., Gough, A.C., and Primrose, J.N. (1999). Optimisation of DNA and RNA extraction from archival formalin-fixed tissue. Nucleic Acids Res. *27*, e12.

Cusi, M.G., Valassina, M., and Valensin, P.E. (1994). Comparison of M-MLV reverse transcriptase and Tth polymerase activity in RT-PCR of samples with low virus burden. Biotechniques *17*, 1034–1036.

Dheda, K., Huggett, J.F., Bustin, S.A., Johnson, M.A., Rook, G., and Zumla, A. (2004). Validation of housekeeping genes for normalizing RNA expression in real-time PCR. Biotechniques *37*, 112–119.

Dheda, K., Huggett, J.F., Chang, J.S., Kim, L.U., Bustin, S.A., Johnson, M.A., Rook, G.A., and Zumla, A. (2005). The implications of using an inappropriate reference gene for real-time reverse transcription PCR data normalization. Anal. Biochem. *344*, 141–143.

Diercks, A., Kostner, H., and Ozinsky, A. (2009). Resolving cell population heterogeneity: real-time PCR for simultaneous multiplexed gene detection in multiple single-cell samples. PLoS One *4*, e6326.

Dixon, J.M., Lubomirski, M., Amaratunga, D., Morrison, T.B., Brenan, C.J., and Ilyin, S.E. (2009). Nanoliter high-throughput RT-qPCR: a statistical analysis and assessment. Biotechniques *46*, ii-viii.

Doleshal, M., Magotra, A.A., Choudhury, B., Cannon, B.D., Labourier, E., and Szafranska, A.E. (2008). Evaluation and validation of total RNA extraction methods for microRNA expression analyses in formalin-fixed, paraffin-embedded tissues. J. Mol. Diagn. *10*, 203–211.

Easton, L.A., Vilcek, S., and Nettleton, P.F. (1994). Evaluation of a 'one tube' reverse transcription-polymerase chain reaction for the detection of ruminant pestiviruses. J. Virol. Methods *50*, 343–348.

Emmert-Buck, M.R., Bonner, R.F., Smith, P.D., Chuaqui, R.F., Zhuang, Z., Goldstein, S.R., Weiss, R.A., and Liotta, L.A. (1996). Laser capture microdissection. Science *274*, 998–1001.

Fink, L., Kohlhoff, S., Stein, M.M., Hanze, J., Weissmann, N., Rose, F., Akkayagil, E., Manz, D., Grimminger, F., Seeger, W., *et al.* (2002). cDNA array hybridization after laser-assisted microdissection from nonneoplastic tissue. Am. J. Pathol. *160*, 81–90.

Fleige, S., and Pfaffl, M.W. (2006). RNA integrity and the effect on the real-time qRT-PCR performance. Mol. Aspects Med. *27*, 126–139.

Fleige, S., Walf, V., Huch, S., Prgomet, C., Sehm, J., and Pfaffl, M.W. (2006). Comparison of relative mRNA quantification models and the impact of RNA integrity in quantitative real-time RT-PCR. Biotechnol. Lett. *28*, 1601–1613.

Fredricks, D.N., and Relman, D.A. (1998). Improved amplification of microbial DNA from blood cultures by removal of the PCR inhibitor sodium polyanetholesulfonate. J. Clin. Microbiol. *36*, 2810–2816.

Fu, W.J., Hu, J., Spencer, T., Carroll, R., and Wu, G. (2006). Statistical models in assessing fold change of gene expression in real-time RT-PCR experiments. Comput. Biol. Chem. *30*, 21–26.

Gaillard, C., Flavin, M., Woisard, A., and Strauss, F. (1999). Association of double-stranded DNA fragments into multistranded DNA structures. *50*, 679–689.

Gallup, J.M., and Ackermann, M.R. (2008). The 'PREXCEL-Q Method' for qPCR. Int. J. Biomed. Sci. *4*, 273–293.

Gibson, U.E., Heid, C.A., and Williams, P.M. (1996). A novel method for real time quantitative RT-PCR. Genome Res. *6*, 995–1001.

Ginzinger, D.G. (2002). Gene quantification using real-time quantitative PCR: an emerging technology hits the mainstream. Exp. Hematol. *30*, 503–512.

Git, A., Dvinge, H., Salmon-Divon, M., Osborne, M., Kutter, C., Hadfield, J., Bertone, P., and Caldas, C. (2010). Systematic comparison of microarray profiling, real-time PCR, and next-generation sequencing technologies for measuring differential microRNA expression. RNA *16*, 991–1006.

Godfrey, T.E., Kim, S.H., Chavira, M., Ruff, D.W., Warren, R.S., Gray, J.W., and Jensen, R.H. (2000). Quantitative mRNA expression analysis from formalin-fixed, paraffin-embedded tissues using 5' nuclease quantitative reverse transcription polymerase chain reaction. J. Mol. Diagn. *2*, 84–91.

Goldsworthy, S.M., Stockton, P.S., Trempus, C.S., Foley, J.F., and Maronpot, R.R. (1999). Effects of fixation on RNA extraction and amplification from laser capture microdissected tissue. Mol. Carcinog. *25*, 86–91.

Goossens, K., Van Poucke, M., Van Soom, A., Vandesompele, J., Van Zeveren, A., and Peelman, L.J. (2005). Selection of reference genes for quantitative real-time PCR in bovine preimplantation embryos. BMC Dev. Biol. 5, 27.

Grisold, A.J., and Kessler, H.H. (2006). Use of hybridization probes in a real-time PCR assay on the LightCycler for the detection of methicillin-resistant Staphylococcus aureus. Methods Mol. Biol. *345*, 79–89.

Grotzer, M.A., Patti, R., Geoerger, B., Eggert, A., Chou, T.T., and Phillips, P.C. (2000). Biological stability of RNA isolated from RNAlater-treated brain tumor and neuroblastoma xenografts. *34*, 438–442.

Guenin, S., Mauriat, M., Pelloux, J., Van Wuytswinkel, O., Bellini, C., and Gutierrez, L. (2009). Normalization of qRT-PCR data: the necessity of adopting a systematic, experimental conditions-specific, validation of references. J. Exp. Bot. *60*, 487–493.

Gunson, R.N., Bennett, S., Maclean, A., and Carman, W.F. (2008). Using multiplex real time PCR in order to streamline a routine diagnostic service. J. Clin. Virol. *43*, 372–375.

Guy, R.A., Payment, P., Krull, U.J., and Horgen, P.A. (2003). Real-time PCR for quantification of Giardia and Cryptosporidium in environmental water samples and sewage. Appl. Environ. Microbiol. *69*, 5178–5185.

Haller, F., Kulle, B., Schwager, S., Gunawan, B., von Heydebreck, A., Sultmann, H., and Fuzesi, L. (2004). Equivalence test in quantitative reverse transcription polymerase chain reaction: confirmation of reference genes suitable for normalization. Anal. Biochem. *335*, 1–9.

Hartshorn, C., Anshelevich, A., and Wangh, L.J. (2005). Rapid, single-tube method for quantitative preparation and analysis of RNA and DNA in samples as small as one cell. BMC Biotechnol. *5*, 2.

Haupt, C., Tolner, E.A., Heinemann, U., Witte, O.W., and Frahm, C. (2006). The combined use of non-radioactive in situ hybridization and real-time RT-PCR to assess gene expression in cryosections. Brain Res. *1118*, 232–238.

Hellemans, J., Mortier, G., De Paepe, A., Speleman, F., and Vandesompele, J. (2007). qBase relative quantification framework and software for management and automated analysis of real-time quantitative PCR data. Genome Biol. *8*, R19.

Hilscher, C., Vahrson, W., and Dittmer, D.P. (2005). Faster quantitative real-time PCR protocols may lose sensitivity and show increased variability. Nucleic Acids Res. *33*, e182.

Huggett, J., and Bustin, S.A. (2011). Standardisation and reporting for nucleic acid quantification. Accredit. Qual. Assur. *16*, 399–405.

Huggett, J., Dheda, K., Bustin, S., and Zumla, A. (2005). Real-time RT-PCR normalisation; strategies and considerations. Genes Immun. *6*, 279–284.

Huggett, J.F., Novak, T., Garson, J.A., Green, C., Morris-Jones, S.D., Miller, R.F., and Zumla, A. (2008). Differential susceptibility of PCR reactions to inhibitors: an important and unrecognised phenomenon. BMC Res. Notes *1*, 70.

Jiang, J., Alderisio, K.A., Singh, A., and Xiao, L. (2005). Development of procedures for direct extraction of Cryptosporidium DNA from water concentrates and for relief of PCR inhibitors. Appl. Environ. Microbiol. *71*, 1135–1141.

Jin, N., He, K., and Liu, L. (2006). qPCR-DAMS: a database tool to analyze, manage, and store both relative and absolute quantitative real-time PCR data. Physiol. Genomics *25*, 525–527.

Kanno, J., Aisaki, K., Igarashi, K., Nakatsu, N., Ono, A., Kodama, Y., and Nagao, T. (2006). 'Per cell' normalization method for mRNA measurement by quantitative PCR and microarrays. BMC Genomics *7*, 64.

Keilholz, U. (1998). New prognostic factors in melanoma: mRNA tumour markers. Eur. Cancer, J. *34(Suppl. 3)*, S37-S41.

Klimecki, W.T., Futscher, B.W., and Dalton, W.S. (1994). Effects of ethanol and paraformaldehyde on RNA yield and quality. Biotechniques *16*, 1021–1023.

Koch, I., Slotta-Huspenina, J., Hollweck, R., Anastasov, N., Hofler, H., Quintanilla-Martinez, L., and Fend, F. (2006). Real-time quantitative RT-PCR shows variable, assay-dependent sensitivity to formalin fixation: implications for direct comparison of transcript levels in paraffin-embedded tissues. Diagn. Mol. Pathol. *15*, 149–156.

de Kok, J.B., Roelofs, R.W., Giesendorf, B.A., Pennings, J.L., Waas, E.T., Feuth, T., Swinkels, D.W., and Span, P.N. (2005). Normalization of gene expression measurements in tumor tissues: comparison of 13 endogenous control genes. Lab Invest *85*, 154–159.

Kubista, M., Andrade, J.M., Bengtsson, M., Forootan, A., Jonak, J., Lind, K., Sindelka, R., Sjoback, R., Sjogreen, B., Strombom, L., *et al.* (2006). The real-time polymerase chain reaction. Mol. Aspects Med. *27*, 95–125.

Lacey, H.A., Nolan, T., Greenwood, S.L., Glazier, J.D., and Sibley, C.P. (2005). Gestational profile of Na+/H+ exchanger and Cl-/HCO3- anion exchanger mRNA expression in placenta using real-time QPCR. Placenta *26*, 93–98.

Larionov, A., Krause, A., and Miller, W. (2005). A standard curve based method for relative real time PCR data processing. BMC Bioinformatics *6*, 62.

Lefevre, J., Hankins, C., Pourreaux, K., Voyer, H., and Coutlee, F. (2004). Prevalence of selective inhibition of HPV-16 DNA amplification in cervicovaginal lavages. J. Med. Virol. *72*, 132–137.

Lekanne Deprez, R.H., Fijnvandraat, A.C., Ruijter, J.M., and Moorman, A.F. (2002). Sensitivity and accuracy of quantitative real-time polymerase chain reaction using SYBR green I depends on cDNA synthesis conditions. Anal. Biochem. *307*, 63–69.

Lewis, F., and Maughan, N.J. (2004). Extraction of total RNA from formalin-fixed paraffin-embedded tissue. In A-Z of Quantitative PCR, Bustin, S.A., ed. (IUL, La Jolla, CA), pp. 591–603.

Li, J., Smyth, P., Cahill, S., Denning, K., Flavin, R., Aherne, S., Pirotta, M., Guenther, S.M., O'Leary, J.J., and Sheils, O. (2008). Improved RNA quality and TaqMan Pre-amplification method (PreAmp) to enhance expression analysis from formalin fixed paraffin embedded (FFPE) materials. BMC Biotechnol. *8*, 10.

Liss, B. (2002). Improved quantitative real-time RT-PCR for expression profiling of individual cells. *30*, E89.

Liu, J., Enzelberger, M., and Quake, S. (2002). A nanoliter rotary device for polymerase chain reaction. Electrophoresis *23*, 1531–1536.

Liu, W., and Saint, D.A. (2002a). A new quantitative method of real time reverse transcription polymerase chain reaction assay based on simulation of polymerase chain reaction kinetics. Anal. Biochem. *302*, 52–59.

Liu, W., and Saint, D.A. (2002b). Validation of a quantitative method for real time PCR kinetics. Biochem. Biophys. Res. Commun. *294*, 347–353.

Macabeo-Ong, M., Ginzinger, D.G., Dekker, N., McMillan, A., Regezi, J.A., Wong, D.T., and Jordan, R.C. (2002). Effect of duration of fixation on quantitative reverse transcription polymerase chain reaction analyses. *15*, 979–987.

McCaughan, F., and Dear, P.H. (2010). Single-molecule genomics. J. Pathol. *220*, 297–306.

Madejon, A., Manzano, M.L., Arocena, C., Castillo, I., and Carreno, V. (2000). Effects of delayed freezing of liver biopsies on the detection of hepatitis C virus RNA strands. J. Hepatol. *32*, 1019–1025.

Mader, R.M., Schmidt, W.M., Sedivy, R., Rizovski, B., Braun, J., Kalipciyan, M., Exner, M., Steger, G.G., and Mueller, M.W. (2001). Reverse transcriptase template switching during reverse transcriptase-polymerase chain reaction: artificial generation of deletions in ribonucleotide reductase mRNA. J. Lab. Clin. Med. *137*, 422–428.

Marcus, J.S., Anderson, W.F., and Quake, S.R. (2006). Parallel picoliter rt-PCR assays using microfluidics. Anal. Chem. *78*, 956–958.

Marshall, R., Chernesky, M., Jang, D., Hook, E.W., Cartwright, C.P., Howell-Adams, B., Ho, S., Welk, J., Lai-Zhang, J., Brashear, J., *et al.* (2007). Characteristics of the m2000 automated sample preparation and multiplex real-time PCR system for detection of *Chlamydia trachomatis* and *Neisseria gonorrhoeae*. J. Clin. Microbiol. *45*, 747–751.

Marubini, E., Verderio, P., Raggi, C.C., Pazzagli, M., and Orlando, C. (2004). Statistical diagnostics emerging from external quality control of real-time PCR. Int. J. Biol. Markers *19*, 141–146.

Marullo, M., Zuccato, C., Mariotti, C., Lahiri, N., Tabrizi, S.J., Di Donato, S., and Cattaneo, E. (2010). Expressed Alu repeats as a novel, reliable tool for normalization of real-time quantitative RT-PCR data. Genome Biol. *11*, R9.

Masuda, N., Ohnishi, T., Kawamoto, S., Monden, M., and Okubo, K. (1999). Analysis of chemical modification of RNA from formalin-fixed samples and optimization of molecular biology applications for such samples. Nucleic Acids Res. *27*, 4436–4443.

Mestdagh, P., Van Vlierberghe, P., De Weer, A., Muth, D., Westermann, F., Speleman, F., and Vandesompele, J. (2009). A novel and universal method for microRNA RT-qPCR data normalization. Genome Biol. *10*, R64.

Mizuno, T., Nagamura, H., Iwamoto, K.S., Ito, T., Fukuhara, T., Tokunaga, M., Tokuoka, S., Mabuchi, K., and Seyama, T. (1998). RNA from decades-old archival tissue blocks for retrospective studies. Diagn. Mol. Pathol. *7*, 202–208.

Morrison, T., Hurley, J., Garcia, J., Yoder, K., Katz, A., Roberts, D., Cho, J., Kanigan, T., Ilyin, S.E., Horowitz, D., *et al.* (2006). Nanoliter high throughput quantitative PCR. Nucleic Acids Res. *34*, e123.

Muller, M.C., Hordt, T., Paschka, P., Merx, K., La Rosee, P., Hehlmann, R., and Hochhaus, A. (2004). Standardization of preanalytical factors for minimal residual disease analysis in chronic myelogenous leukemia. Acta. Haematol. *112*, 30–33.

Neuzil, P., Zhang, C., Pipper, J., Oh, S., and Zhuo, L. (2006). Ultra fast miniaturized real-time PCR: 40 cycles in less than six minutes. Nucleic Acids Res. *34*, e77.

Nolan, T., Hands, R.E., and Bustin, S.A. (2006a). Quantification of mRNA using real-time RT-PCR. Nat. Protocols *1*, 1559–1582.

Nolan, T., Hands, R.E., Ogunkolade, B.W., and Bustin, S.A. (2006b). SPUD: a qPCR assay for the detection of inhibitors in nucleic acid preparations. Anal. Biochem. *351*, 308–310.

Nordgard, O., Kvaloy, J.T., Farmen, R.K., and Heikkila, R. (2006). Error propagation in relative real-time reverse transcription polymerase chain reaction quantification models: the balance between accuracy and precision. Anal. Biochem. *356*, 182–193.

O'Connor, W., and Runquist, E.A. (2008). Error minimization algorithm for comparative quantitative PCR analysis: Q-Anal. Anal. Biochem. *378*, 96–98.

Oji, Y., Yamamoto, H., Nomura, M., Nakano, Y., Ikeba, A., Nakatsuka, S., Abeno, S., Kiyotoh, E., Jomgeow, T., Sekimoto, M., *et al.* (2003). Overexpression of the Wilms' tumor gene WT1 in colorectal adenocarcinoma. Cancer Sci. *94*, 712–717.

Pasloske, B.L., Walkerpeach, C.R., Obermoeller, R.D., Winkler, M., and DuBois, D.B. (1998). Armored RNA technology for production of ribonuclease-resistant viral RNA controls and standards. J. Clin. Microbiol. *36*, 3590–3594.

Pattyn, F., Speleman, F., De Paepe, A., and Vandesompele, J. (2003). RTPrimerDB: the real-time PCR primer and probe database. Nucleic Acids Res. *31*, 122–123.

Pattyn, F., Robbrecht, P., De Paepe, A., Speleman, F., and Vandesompele, J. (2006). RTPrimerDB: the real-time PCR primer and probe database, major update 2006. Nucleic Acids Res. *34*, D684–D688.

Peixoto, A., Monteiro, M., Rocha, B., and Veiga-Fernandes, H. (2004). Quantification of multiple gene expression in individual cells. Genome Res. *14*, 1938–1947.

Peltier, H.J., and Latham, G.J. (2008). Normalization of microRNA expression levels in quantitative RT-PCR assays: identification of suitable reference RNA targets in normal and cancerous human solid tissues. RNA *14*, 844–852.

Perch-Nielsen, I.R., Bang, D.D., Poulsen, C.R., El-Ali, J., and Wolff, A. (2003). Removal of PCR inhibitors using dielectrophoresis as a selective filter in a microsystem. Lab. Chip *3*, 212–216.

Perez-Novo, C.A., Claeys, C., Speleman, F., Van Cauwenberge, P., Bachert, C., and Vandesompele, J. (2005). Impact of RNA quality on reference gene expression stability. Biotechniques *39*, 52, 54, 56.

Pfaffl, M.W., Horgan, G.W., and Dempfle, L. (2002). Relative expression software tool (REST) for group-wise comparison and statistical analysis of relative expression results in real-time PCR. Nucleic Acids Res. *30*, e36.

Pfaffl, M.W., Tichopad, A., Prgomet, C., and Neuvians, T.P. (2004). Determination of stable housekeeping genes, differentially regulated target genes and sample integrity: BestKeeper – Excel-based tool using pair-wise correlations. Biotechnol. Lett. *26*, 509–515.

Picard, C., Silvy, M., and Gabert, J. (2006). Overview of real-time RT-PCR strategies for quantification of gene rearrangements in the myeloid malignancies. Methods Mol. Med. *125*, 27–68.

Pierce, K.E., and Wangh, L.J. (2011). LATE-PCR and allied technologies: real-time detection strategies for rapid, reliable diagnosis from single cells. Methods Mol. Biol. *688*, 47–66.

Platts, A.E., Johnson, G.D., Linnemann, A.K., and Krawetz, S.A. (2008). Real-time PCR quantification using a variable reaction efficiency model. Anal. Biochem. *380*, 315–322.

Poddar, S.K., Sawyer, M.H., and Connor, J.D. (1998). Effect of inhibitors in clinical specimens on Taq and Tth DNA polymerase-based PCR amplification of influenza A virus. J. Med. Microbiol. *47*, 1131–1135.

Port, M., Schmelz, H.U., Stassen, T., Mueller, K., Stockinger, M., Obermair, R., and Abend, M. (2007). Correcting false gene expression measurements from degraded RNA using RTQ-PCR. Diagn. Mol. Pathol. *16*, 38–49.

Pushkarev, D., Neff, N.F., and Quake, S.R. (2009). Single-molecule sequencing of an individual human genome. Nat. Biotechnol. *27*, 847–852.

Radstrom, P., Knutsson, R., Wolffs, P., Lovenklev, M., and Lofstrom, C. (2004). Pre-PCR processing: strategies to generate PCR-compatible samples. Mol. Biotechnol. *26*, 133–146.

Raengsakulrach, B., Nisalak, A., Maneekarn, N., Yenchitsomanus, P.T., Limsomwong, C., Jairungsri, A., Thirawuth, V., Green, S., Kalayanarooj, S., Suntayakorn, S., *et al.* (2002). Comparison of four reverse transcription-polymerase chain reaction procedures for the detection of dengue virus in clinical specimens. J. Virol. Methods *105*, 219–232.

Raggi, C.C., Verderio, P., Pazzagli, M., Marubini, E., Simi, L., Pinzani, P., Paradiso, A., and Orlando, C. (2005). An Italian program of external quality control for quantitative assays based on real-time PCR with Taq-Man probes. Clin. Chem. Lab. Med. *43*, 542–548.

Rainen, L., Oelmueller, U., Jurgensen, S., Wyrich, R., Ballas, C., Schram, J., Herdman, C., Bankaitis-Davis, D., Nicholls, N., Trollinger, D., *et al.* (2002). Stabilization of mRNA expression in whole blood samples. Clin. Chem. *48*, 1883–1890.

Ramakers, C., Ruijter, J.M., Deprez, R.H., and Moorman, A.F. (2003). Assumption-free analysis of quantitative real-time polymerase chain reaction (PCR) data. Neurosci. Lett. *339*, 62–66.

Ramsden, S.C., Daly, S., Geilenkeuser, W.J., Duncan, G., Hermitte, F., Marubini, E., Neumaier, M., Orlando, C., Palicka, V., Paradiso, A., *et al.* (2006). EQUAL-quant: an international external quality assessment scheme for real-time PCR. Clin. Chem. *52*, 1584–1591.

Raymaekers, M., Smets, R., Maes, B., and Cartuyvels, R. (2009). Checklist for optimization and validation of real-time PCR assays. J. Clin. Lab. Anal. *23*, 145–151.

Renfrow, J.J., Scheck, A.C., Dhawan, N.S., Lukac, P.J., Vogel, H., Chandler, J.P., Raizer, J.J., Harsh, G.R., Chakravarti, A., and Bredel, M. (2011). Gene-protein correlation in single cells. Neuro. Oncol. *13*, 880–885.

Rubie, C., Kempf, K., Hans, J., Su, T., Tilton, B., Georg, T., Brittner, B., Ludwig, B., and Schilling, M. (2005). Housekeeping gene variability in normal and cancerous colorectal, pancreatic, esophageal, gastric and hepatic tissues. Mol. Cell. Probes *19*, 101–109.

Rutledge, R.G. (2004). Sigmoidal curve-fitting redefines quantitative real-time PCR with the prospective of developing automated high-throughput applications. Nucleic Acids Res. *32*, e178.

Rutledge, R.G., and Stewart, D. (2008a). A kinetic-based sigmoidal model for the polymerase chain reaction and its application to high-capacity absolute quantitative real-time PCR. BMC Biotechnol. *8*, 47.

Rutledge, R.G., and Stewart, D. (2008b). Critical evaluation of methods used to determine amplification efficiency refutes the exponential character of real-time PCR. BMC Mol. Biol. *9*, 96.

Schroeder, A., Mueller, O., Stocker, S., Salowsky, R., Leiber, M., Gassmann, M., Lightfoot, S., Menzel, W., Granzow, M., and Ragg, T. (2006). The RIN: an RNA integrity number for assigning integrity values to RNA measurements. BMC Mol. Biol. *7*, 3.

Shanmugam, V., Sell, K.W., and Saha, B.K. (1993). Mistyping ACE heterozygotes. PCR Methods Appl. *3*, 120–121.

Sindelka, R., Jonak, J., Hands, R., Bustin, S.A., and Kubista, M. (2008). Intracellular expression profiles measured by real-time PCR tomography in the Xenopus laevis oocyte. Nucleic Acids Res. *36*, 387–392.

Smith, M.V., Miller, C.R., Kohn, M., Walker, N.J., and Portier, C.J. (2007). Absolute estimation of initial concentrations of amplicon in a real-time RT-PCR process. BMC Bioinformatics *8*, 409.

Smith, R.D., Brown, B., Ikonomi, P., and Schechter, A.N. (2003). Exogenous reference RNA for normalization of real-time quantitative PCR. Biotechniques *34*, 88–91.

von Smolinski, D., Leverkoehne, I., von Samson-Himmelstjerna, G., and Gruber, A.D. (2005). Impact of formalin-fixation and paraffin-embedding on the ratio between mRNA copy numbers of differently expressed genes. Histochem. Cell Biol. *124*, 177–188.

Specht, K., Richter, T., Muller, U., Walch, A., Werner, M., and Hofler, H. (2001). Quantitative gene expression analysis in microdissected archival formalin-fixed and paraffin-embedded tumor tissue. Am. J. Pathol. *158*, 419–429.

Spiess, A.N., Feig, C., and Ritz, C. (2008). Highly accurate sigmoidal fitting of real-time PCR data by introducing a parameter for asymmetry. BMC Bioinformatics *9*, 221.

Stahlberg, A., and Bengtsson, M. (2010). Single-cell gene expression profiling using reverse transcription quantitative real-time PCR. Methods *50*, 282–288.

Stahlberg, A., Aman, P., Ridell, B., Mostad, P., and Kubista, M. (2003). Quantitative real-time PCR method for detection of B-lymphocyte monoclonality by comparison of kappa and lambda immunoglobulin light chain expression. Clin. Chem. *49*, 51–59.

Stahlberg, A., Hakansson, J., Xian, X., Semb, H., and Kubista, M. (2004a). Properties of the reverse transcription reaction in mRNA quantification. Clin. Chem. *50*, 509–515.

Stahlberg, A., Kubista, M., and Pfaffl, M. (2004b). Comparison of reverse transcriptases in gene expression analysis. Clin. Chem. *50*, 1678–1680.

Stahlberg, A., Zoric, N., Aman, P., and Kubista, M. (2005). Quantitative real-time PCR for cancer detection: the lymphoma case. Expert Rev. Mol. Diagn. *5*, 221–230.

Stahlberg, A., Elbing, K., Andrade-Garda, J.M., Sjogreen, B., Forootan, A., and Kubista, M. (2008). Multiway real-time PCR gene expression profiling in yeast Saccharomyces cerevisiae reveals altered transcriptional response of ADH-genes to glucose stimuli. BMC Genomics *9*, 170.

Stangegaard, M., Dufva, I.H., and Dufva, M. (2006). Reverse transcription using random pentadecamer primers increases yield and quality of resulting cDNA. Biotechniques *40*, 649–657.

Stanley, K.K., and Szewczuk, E. (2005). Multiplexed tandem PCR: gene profiling from small amounts of RNA using SYBR Green detection. Nucleic Acids Res. *33*, e180.

Sunen, E., Casas, N., Moreno, B., and Zigorraga, C. (2004). Comparison of two methods for the detection of hepatitis A virus in clam samples (Tapes spp.) by reverse transcription-nested PCR. Int. J. Food Microbiol. *91*, 147–154.

Suslov, O., and Steindler, D.A. (2005). PCR inhibition by reverse transcriptase leads to an overestimation of amplification efficiency. Nucleic Acids Res. *33*, e181.

Swartzman, E., Shannon, M., Lieu, P., Chen, S.M., Mooney, C., Wei, E., Kuykendall, J., Tan, R., Settineri, T., Egry, L., *et al.* (2010). Expanding applications of protein analysis using proximity ligation and qPCR. Methods *50*, S23–6.

Szabo, A., Boucher, K., Carroll, W.L., Klebanov, L.B., Tsodikov, A.D., and Yakovlev, A.Y. (2002). Variable selection and pattern recognition with gene expression data generated by the microarray technology. Math. Biosci. *176*, 71–98.

Tichopad, A., Dilger, M., Schwarz, G., and Pfaffl, M.W. (2003). Standardized determination of real-time PCR efficiency from a single reaction set-up. Nucleic Acids Res. *31*, e122.

Tichopad, A., Didier, A., and Pfaffl, M.W. (2004). Inhibition of real-time RT-PCR quantification due to tissue-specific contaminants. Mol. Cell. Probes *18*, 45–50.

Tichopad, A., Kitchen, R., Riedmaier, I., Becker, C., Stahlberg, A., and Kubista, M. (2009). Design and optimization of reverse-transcription quantitative PCR experiments. Clin. Chem. *55*, 1816–1823.

Tricarico, C., Pinzani, P., Bianchi, S., Paglierani, M., Distante, V., Pazzagli, M., Bustin, S.A., and Orlando, C. (2002). Quantitative real-time reverse transcription polymerase chain reaction: normalization to rRNA or single housekeeping genes is inappropriate for human tissue biopsies. Anal. Biochem. *309*, 293–300.

Valasek, M.A., and Repa, J.J. (2005). The power of real-time PCR. Adv. Physiol. Educ. *29*, 151–159.

Vandesompele, J., De Preter, K., Pattyn, F., Poppe, B., Van Roy, N., De Paepe, A., and Speleman, F. (2002a). Accurate normalization of real-time quantitative RT-PCR data by geometric averaging of multiple internal control genes. Genome Biol. *3*, 0034.1–0034.11.

Vandesompele, J., De Paepe, A., and Speleman, F. (2002b). Elimination of primer-dimer artifacts and genomic coamplification using a two-step SYBR green I real-time RT-PCR. Anal. Biochem. *303*, 95–98.

Verderio, P., Pizzamiglio, S., Gallo, F., and Ramsden, S.C. (2008a). FCI: an R-based algorithm for evaluating uncertainty of absolute real-time PCR quantification. BMC Bioinformatics *9*, 13.

Verderio, P., Ramsden, S.C., Orlando, C., Pizzamiglio, S., Paradiso, A., Neumaier, M., Pazzagli, M., and Marubini, E. (2008b). External quality assessment schemes for real-time PCR: a statistical procedure to corrective actions. Clin. Chem. Lab. Med. *46*, 717–721.

Vermeulen, J., De Preter, K., Lefever, S., Nuytens, J., De Vloed, F., Derveaux, S., Hellemans, J., Speleman, F., and Vandesompele, J. (2011). Measurable impact of RNA quality on gene expression results from quantitative PCR. Nucleic Acids Res. *39*, e63.

Vlems, F., Soong, R., Diepstra, H., Punt, C., Wobbes, T., Tabiti, K., and van Muijen, G. (2002a). Effect of blood sample handling and reverse transcriptase-polymerase chain reaction assay sensitivity on detection of CK20 expression in healthy donor blood. Diagn. Mol. Pathol. *11*, 90–97.

Vlems, F.A., Diepstra, J.H., Cornelissen, I.M., Ruers, T.J., Ligtenberg, M.J., Punt, C.J., van Krieken, J.H., Wobbes, T., and van Muijen, G.N. (2002b). Limitations of cytokeratin 20 RT-PCR to detect disseminated tumour cells in blood and bone marrow of patients with colorectal cancer: expression in controls and downregulation in tumour tissue. Mol. Pathol. *55*, 156–163.

de Vries, T.J., Fourkour, A., Punt, C.J., Ruiter, D.J., and van Muijen, G.N. (2000). Analysis of melanoma cells in peripheral blood by reverse transcription-polymerase chain reaction for tyrosinase and MART-1 after mononuclear cell collection with cell preparation tubes: a comparison with the whole blood guanidinium isothiocyanate RNA isolation method. Melanoma Res. *10*, 119–126.

Wagner, B.K., and Arany, Z. (2009). High-throughput real-time PCR for detection of gene-expression levels. Methods Mol. Biol. *486*, 167–175.

Wang, Z., Dong, D., Ru, B., Young, R.L., Han, N., Guo, T., and Zhang, S. (2010). Digital gene expression tag profiling of bat digits provides robust candidates contributing to wing formation. BMC Genomics *11*, 619.

Weissensteiner, T., and Lanchbury, J.S. (1996). Strategy for controlling preferential amplification and avoiding false-negatives in PCR typing. Biotechniques *21*, 1102–1108.

Westh, H., Lisby, G., Breysse, F., Boddinghaus, B., Chomarat, M., Gant, V., Goglio, A., Raglio, A., Schuster, H., Stuber, F., *et al.* (2009). Multiplex real-time PCR and blood culture for identification of bloodstream pathogens in patients with suspected sepsis. Clin. Microbiol. Infect. *15*, 544–551.

Whitcombe, D., Theaker, J., Guy, S.P., Brown, T., and Little, S. (1999). Detection of PCR products using self-probing amplicons and fluorescence. Nat. Biotechnol. *17*, 804–807.

Wilkes, T.M., Devonshire, A.S., Ellison, S.L., and Foy, C.A. (2010). Evaluation of a novel approach for the measurement of RNA quality. BMC Res. Notes 3, 89.

Yang, W., Maqsodi, B., Ma, Y., Bui, S., Crawford, K.L., McMaster, G.K., Witney, F., and Luo, Y. (2006). Direct quantification of gene expression in homogenates of formalin-fixed, paraffin-embedded tissues. Biotechniques *40*, 481–486.

Zaayman, D., Human, S., and Venter, M. (2009). A highly sensitive method for the detection and genotyping of West Nile virus by real-time PCR. J. Virol. Methods *157*, 155–160.

Zhang, C., and Xing, D. (2007). Miniaturized PCR chips for nucleic acid amplification and analysis: latest advances and future trends. Nucleic Acids Res. *35*, 4223–4237.

Detection of Biothreats by Real-time PCR

4

Christina Egan and Cassandra D. Kelly-Cirino

The world is a dangerous place to live, not just because of the people who are evil, but because of the people who don't do anything about it.

Albert Einstein

Abstract

With the public's reawakened concern regarding use of biological agents as weapons, the rapid detection, discrimination, and identification of pathogenic organisms and toxins has become a priority for state and federal government agencies. High-confidence, cost-effective and near real-time diagnostic methods are essential to protecting national health security whether the target is public health, agriculture, commodities, or water supply infrastructures. While culture-based methods have been, and will likely remain, the gold standard for microbiological diagnostics, PCR-based tests offer significant advantages in sensitivity, specificity, speed, and data richness that make them invaluable to diagnostic laboratories. In this chapter, we will describe the application of real-time PCR methods in biodefence. We will discuss the use of real-time PCR in biodefence in terms of general workflow and processing considerations, clinical diagnostic applications, environmental diagnostic applications, and multiplex screening. Real-time PCR assays can be either quantitative (qPCR) or qualitative, depending on whether a standard curve is included with the analytical run. Most diagnostic and biodefence applications utilize the qualitative nature of real-time PCR as a detection platform; this chapter will focus on the benefits of these types of assays. Finally, we will consider the future uses and anticipated advances in real-time PCR applications as related to biodefence.

Introduction

Biodefence is a field that has readily embraced the diagnostic capabilities of real-time PCR, given that the ability to rapidly and specifically detect a pathogen involved in an event of bioterrorism (BT) is of utmost importance. Real-time PCR is the method of choice for initial detection because it is rapid, highly sensitive, specific, and can be performed in a portable laboratory (Edwards *et al.*, 2006). Real-time PCR is utilized in many aspects of biodefence response and preparedness, including testing in the field, surveillance of water or air, environmental sampling of letters and facilities (Higgins *et al.*, 2003; Canter *et al.*, 2005) and testing of clinical specimens. This chapter will describe the application of real-time PCR and highlight the benefits of utilizing the technology in many different areas of biodefence.

The use of real-time PCR is documented by many military, federal, state, and local entities that are involved in the detection of select agents, defined as organisms or toxins that can be used as biological weapons (Table 4.1). One such entity is the Laboratory Response Network (LRN) (Website: http://www.bt.cdc.gov/lrn/) that was created by the CDC to provide validated real-time PCR protocols and reagents to public health laboratories across the country for the detection of potential bioterrorist agents. The reagents used in these advanced diagnostic methods including real-time PCR are provided by the LRN and are restricted to approved laboratories, usually state and county public health laboratories. The rationale to restrict these tests is based on many factors, including safety of laboratory personnel, availability of expertise and dedicated personnel necessary to run and maintain facilities for these tests, access to high-quality reagents, and maintenance of proficiency standards (Synder *et al.*, 2004).

Many of the agents considered the primary threats for a bioterrorism attack, such as *Francisella tularensis, Yersinia pestis,* and *Clostridium botulinum,* are slow-growing and fastidious, and may require specialized growth medium or environmental conditions. Additionally, testing for any of these agents necessitates the availability of specialized facilities at minimally biosafety level (BSL) 2, with many needing to be handled in the more limited BSL-3 and BSL-4 facilities, where gold-standard tests, including culture, must be carried out; testing for some agents may necessitate mouse inoculation. Agents such as *Bacillus anthracis* occur in highly aerosolized forms that are especially conducive to laboratory-acquired infections. Agents such as variola virus (smallpox) and the viral agents of haemorrhagic fever such as Ebola and Marburg viruses are classified as BSL-4 agents and cannot be handled in most state or local public health laboratories. Real-time PCR is an ideal method by which to screen for these agents without amplifying them in culture; in real-time PCR, the analyst first inactivates the agent(s) in a biothreat sample, and then tests for multiple organisms or multiple targets. Thus, this methodology provides a measure of safety, because a portion of a sample can be disinfected or inactivated prior to analysis (Nitsche *et al.*, 2004).

Real-time PCR assays have been developed for some biothreat-classified agents such as *B. anthracis, Y. pestis, F. tularensis, C. botulinum* toxins, orthopoxviruses, and viral haemorrhagic fever viruses. To date, the number of real-time PCR detection assays applied to the diagnosis of infectious diseases, is relatively small in comparison to the published literature utilizing this technology. In the last few years, however, the number of real-time PCR assays for detection of biothreat agents has grown rapidly (Fig. 4.1a). As expected, there was an increase after 2001 in publications describing assays to detect *B. anthracis*. In the last 2–3 years, assays have been published that can identify other BT agents such as *Y. pestis* or orthopoxvirus (Fig. 4.1b). Most promising is the small but growing number of assays that can detect multiple agents (Tong *et al.*, 2008; Elsholz *et al.*, 2009; Janse *et al.*, 2010; Hong-Geller *et al.*, 2010). Since the threat of bioagents other than *B. anthracis* exists, assays that can detect multiple agents are critical to a rapid public health response, if potential exposures are to be minimized. Multiplex real-time PCR is a technology that can be utilized to detect more than one agent in a single assay. The use of multiplex real-time PCR in biodefence will be discussed in this chapter.

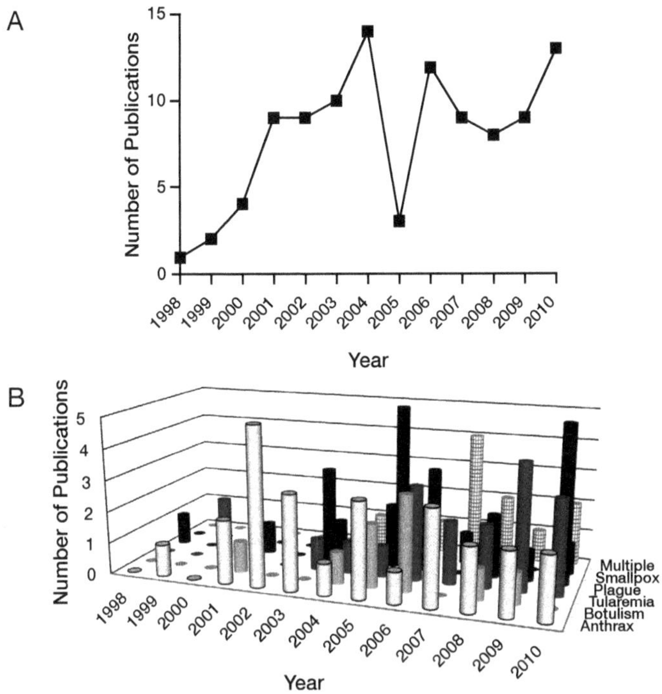

Figure 4.1 Recent literature published on biothreat agent detection by real-time PCR. (A) Publications on real-time PCR designed for the detection of selects agents; 1998–2010. (B) Publications on real-time PCR designed for the detection of select agents by year and agent; 1998–2010. This analysis was performed by searching the http://www.ncbi.nlm.nih.gov Website using the PubMed database. Queries were performed by combining the keywords: [real-time PCR OR qPCR OR 5′ nuclease OR lightcycler OR beacon OR scorpion OR taqman] and the bacterial genus and species and the disease caused by the agent [*Bacillus anthracis* OR anthrax; *Yersinia pestis* OR plague; *Francisella tularensis* OR tularaemia; *Clostridium botulinum* OR botulism; variola OR orthopox OR smallpox]. The information gathered by the authors and then evaluated by reading the abstract for each publication listed and either including or excluding the article from the data. Publications that described the detection of more than one select agent were included only in the multiple agent category.

Assay validation

The validation of protocols is a critical area of method development for real-time PCR assays in the field of biodefence. Assay validation should take into account any applicable state and federal regulations as well as sample considerations such as the range of concentrations of bacteria or DNA expected in a particular clinical or environmental sample (Cirino *et al.*, 2006). The Select Agent Rule, designed to regulate laboratories that work with agents with biothreat potential, has imposed additional restrictions on laboratories that develop and/or use real-time PCR assays. Stringent record keeping for the use, storage, and destruction of culture stocks of agents must be maintained by the laboratory; thereby adding an additional layer of complexity when select agent assays are being developed, validated and used. Certain exemption limits may apply when working with select agent toxins; however, these exemption limits do not apply to the bacterial or viral organisms themselves (Tables 4.1 and 4.2).

Table 4.1A List of select agents (current as of 15 November 2011): health and human services select agents and toxins

Abrin
Botulinum neurotoxins
Botulinum neurotoxin producing species of *Clostridium*
Cercopithecine herpesvirus 1 (Herpes B virus)
Clostridium perfringens epsilon toxin
Coccidioides posadasii/Coccidioides immitis
Conotoxins
Coxiella burnetii
Crimean-Congo haemorrhagic fever virus
Diacetoxyscirpenol
Eastern equine encephalitis virus
Ebola virus
Francisella tularensis
Lassa fever virus
Marburg virus
Monkeypox virus
Reconstructed replication competent forms of the 1918 pandemic influenza virus containing any portion of the coding regions of all eight gene segments (Reconstructed1918 Influenza virus)
Ricin
Rickettsia prowazekii
Rickettsia rickettsii
Saxitoxin
Shiga-like ribosome-inactivating proteins
Shigatoxin
South American haemorrhagic fever viruses
Flexal
Guanarito
Junin
Machupo
Sabia
Staphylococcal enterotoxins
T-2 toxin
Tetrodotoxin
Tick-borne encephalitis complex (flavi) viruses
Central European tick-borne encephalitis
Far Eastern tick-borne encephalitis
Kyasanur Forest disease
Omsk haemorrhagic fever
Russian spring and summer encephalitis
Variola major virus (smallpox virus)
Variola minor virus (Alastrim)
Yersinia pestis

Table 4.1B List of select agents (current as of 15 November 2011): overlap select agents and toxins

Bacillus anthracis
Brucella abortus
Brucella melitensis
Brucella suis
Burkholderia mallei (formerly *Pseudomonas mallei*)
Burkholderia pseudomallei (formerly *Pseudomonas pseudomallei*)
Hendra virus
Nipah virus
Rift Valley fever virus
Venezuelan equine encephalitis virus

Table 4.1C List of select agents (current as of 15 November 2011): USDA select agents and toxins

African horse sickness virus
African swine fever virus
Akabane virus
Avian influenza virus (highly pathogenic)
Bluetongue virus (exotic)
Bovine spongiform encephalopathy agent
Camel pox virus
Classical swine fever virus
Ehrlichia ruminantium (Heartwater)
Foot-and-mouth disease virus
Goat pox virus
Japanese encephalitis virus
Lumpy skin disease virus
Malignant catarrhal fever virus (Alcelaphine herpesvirus type 1)
Menangle virus
Mycoplasma capricolum subspecies *capripneumoniae* (contagious caprine pleuropneumonia)
Mycoplasma mycoides subspecies *mycoides* small colony (*Mmm* SC) (contagious bovine pleuropneumonia)
Peste des petits ruminants virus
Rinderpest virus
Sheep pox virus
Swine vesicular disease virus
Vesicular stomatitis virus (exotic): Indiana subtypes VSV-IN2, VSV-IN3
Virulent Newcastle disease virus

Table 4.1D List of select agents (current as of 15 November 2011): USDA plant select agents and toxins

Peronosclerospora philippinensis (*Peronosclerospora sacchari*)
Phoma glycinicola (formerly *Pyrenochaeta glycines*)
Ralstonia solanacearum race 3, biovar 2
Rathayibacter toxicus
Sclerophthora rayssiae var. *zeae*
Synchytrium endobioticum
Xanthomonas oryzae
Xylella fastidiosa (citrus variegated chlorosis strain)

Table 4.2 Permissible toxin amounts

HHS toxins [§73.3(d)(3)]	Amount
Abrin	100mg
Botulinum neurotoxins	0.5mg
Clostridium perfringens epsilon toxin	100mg
Conotoxin	100mg
Diacetoxyscirpenol (DAS)	1000mg
Ricin	100mg
Saxitoxin	100mg
Shiga-like ribosome-inactivating proteins	100mg
Shigatoxin	100mg
Staphylococcal enterotoxins	5mg
T-2 toxin	1000mg
Tetrodotoxin	100mg

The above toxins are not regulated if the amount under the control of a principal investigator, treating physician or veterinarian, or commercial manufacturer or distributor does not exceed, at any time, the amounts indicated in the table above.

Because there are many real-time PCR chemistries, instruments, and probe dyes available, users must take care to fully validate an existing protocol for their particular facility's equipment and the manner in which the assay is used such as sample types that may be tested. Validation studies must include sensitivity, specificity, and reproducibility data for the assay, using the specific matrix that will be tested. The validation process, while important for any real-time PCR assay, is crucial for biothreat agent diagnostics, because many decisions regarding vaccination, quarantine, and treatment will potentially be made on the basis of this initial information, prior to full confirmatory work involving culture, biochemical characterization, and antibiotic susceptibility profiles being completed.

Other elements important to the success of real-time PCR for biodefence applications are sampling and collection protocols appropriate for the event or agent of interest. Sample processing can be critical to effective diagnostics. This is especially true for any PCR-based method. DNA extraction is an important, initial step in the evaluation of the performance

Table 4.3 Available commercial real-time PCR kits for biothreat detection

Real-time PCR kit	Manufacturer	Agent(s) detected	Rt PCR platform	No. of targets	Specimen type	Notes
BA 4-Plex Assay	Cepheid	*B. anthracis*	GeneXpert	2+2 controls	Environmental	
CycleavePCR *Bacillus anthracis* Detection Kit	TaKaRa	*B. anthracis*	Smart Cycler	2	Environmental, clinical	Uses cycling probe technology
LightCycler *Bacillus anthracis* Detection Kit	Roche	*B. anthracis*	LightCycler	2	Environmental, clinical	
PathAlert™ Detection Kit	Invitrogen	*B. anthracis*, *Y. pestis*, Orthopoxvirus, F. tularensis	Agilent 2100 Bioanalyzer, 5100 Automater Lab on a Chip Platform	4 multiplex kits	Environmental (Air, food, water)	Combines Multiplex real-time PCR with a microfluidic chip
Certified Lux™ Primer Sets for infectious agents	Invitrogen	*B. anthracis*, *Y. pestis*, variola major, *F. tularensis*, C. botulinum toxin Types B, E	Compatible with most real-time PCR instruments	Ba (3), Yp (3), Var (1), Ft (2), CbotB (1), CbotE (1)	Linear DNA	
artus™ *B. anthracis* PCR kit, artus™ Orthopox PCR kit	Qiagen	*B. anthracis*, orthopoxvirus, *C. burnetii*, *Y. pestis*, *F. tularensis*, filoviruses	LightCycler, ABI Prism 7000,7900HT, Rotor-gene	2+2 controls	Spores and vegetative cells from diverse sample types (tissues, blood, stool, soil)	
BioThreat Screening Kit	Idaho Technology	*B. anthracis*, *F. tularensis*, *Y. pestis*, *Brucella* spp.	LightCycler, Razor, RAPID	1 target each	Environmental	Reagents are freeze-dried
JBAIDS Detection Kits for *B. anthracis* (Ba), *F. tularensis* (Ft), and *Y. pestis* (Yp)	Idaho Technology	*B. anthracis*, *F. tularensis*, *Y. pestis*	RAPID	2 targets each	Clinical specimens	Reagents are freeze dried, 510 cleared for use in whole blood and isolates for Ba and isolates, blood and sputum for Ft and Yp
Pathogen BioReagents	Idaho Technology	*B. anthracis*, *F. tularensis*, *Y. pestis*, *Brucella* spp., variola, *C. botulinum* Type A	Razor, R.A.P.I.D.	BA (3), Ft (2), Yp (2), Br (1) Va (1), Cbot (1)	Environmental	Reagents are freeze-dried
BioSeeq PLUS	Smiths Detection	*B. anthracis*. *F. tularensis*, *Y. pestis*, orthopox	BioSeeq	N.A.	Environmental (powders)	Hand-Held Unit, LATE PCR technology

Table 4.4 Real-time PCR assays published for clinical specimen types

Agent	Specimen type	Target	Chemistry	Sensitivity	Instrument	Reference
Bacillus anthracis	Serum, tissue, pleural fluid, sputum	PX01, PXO2, Chromosome	Taqman	167 cells (1pg DNA)	Light Cycler, ABI 7700, Smart Cycler	Bell (2002), Drago *et al*. (2002), Hoffmaster *et al*. (2002b)
C. botulinum toxin types A, B, E	Serum, wound swabs aspirates, pus	Toxin gene	Taqman	16–141 GE	Light Cycler, ABI 7700	Akbulut (2004)
Brucella spp.	Serum	Bcsp31	Light Cycler Hyb probes	6 gc	Light Cycler	Debeaumont (2005)
Yersinia pestis	Sputum	Pla gene	MGB Taqman	1.5cfu/ml	ABI7000	Loiez (2003), Chase (2005)
Francisella tularensis	Lymph node	rRNA, LpnA	Sybr Green	10–100cfu, 100–150 GE	Icycler	Sjostedt (1997)
Orthopoxviruses		Haemagglutinin gene	Taqman	25 gene copies	Light Cycler, Smart Cycler	Panning (2004)
Burkholderia pseudomallei	Blood, sputum, pus, urine, other body fluids	TTS1 gene (orf2)	Taqman	8.4×103cfu/ml	Rotorgene	Novak *et al*. (2006)
Coxiella burnetii	Serum	htpAB	Sybr Green	1 copy	Light Cycler	Fournier and Renaoult (2003)
Ebola/Marburg virus	Serum	GP gene	Sybr Green	8.6–16 RNA copies	Light Cycler	Drosten *et al*. (2002)
Lassa virus	Serum	S RNA and Glycoprotein gene	Sybr Green	8.6–16 RNA copies	Light Cycler	Drosten *et al*. (2002)

of real-time PCR on biothreat samples. Fortunately, many commercially available reaction components (e.g. mastermixes) for real-time PCR include additives that reduce the effects of potential PCR-specific inhibitors that are commonly found in environmental and clinical sample matrices (Halse *et al.*, 2006) typical of biothreat events.

As for any technique applied to the diagnosis of pathogens, the capability to operate using only limited sample volume, and the ability to assess more than one pathogen or gene in that volume (i.e. multiplex capability) are essential. Real-time PCR requires minimal sample volume and can be used on sample types commonly analysed in laboratories, including clinical or environmental samples (Ryu *et al.*, 2003). The utility of traditional diagnostic methods, such as microscopy and culture, can be limited by such factors as the lack of sensitivity and specificity and the long turn-around-time for final results (Whelen and Persing, 1996; Mackay, 2004). Real-time PCR, as compared to gold-standard methods such as culture, can achieve a very high level of sensitivity, down to detection of a single bacterial organism. Additionally, real-time PCR is effective for the detection of fastidious or non-cultivable organisms, and for application to non-viable or mixed samples for which culture or other conventional methods are not possible. Melo and co-workers found multiplex PCR to be more sensitive than culture for the diagnosis of plague, for both older retrospective and more recent samples (Melo *et al.*, 2003). It has also been shown that real-time PCR is significantly more sensitive than conventional PCR methods, and significantly faster (Cockerill, 2003). The coupling of the high sensitivity and the short time to final results makes real-time PCR an invaluable tool for the investigation of biothreat agents in clinical and environmental samples.

Laboratories performing real-time PCR on potential biothreat agents must also take into account many factors that may not be necessary for other applications of real-time PCR. These include chain-of-custody protocols and biosecurity measures such as secure facilities, surveillance cameras, and locked incubators and freezers with minimal access, consistent with federal guidelines such as the Select Agent Rule. Additionally, laboratorians must have knowledge of the appropriate protocols expected to be performed and the location where a specimen will need to be forwarded for final confirmatory testing. Finally, a laboratory must have in place appropriate protocols, equipment and reagents for testing specimens by real-time PCR in the facility in which the specimen is processed (i.e. BSL-3, BSL-4) or approved protocols for removing DNA extracts from this facility following sterility determinations. The latter option is of interest to many laboratories that have only limited number of real-time PCR instruments and need to maximize their diagnostic capabilities for other agents processed in BSL-2 laboratories to avoid having real-time PCR platforms, reagent clean facilities, and additional personnel in high containment space.

As with any validated assay, a proficiency testing program must be established to maintain the competency of the laboratory staff. Unlike most traditional infectious agents, there are no commercially available Proficiency Testing programs available for the biothreat agents (Synder *et al.*, 2004). In light of this challenge, it becomes each laboratory's responsibility to establish a robust Competency/Proficiency Program to assess the competency of each analyst on a routine basis.

PCR workflow

As with any area of microbiology, laboratory design and workflow are important to the success of the testing. The use of proper PCR work-flow, as in standard PCR applications, is a necessity, as is the use of proper extraction and amplification controls. Although real-time PCR is a closed-system approach, PCR workflow should still be carefully thought-out. It is necessary to maintain separate areas for reagents and mastermix considered 'PCR Clean', as well as areas for specimen processing and extraction, amplification, and analysis (Persing, 1991; Mitchell *et al.*, 2004). Laboratories should maintain this PCR flow as unidirectional, such that staff are not allowed to enter a 'PCR Clean' area once they have entered PCR amplification areas on a given day. Many large laboratories have dedicated personnel for each of the different areas of PCR analysis. Published literature contains laboratory flow schemes that are excellent references for those contemplating the addition of real-time PCR into their laboratory testing algorithms (Millar *et al.*, 2002, 2004). Laboratories should consider, when possible, dedicated equipment such as pipettes, centrifuges, vortex mixers, refrigerators and freezers in each area as well as protocols for the use of each area. Protocols should include proper cleaning of the area and equipment with 10% bleach to reduce the potential for amplicon or DNA contamination. Additionally, racks utilized in the real-time PCR process can be soaked in 10% bleach (Mitchell *et al.*, 2004). The procedure for wearing gloves, shoe covers, lab coats, and other personal protective equipment should also be established. At the very least, dedicated pipettes and adequate glove use should be maintained.

Routine environmental sampling of rooms and/or areas should also be considered. This testing process can involve the swabbing of areas such as doorknobs to rooms and refrigerators/freezers, centrifuges, racks, and pipettes. A swab can be moistened in sterile water or buffer and used to sample an area, after which it is returned to a microfuge tube containing the water or buffer and properly labelled. This tube can then undergo DNA extraction or testing directly in a PCR assay either for a specific target or a broad-range target like 16S rDNA that would detect all bacteria.

It is critical in the diagnosis of biothreat agents that false-positives and false-negatives minimized. To decrease the risk of false-positives, nested primers should be avoided, whenever possible, since these are a major contributor to contamination (Wannamaker *et al.*, 1989). Additionally, proper training of analysts is crucial to the success of this method. Real-time PCR instrumentation is extremely complex and requires routine maintenance for background determination, for calibration of dyes, and for ensuring that heating blocks and elements are functioning properly. The occurrence of false-negatives can be greatly decreased by timely attention to these details.

Clinical assays

Test results for specimens from the clinically ill must be obtained as rapidly as possible, to provide the physician with beneficial information crucial for patient treatment. This is especially true in the field of infectious disease where treatment must be given as soon as possible. PCR fills an important gap in the testing for infectious agents in that culture-based testing may take days to produce a result, whereas PCR requires a few hours. Real-time PCR assays have vastly improved clinical diagnostics for the detection of biothreat agents. Real-time PCR has several added benefits over traditional PCR analysis. Real-time PCR provides an extra layer of specificity by the inclusion of a specific oligonucleotide probe that is not present in a traditional PCR assay. Quick turnaround time for test results is another benefit of real-time PCR as results may be available within 1–2 hours as compared with traditional PCR, which generally requires 5–6 hours. Rapid turnaround times are essential in a clinical setting when a biothreat agent such as *Bacillus anthracis* or variola major is suspected because the initiation of an epidemiological or criminal investigation hinges on the rapid identification of the agent.

Many naturally occurring infections are caused by organisms that are classified as biothreat agents, such as *Brucella* spp. or *F. tularensis.* Through an epidemiological investigation it can be determined whether an infection was acquired in an area endemic for a particular pathogen, from a contaminated food source, from an accidental infection in a laboratory setting, or from an actual biothreat event. Real-time PCR has enabled rapid detection of agents that may be used in biothreat events, and it is these rapid determinations that initiate epidemiologic investigation to either rule in or rule out bioterrorism. In a biothreat event, rapid determination is essential to begin both criminal and public health investigations so as to minimize the number of individuals exposed to the agents and begin prophylaxis for those already exposed.

The power of both conventional and real-time PCR methods is the exquisite specificity that may be obtained. This is a critical component in testing for biothreat agents. Many biothreat agents are genetically similar to organisms that pose a low risk to human health. For example, *B. anthracis* shows > 99% genetic homology with *B. thuringiensis* and *B. cereus* within certain areas of the genome (Ash *et al.*, 1991; Ash and Collins, 1992). For a properly designed real-time PCR assay, the addition of an additional fluorescent oligonucleotide sequence referred to as a probe adds significant specificity to the assay. This type of PCR assay was utilized by the LRN in the investigation of the intentional releases of anthrax in the US postal system in 2001. The CDC had developed real-time PCR assays for *B. anthracis* and utilized these assays to analyse large numbers of both clinical and environmental samples (Hoffmaster *et al.*, 2002b). These real-time PCR assays were extremely useful when laboratories received hundreds of items a day and a mechanism was needed to quickly rule out *B. anthracis* in these samples. This allowed the laboratories to focus further additional molecular and culture testing only on samples that produced positive results in the initial real-time PCR assay. Valuable information was quickly reported back to law enforcement personnel and epidemiologists to aid in their investigation. As reported by Hoffmaster and coworkers (2002), every clinical specimen that tested positive at CDC in the real-time PCR assays was subsequently confirmed positive by culture (Hoffmaster *et al.*, 2002a). Due to the highly sensitive and specific nature of the real-time PCR assays, it proved to be an invaluable tool in the 2001 anthrax investigations. Future bioterrorism-related investigations will also rely on real-time PCR for its invaluable capacity as a screening tool.

Real-time PCR has also been combined with melting point analysis, in order to aid in the diagnosis of closely related organisms. Bystrom and coworkers described the use of real-time PCR to distinguish among closely related *F. tularensis* strains, in order to detect tularaemia from lymph node tissue (Bystrom *et al.*, 2005). Utilizing melting point analysis after real-time PCR, they obtained subspecies information that could be extremely useful in a bioterrorism event for determining whether an intentional release occurred; further, it could provide epidemiological evidence to link a potential source to infected patients, and identify where the strain originated.

Real-time PCR assays for biothreat agents other than *Bacillus anthracis* have been developed and utilized in recent years (Newby *et al.*, 2003; Tomaso *et al.*, 2003; Debeaumont *et al.*, 2005; Jones *et al.*, 2005; Christensen *et al.*, 2006). While many of these assays were designed for diagnosis of agents such as *Brucella* spp., *F. tularensis*, and *Yersinia pestis* in a biothreat situation, they have been particularly useful for the diagnosis of naturally occurring infections in areas endemic for these bacteria. After a primary clinical diagnosis of infection with a biothreat agent, it may be difficult to determine whether the source of infection is a true intentional bioterrorist event. The following case highlights the importance of a rapid diagnosis in order to confirm a clinical diagnosis and begin an epidemiological investigation. In 2003, in New York City, a male and female presented at in an emergency room with symptoms consistent with bubonic plague (CDC, 2003). Specimens were collected, and *Y. pestis* was quickly identified by the real-time PCR assay that is available to public health laboratories that are members of the LRN. Because plague is not endemic in New York State, there was a suspicion of bioterrorism. An epidemiological investigation quickly determined that the infection was of natural origin. The ill couple had just recently arrived in New York from New Mexico, where plague is endemic. *Yersinia pestis* had been isolated from fleas trapped in the couple's backyard in New Mexico. Additional detailed genetic analyses were used to determine that the pathogen that caused the infection in the couple had originated in the flea population behind their home.

Commercial kits and reagents have become available for use in biothreat detection (Table 4.3). There are a few kits that have been developed to test clinical specimens for *B. anthracis*. However, the majority of the available kits and assays were designed to be used with environmental samples and have not been validated on clinical specimens, leaving a critical gap in diagnostic availability of reagents in a biothreat event.

Several real-time PCR assays have been reported in the literature for use with clinical specimen types for biothreat detection (Table 4.4). For pathogens thought to be viable biothreat agents, there is at least one published real-time PCR assay evaluated in at least one clinical specimen type. One example is the detection of *Y. pestis* in sputum (Loiez *et al.*, 2003). Sputum is known to contain substances that inhibit the PCR reaction, a fact which caused the authors to modify their extraction procedure. This and other published assays highlight the importance of optimization of real-time PCR assays as well as the inclusion of a step to monitor inhibition in each sample tested.

One pathogen clearly highlights the need for rapid diagnosis within hours. Intoxication by *Clostridium botulinum* toxin causes a descending paralysis that can occur extremely quickly, especially in infants. Akbulant and colleagues recently reported the development of a real-time PCR assay for the detection *C. botulinum* toxins A, B, and E from serum, tissue samples, or wound swabs (Akbulut *et al.*, 2004). The use of real-time PCR is essential when *C. botulinum* toxin is suspected, because the gold-standard testing method, the mouse bioassay, can require several days to produce a result. Further, the laboratory may not be supplied with the appropriate amount of specimen required for the mouse bioassay, particularly for infant cases. Real-time PCR can serve as a screen for presumptive toxin identification, as well as a screen for associated environmental samples to determine the source of the infection. Our laboratory has determined that real-time PCR analysis for *C. botulinum* toxin is invaluable in this regard, given that many sample types, such as stool, food, water, and soil samples, can contain competing organisms and other *Clostridium* species. Real-time PCR can be used as a screening tool, and positive isolates can then be identified for testing in the mouse bioassay.

This strategy significantly reduces the number of mice needed for the bioassay, as well as providing a cost savings, since only a subset of the original bacterial colonies must be tested.

While the number of published real-time PCR assays for biothreats has remained steady in the last few years, there continues to be very few papers published on the use of these real-time PCR assays in different clinical specimen types. Often, significant research needs to be done in order to develop PCR protocols for extraction of DNA from various specimen types which can hinder the number of laboratories that are interested in pursuing this avenue of research. The continued recognition of the importance of developing assays for real-time PCR is a positive sign. However, there need to be initiatives for supporting research into the validation of these assays for use with clinical specimens. In addition, support must be available to enable the publication of this data so that other laboratories may adopt well designed and validated assays for routine use. The paucity of nationally validated assays could have serious consequences if and when another biothreat attack occurs. Often, the development of such protocols and the validation of appropriate number of specimens requires significant allocation of resources that would be difficult to free up during a biothreat event, when there will be large numbers of specimens that must be analysed.

Environmental diagnostic applications

The processing and analysis of environmental samples are multifaceted tasks. While there exists a limited number of typical clinical matrices (e.g. blood, serum, urine, sputum, tissue), there are a myriad of highly diverse, complex matrix types in the environment. From the simplest matrix (potable water) to the most complex matrices (raw sewage, soil, or foods) the primary sample processing and extraction will be critical to the success of real-time PCR analyses. Each matrix type has particular characteristics that necessitate specific protocols to achieve efficient recovery of nucleic acids, while excluding confounding materials (Lewis *et al.*, 2000). For example, humic and fulvic acids commonly found in soil will hinder PCR amplification (Kreader, 1996; Miller, 2001) as will metal chelators and heavy metals in certain foods (Ijzerman *et al.*, 1997; Kreader, 1996) and haem in blood (Akane *et al.*, 1994). Wilson provides an excellent review of the factors that are known to inhibit PCR amplification (Wilson, 1997). Unfortunately, it is impossible to validate extraction procedures on every possible environmental sample matrix, so most kits or methods are marketed for one or two general sample matrix types like soil, blood, or stool. Therefore, when new or atypical matrices are being processed, it is essential to include positive and internal controls in each extraction procedure.

Each class of these PCR-confounding materials (e.g. polymerase inhibitors, nucleases, extreme pH modifiers) requires implementation of defined steps to isolate the amplifiable nucleic acids from the inhibitory compounds (Tsai and Olson, 1992; de Franchis *et al.*, 1988; Kuske *et al.*, 1998; Filion *et al.*, 2003; Sutlovic *et al.*, 2005). Fortunately, the basic methods for nucleic acid extraction have been well characterized, and solid-phase extraction procedures are used routinely in most research and diagnostic laboratories (Yamamoto, 2002). The many extraction kits available for nucleic acid processing provide differing levels of PCR-inhibitor mitigation. In addition, certain pathogens, like spore-forming bacteria, require harsher or additional processing steps to release genetic material for PCR-based detection. For all of these reasons, the processing and PCR analysis of environmental samples require significant preliminary testing and evaluation, in addition to inclusion of processing controls.

To facilitate better understanding by the reader of the various applications of real-time PCR, this section is organized according to end-product applications: liquid testing, food testing, solids testing (e.g. powders, soils), and air testing. In addition, the areas of forensic PCR and automated PCR for environmental monitoring will be discussed.

Real-time PCR analysis of liquids and foods

Water quality testing is the most common environmental testing enterprise in developed countries and one of the highest priority public health concerns in underdeveloped countries (Pedley and Pond, 2003). Comprehensive assessment of water quality is complex, and many diagnostic methods are required to ensure a safe potable water supply. Because water quality testing is a monitoring/surveillance activity meant to ensure integrity of product, it is not routinely performed in response to a defined incident. PCR-based analysis constitutes only a small proportion of the diagnostic tests used to assess water quality. Determinations of turbidity, pH, metal ion levels, and general bioburden far outweigh analysis by PCR for specific pathogens in the water (Shepard *et al.*, 2006). Initial application of PCR to determine bioburden was found to be less cost-effective than the use of classical methods, so use of PCR for this assessment has not been extensive. Instead of generic bioburden determination, levels of certain indicator organisms have been shown to be predictive of water quality, and PCR screens for these indicator organisms have been developed (Lebuhn *et al.*, 2004). Because of dilution factors in municipal water supplies, and because of the fact that many of the select agents can be found naturally in the environment somewhere in the U.S., PCR-based analysis for many select agents (Table 4.1) has not been performed routinely, even though real-time PCR methods are available for these targets (Ivnitski *et al.*, 2003; McAvin *et al.*, 2003; Bode *et al.*, 2004; McAvin *et al.*, 2004 ; Christensen *et al.*, 2006); (Tomaso *et al.*, 2003; Chase *et al.*, 2005). The most common and widespread application of PCR to water testing is PCR monitoring for trace levels of enteropathogenic *Escherichia coli* O157:H7 either at the water source or after post-processing at waste treatment facilities (Campbell *et al.*, 2001; Ibekwe *et al.*, 2002; Ibekwe *et al.*, 2004). Additional applications to detect viruses (Carducci *et al.*, 2003; Tong and Lu, 2011) and parasites (Fontaine and Guillot, 2002) are becoming more common in public water sources and recreational waters, as is screening of livestock run-off wastewater for bacteria carrying antibiotic resistance genes (Volkmann *et al.*, 2004).

Solid (e.g. beef, chicken, fresh produce) and liquid (e.g. milk, carbonated beverages, baby formula) foods are also amenable to diagnosis by real-time PCR, although the method is not routinely employed for them, because of cost considerations (Perelle *et al.*, 2004). Similar to water testing, the most common application of real-time PCR to food testing is the detection of enteropathogenic *E. coli* O157:H7 (Demarco and Lim, 2002; Nguyen *et al.*, 2004), typically in beef. Additional tests for *Listeria* (Nguyen *et al.*, 2004) and *Staphylococcus aureus* in beef (Alarcon *et al.*, 2006) and *Salmonella* in meat products (Perelle *et al.*, 2004), have been reported. These directed diagnostics were developed because the pathogens of interest are commonly associated with certain livestock or dairy products and are known to cause some level of morbidity or mortality in consumers. Clearly, the ability to screen food for pathogenic bacteria is a major focus of food biosecurity (McKillip and Drake, 2004) as is the ability to detect plant pathogens on crops (Schaad *et al.*, 2003).

Botulism, one form of food poisoning, is caused by ingestion of a toxin secreted by the bacterium *Clostridium botulinum*; this toxin is one of the most toxic substances known.

Botulism is typically associated with home-canned foods, although the bacteria can occur naturally in raw honey and the soil. Because of its extreme toxicity, ease of acquisition, and production, botulinum toxin is classed as a select agent. Therefore, detection of the toxin or the bacterial DNA in solid (Yoon *et al.*, 2005) or liquid (Perdue, 2003) foods is considered a high priority for public health and national security. Several real-time PCR methods exist to detect these signatures in food and environmental matrices (Szabo *et al.*, 1994; Lindberg, 2010; Sachdeva *et al.*, 2010).

Real-time PCR analysis of soils and powders

Whereas real-time PCR analysis of water, and of liquid and solid foods, is done for surveillance, analysis of soil or powders is typically done in response to a suspected or known event, be it a natural outbreak or intentional contamination. Hence, many of the real-time PCR assays that have been developed for detection of biothreat agents are employed for soil and/or powder matrices (Ivnitski *et al.*, 2003). Not surprisingly, identification of *B. anthracis* from powders associated with mail has become a routine procedure for many public health laboratories since 2001 (Teshale *et al.*, 2002). Real-time PCR assays for *B. anthracis* have also been validated for samples in soil matrices (Beyer *et al.*, 1995; Alam *et al.*, 2003; Cheun *et al.*, 2003; Ryu *et al.*, 2003), since *B. anthracis* spores can survive in soils for decades.

Since the intentional release of *B. anthracis* through the US Postal System, there was an effort by a number of laboratories to develop and validated methods that could be used in the event of another anthrax investigation. In the years since this bioterror event, there have been several cases of anthrax in humans in NYC, CT and most recently, New Hampshire that have been associated with unintentional infections (CDC, 2010; Guh, 2010; Mayo *et al.*, 2010; Nguyen, 2010). Each of these cases involved intensive investigations of the dwellings and other associated areas of the infected individuals. Numerous environmental samples were tested, mainly by LRN reference laboratories. Environmental wipes and swabs were tested to determine the extent of contamination. The methods used in these investigations were developed and validated through the Laboratory Response Network and included both culture and PCR-based testing. Following any future bioterrorism event, reclamation of buildings and the surrounding environment is likely to entail a process of repeated testings for remaining viable spores. Other assays for detection of biothreats in soils/powders include detection of *Burkholderia* spp. (Miller *et al.*, 2002; Salles *et al.*, 2002), *Y. pestis* (McAvin *et al.*, 2003), *F. tularensis* (Versage *et al.*, 2003), and *E. coli* O157:H7 (Campbell *et al.*, 2001; Ibekwe *et al.*, 2004).

Real-time PCR analysis of air samples

Prior to 2001, air monitoring was performed to detect chemical contaminants or pollution, but was rarely employed to test for biological organisms. Since that time, however, air monitoring for biothreats has become a national biodefence focus (Lim *et al.*, 2005). Now, operating throughout the US Postal Service system is the Biohazard Detection System, or BDS, which currently screens all mailed letters for *B. anthracis*. The BDS system collects air from 'pinch-points' in the mail processing stream and subjects the collected material to real-time PCR analysis. The system has been operating in postal facilities since 2003, with exemplary results (Jaffer, 2004). Additionally, the BioWatch system is operating in many major US metropolitan areas. BioWatch collects air samples and analyses them for several biothreat agents. Both BioWatch and BDS operate in near-real time and have emergency response protocols incorporated in case of biothreat detection. Other automated air sampling systems are in development and are expected to reduce analysis time and operational costs, while expanding the number of pathogens detected. One such system, the autonomous pathogen detection system (APDS) has been deployed for field evaluation and others are also in various stages of development (McBride *et al.*, 2003; Hindson *et al.*, 2005).

Multiplex real-time PCR in biodefense

The foregoing sections have highlighted the value of real-time PCR in situations when a biothreat agent is suspected. In certain cases, the causative agent may be unclearly identified, or it may be one among a list of bacterial or viral agents. In the early stages of the anthrax events of 2001, anthrax could potentially be misdiagnosed in patients as influenza due to similar symptomatology (Anon, 2001; CDC, 2001). Many biothreat agents initially present with symptoms similar to those of the flu. Given this lack of symptomatic specificity, it is likely that a particular specimen will require testing for a number of agents. Multiplex real-time PCR can provide the necessary tool in this situation. Until recently, multiplexed assays could be used to reliably detect up to four or five target analytes in one tube. The latest generation instruments such as the InSyte from Biobank and the Rotorgene 6000 from Corbett research are reported to be able to detect at least six or seven different analytes by means of multiple channels that can distinguish different wavelengths in a single PCR assay (Kubista *et al.*, 2006). Newer technology such as the OpenArray platform from AB Biosystems has the ability to greatly increase the multiplexing ability through PCR but it has yet to be evaluated in the biothreat testing arena.

Multiplexed real-time PCR can be used for detection of multiple agents, to confirm pathogen isolates by using multiple targets, or to differentiate an organism from closely related organisms at the species level. Multiplex PCR allows for the simultaneous detection of multiple genes in one organism such as the two plasmids (pX01 and pX02) necessary for virulence in *B. anthracis* to distinguish it from its closely related near neighbours. Several real-time PCR kits that have been developed that are syndromic in design, (i.e. based on patient symptomatology); however most of the commercially available kits for detection of biothreat agents have been designed and validated using environmental samples (Table 4.2). Multiplex PCR assays have been developed for select agents in clinical specimens such as *B. anthracis* (Ramisse *et al.*, 1996; Bell *et al.*, 2002; Ryu *et al.*, 2003), *Y. pestis* (Melo *et al.*, 2003; Tomaso *et al.*, 2003), *F. tularensis* (Sjostedt *et al.*, 1997; Grunow *et al.*, 2000; Versage *et al.*, 2003), *C. botulinum* toxins (Lindstrom *et al.*, 2001), *Staphylococcus aureus* enterotoxin type B (SEB) (Letertre *et al.*, 2003), Orthopoxviruses (Kulesh *et al.*, 2004; Panning *et al.*, 2004), and viral haemorrhagic fever viruses (Bronzoni *et al.*, 2004). For a review of multiplex assays for biothreat agents, see Cirino *et al.* (2004). The literature published on multiplex real-time PCR assays is small, and few of these assays have been applied to the diagnosis of infectious diseases (Mackay, 2004). Assays have been developed in the branches of clinical microbiology that typically investigate multiple organisms, such as common respiratory pathogens (Templeton *et al.*, 2004) or food-borne bacteria (Fout *et al.*, 2003). In the last few years, however, more assays have been reported, including several for biothreat agents.

Future prospects

A breakthrough technology is defined as one that rapidly displaces other technologies and renders them obsolete. PCR was one such breakthrough technology, for genetic manipulation and nucleic acid detection. The last 5 years have seen an explosive growth in the number of new technologies being developed for diagnostics, and the next 5 years should see some of these nascent technologies becoming widely employed for detection of biothreat agents.

Real-time PCR technology has existed for almost 10 years and is likely to be displaced within the next decade, although the ability to better multiplex this elegant platform may keep it viable for a longer period. Availability of new dyes, better methods for optical discrimination, and more powerful software will be critical to the longevity of real-time PCR. Microsphere arrays, microarrays, and biosensors are still in their early evolutionary phases but will likely become significant technologies during the next 5 years. The limitations of the technology, such as expense and time for development, will decrease as these platforms are increasingly combined with other technologies such as magnetic separation, biofilms, microfluidics, and fibre optics. It is likely that each successful technology will develop and expand into a defined analytical niche.

Advances in reporter technologies will also play a significant role in determining what types of methodologies become widely employed. Novel fluorophores and fluorescent quenchers are being developed, and these will be critical in fluorescent or FRET-based diagnostic platform multiplexing. Similarly, up-conversion of phosphor reporters, which have been already been applied to monitor changes in RNA expression levels (van De Rijke *et al.*, 2001), may prove useful in the detection of biothreat agents. More diverse reporter compounds will improve the sensitivity of various platforms and will offer significant advantages for multiplexing.

Another area in which breakthroughs are likely to occur in the next 5 years is sample processing. The various detection platforms require very diverse pre-analysis processing, a fact which makes a unified diagnostic algorithm difficult to develop. Better software, robotics, and understanding of the field of diagnostics on a global basis should facilitate continued development improvements in sample processing methodologies. Current lab-on-a-chip technology should evolve into a broadly integrated component of diagnostics.

Although real-time PCR was developed over a decade ago, it has only recently evolved to become sufficiently mature and cost-effective for integration into the routine workflow of diagnostic labs and public health laboratories. Real-time PCR has been used for some time in the pharmaceutical industry for drug target quantitation, genetic screening, and tracking of genetically modified organisms (GMO). With recent advances in real-time PCR, the field of biodefence may have the ability to have access to assays in coming years that will be able to determine numbers of spores or cells in a sample, type the biothreat agent present, and even determine the viability of the organisms.

The ongoing improvements in real-time PCR analytical systems and software, combined with reduced cost and the steady increasing ability to multiplex analytical assays, should ensure a key role for real-time PCR in both diagnostic laboratories and in biodefence research and applications for another decade or more.

References

Akane, A., Matsubara, K., Nakamura, H., Takahashi, S., and Kimura, K. (1994). Identification of the heme compound copurified with deoxyribonucleic acid (DNA) from bloodstains, a major inhibitor of polymerase chain reaction (PCR) amplification. J. Forensic Sci. *39*, 362–372.

Akbulut, D., Grant, K.A., and McLauchlin, J. (2004). Development and application of Real-Time PCR assays to detect fragments of the Clostridium botulinum types A, B, and E neurotoxin genes for investigation of human foodborne and infant botulism. Foodborne Pathog. Dis. *1*, 247–257.

Alam, S.I., Agarwal, G.S., Kamboj, D.V., Rai, G.P., and Singh, L. (2003). Detection of spores of *Bacillus anthracis* from environment using polymerase chain reaction. Indian J. Exp. Biol. *41*, 177–180.

Alarcon, B., Vicedo, B., and Aznar, R. (2006). PCR-based procedures for detection and quantification of Staphylococcus aureus and their application in food. J. Appl. Microbiol. *100*, 352–364.

Ash, C., and Collins, M.D. (1992). Comparative analysis of 23S ribosomal RNA gene sequences of *Bacillus anthracis* and emetic *Bacillus cereus* determined by PCR-direct sequencing. FEMS Microbiol. Lett. *73*, 75–80.

Ash, C., Farrow, J.A., Dorsch, M., Stackebrandt, E., and Collins, M.D. (1991). Comparative analysis of *Bacillus anthracis, Bacillus cereus,* and related species on the basis of reverse transcriptase sequencing of 16S rRNA. Int. J. Syst. Bacteriol. *41*, 343–346.

Bell, C.A., Uhl, J.R., Hadfield, T.L., David, J.C., Meyer, R.F., Smith, T.F., and Cockerill, F.R., 3rd. (2002). Detection of *Bacillus anthracis* DNA by LightCycler PCR. J. Clin. Microbiol. *40*, 2897–2902.

Beyer, W., Glockner, P., Otto, J., and Bohm, R. (1995). A nested PCR method for the detection of *Bacillus anthracis* in environmental samples collected from former tannery sites. Microbiol. Res. *150*, 179–186.

Bode, E., Hurtle, W., and Norwood, D. (2004). Real-time PCR assay for a unique chromosomal sequence of *Bacillus anthracis*. J. Clin. Microbiol. *42*, 5825–5831.

Bronzoni, R.V., Moreli, M.L., Cruz, A.C., and Figueiredo, L.T. (2004). Multiplex nested PCR for Brazilian Alphavirus diagnosis. Trans. R. Soc. Trop. Med. Hyg. *98*, 456–461.

Bystrom, M., Bocher, S., Magnusson, A., Prag, J., and Johansson, A. (2005). Tularemia in Denmark: identification of a *Francisella tularensis* subsp. holarctica strain by real-time PCR and high-resolution typing by multiple-locus variable-number tandem repeat analysis. J. Clin. Microbiol. *43*, 5355–5358.

Campbell, G.R., Prosser, J., Glover, A., and Killham, K. (2001). Detection of *Escherichia coli* O157:H7 in soil and water using multiplex PCR. J. Appl. Microbiol. *91*, 1004–1010.

Canter, D.A., Gunning, D., Rodgers, P., O'Connor, L., Traunero, C., and Kempter, C.J. (2005). Remediation of *Bacillus anthracis* contamination in the U.S. Department of Justice mail facility. Biosecur. Bioterror. *3*, 119–127.

Carducci, A., Casini, B., Bani, A., Rovini, E., Verani, M., Mazzoni, F., and Giuntini, A. (2003). Virological control of groundwater quality using biomolecular tests. Water Sci. Technol. *47*, 261–266.

C.D.C. (2001). Considerations for distinguishing influenza-like illness from inhalational anthrax. MMWR Morb. Mortal. Wkly Rep. *50*, 984–986.

C.D.C. (2003). Imported plague – New York City, 2002. MMWR Morb. Mortal. Wkly Rep. *52*, 725–728.

C.D.C. (2010). Gastrointestinal anthrax after an animal-hide drumming event – New Hampshire and Massachusetts, 2009. MMWR Morb. Mortal. Wkly Rep. 2010 *59*, 872–877.

Chase, C.J., Ulrich, M.P., Wasieloski, L.P., Jr., Kondig, J.P., Garrison, J., Lindler, L.E., and Kulesh, D.A. (2005). Real-time PCR assays targeting a unique chromosomal sequence of *Yersinia pestis*. Clin. Chem. *51*, 1778–1785.

Cheun, H.I., Makino, S.I., Watarai, M., Erdenebaatar, J., Kawamoto, K., and Uchida, I. (2003). Rapid and effective detection of anthrax spores in soil by PCR. J. Appl. Microbiol. *95*, 728–733.

Christensen, D.R., Hartman, L.J., Loveless, B.M., Frye, M.S., Shipley, M.A., Bridge, D.L., Richards, M.J., Kaplan, R.S., Garrison, J., Baldwin, C.D., *et al.* (2006). Detection of biological threat agents by real-time PCR: comparison of assay performance on the R.A.P.I.D., the LightCycler, and the Smart Cycler platforms. Clin. Chem. *52*, 141–145.

Cirino, N.M., Musser, K.A., and Egan, C. (2004). Multiplex diagnostic platforms for detection of biothreat agents. Expert Rev. Mol. Diagn. *4*, 841–857.

Cirino, N.M., Shail, M., and Egan, C. (2006). Biodefense: New Research (Nova Science Publishing, Inc., New York).

Cockerill, F.R., 3rd. (2003). Application of rapid-cycle real-time polymerase chain reaction for diagnostic testing in the clinical microbiology laboratory. Arch. Pathol. Lab. Med. *127*, 1112–1120.

Debeaumont, C., Falconnet, P.A., and Maurin, M. (2005). Real-time PCR for detection of *Brucella* spp. DNA in human serum samples. Eur. J. Clin. Microbiol. Infect. Dis. *24*, 842–845.

Demarco, D.R., and Lim, D.V. (2002). Detection of *Escherichia coli* O157:H7 in 10- and 25-gram ground beef samples with an evanescent-wave biosensor with silica and polystyrene waveguides. J. Food Prot. *65*, 596–602.

Drago L., Lombardi, A., Vecchi, E.D., Gismondo, M.R. (2002). Real-time PCR assay for rapid detection of *Bacillus anthracis* spores in clinical samples. J. Clin. Microbiol. *40*, 4399.

Drosten C., Göttig, S., Schilling, S., Asper, M., Panning, M., Schmitz, H., and Günther, S. (2002). Rapid detection and quantification of RNA of Ebola and Marburg viruses, Lassa virus, Crimean-Congo hemorrhagic fever virus, Rift Valley fever virus, dengue virus, and yellow fever virus by real-time reverse transcription-PCR. J Clin Microbiol. *40*, 2323–30.

Edwards, K.A., Clancy, H.A., and Baeumner, A.J. (2006). *Bacillus anthracis*: toxicology, epidemiology and current rapid-detection methods. Anal. Bioanal. Chem. *384*, 73–84.

Elsholz, B., Nitche, A., Achenbach, J., Ellerbrok, H., Blohm, L., Albers, J., Pauli, G., Hintsche, R., and Worl, R. (2009). Electrical microarrays for highly sensitive detection of multiplex PCR products from biological agents. Biosens. Bioelectron. *24*, 1737–1743.

Filion, M., St-Arnaud, M., and Jabaji-Hare, S.H. (2003). Direct quantification of fungal DNA from soil substrate using real-time PCR. J. Microbiol. Methods *53*, 67–76.

Fontaine, M., and Guillot, E. (2002). Development of a TaqMan quantitative PCR assay specific for *Cryptosporidium parvum*. FEMS Microbiol. Lett. *214*, 13–17.

Fournier, P.E., and Raoult, D. (2003). Comparison of PCR and serology assays for early diagnosis of acute Q fever. J. Clin. Microbiol. *41*, 5094–5098.

Fout, G.S., Martinson, B.C., Moyer, M.W., and Dahling, D.R. (2003). A multiplex reverse transcription-PCR method for detection of human enteric viruses in groundwater. Appl. Environ. Microbiol. *69*, 3158–3164.

de Franchis, R., Cross, N.C., Foulkes, N.S., and Cox, T.M. (1988). A potent inhibitor of Taq polymerase copurifies with human genomic DNA. Nucleic Acids Res. *16*, 10355.

Grunow, R., Splettstoesser, W., McDonald, S., Otterbein, C., O'Brien, T., Morgan, C., Aldrich, J., Hofer, E., Finke, E.J., and Meyer, H. (2000). Detection of *Francisella tularensis* in biological specimens using a capture enzyme-linked immunosorbent assay, an immunochromatographic handheld assay, and a PCR. Clin. Diagn. Lab. Immunol. *7*, 86–90.

Guh A., Heyman, M.L., Barden, D., Fontana, J., and Hadler, J.L. (2010). Lessons learned from the investigation of a cluster of cutaneous anthrax cases in Connecticut. J. Public Health Manag. Pract. *16*, 201–210.

Halse, T.A., Musser, K.A., and Limberger, R.J. (2006). A multiplexed real-time PCR assay for rapid detection of chlamydia trachomatis and identification of serovar L-2, the major cause of lymphogranuloma venereum in New York. Mol. Cell. Probes *20*, 290–297.

Higgins, J.A., Cooper, M., Schroeder-Tucker, L., Black, S., Miller, D., Karns, J.S., Manthey, E., Breeze, R., and Perdue, M.L. (2003). A field investigation of *Bacillus anthracis* contamination of U.S. Department of Agriculture and other Washington, D.C., buildings during the anthrax attack of October 2001. Appl. Environ. Microbiol. *69*, 593–599.

Hindson, B.J., Makarewicz, A.J., Setlur, U.S., Henderer, B.D., McBride, M.T., and Dzenitis, J.M. (2005). APDS: the autonomous pathogen detection system. Biosens. Bioelectron. *20*, 1925–1931.

Hoffmaster, A.R., Fitzgerald, C.C., Ribot, E., Mayer, L.W., and Popovic, T. (2002a). Molecular subtyping of *Bacillus anthracis* and the 2001 bioterrorism-associated anthrax outbreak, United States. Emerg. Infect. Dis. *8*, 1111–1116.

Hoffmaster, A.R., Meyer, R.F., Bowen, M.D., Marston, C.K., Weyant, R.S., Thurman, K., Messenger, S.L., Minor, E.E., Winchell, J.M., Rassmussen, M.V., *et al.* (2002b). Evaluation and validation of a real-time polymerase chain reaction assay for rapid identification of *Bacillus anthracis*. Emerg. Infect. Dis. *8*, 1178–1182.

Hong-Geller, E., Valdez, Y.E., Shou, Y., Yoshida, T.M., Marrone, B.L., and Dunbar, J.M. (2010). Evaluation of *Bacillus anthracis* and *Yersinia pestis* sample collection from nonporous surfaces by quantitative real-time PCR. Lett Appl Microbiol. *50*, 431–437.

Ibekwe, A.M., Watt, P.M., Grieve, C.M., Sharma, V.K., and Lyons, S.R. (2002). Multiplex fluorogenic real-time PCR for detection and quantification of *Escherichia coli* O157:H7 in dairy wastewater wetlands. Appl. Environ. Microbiol. *68*, 4853–4862.

Ibekwe, A.M., Watt, P.M., Shouse, P.J., and Grieve, C.M. (2004). Fate of *Escherichia coli* O157:H7 in irrigation water on soils and plants as validated by culture method and real-time PCR. Can. J. Microbiol. *50*, 1007–1014.

Ijzerman, M.M., Dahling, D.R., and Fout, G.S. (1997). A method to remove environmental inhibitors prior to the detection of waterborne enteric viruses by reverse transcription-polymerase chain reaction. J. Virol. Methods *63*, 145–153.

Ivnitski, D., O'Neil, D.J., Gattuso, A., Schlicht, R., Calidonna, M., and Fisher, R. (2003). Nucleic acid approaches for detection and identification of biological warfare and infectious disease agents. Biotechniques *35*, 862–869.

Jaffer, A.S. (2004). Biohazard detection system deployment resumes. U. S. Postal Service Postal News *38*.

Janse, I., Hamidjaja, R.A., Bok, J.M., van Rotterdam, B.J. (2010). Reliable detection of *Bacillus anthracis, Francisella tularensis* and *Yersinia pestis* by using multiplex qPCR including internal controls for nucleic acid extraction and amplification. BMC Microbiol. *10*(1), 314.

Jones, S.W., Dobson, M.E., Francesconi, S.C., Schoske, R., and Crawford, R. (2005). DNA assays for detection, identification, and individualization of select agent microorganisms. Croat. Med. J. *46*, 522–529.

Kreader, C.A. (1996). Relief of amplification inhibition in PCR with bovine serum albumin or T4 gene 32 protein. Appl. Environ. Microbiol. *62*, 1102–1106.

Kubista, M., Andrade, J.M., Bengtsson, M., Forootan, A., Jonak, J., Lind, K., Sindelka, R., Sjoback, R., Sjogreen, B., Strombom, L., *et al.* (2006). The real-time polymerase chain reaction. Mol. Aspects Med. *27*, 95–125.

Kulesh, D.A., Baker, R.O., Loveless, B.M., Norwood, D., Zwiers, S.H., Mucker, E., Hartmann, C., Herrera, R., Miller, D., Christensen, D., *et al.* (2004). Smallpox and pan-orthopox virus detection by real-time 3′-minor groove binder TaqMan assays on the roche LightCycler and the Cepheid smart Cycler platforms. J. Clin. Microbiol. *42*, 601–609.

Kuske, C.R., Banton, K.L., Adorada, D.L., Stark, P.C., Hill, K.K., and Jackson, P.J. (1998). Small-scale DNA sample preparation method for field PCR detection of microbial cells and spores in soil. Appl. Environ. Microbiol. *64*, 2463–2472.

Lebuhn, M., Effenberger, M., Garces, G., Gronauer, A., and Wilderer, P.A. (2004). Evaluating real-time PCR for the quantification of distinct pathogens and indicator organisms in environmental samples. Water Sci. Technol. *50*, 263–270.

Letertre, C., Perelle, S., Dilasser, F., and Fach, P. (2003). A strategy based on 5′ nuclease multiplex PCR to detect enterotoxin genes sea to sej of Staphylococcus aureus. Mol. Cell. Probes *17*, 227–235.

Lewis, G.D., Molloy, S.L., Greening, G.E., and Dawson, J. (2000). Influence of environmental factors on virus detection by RT-PCR and cell culture. J. Appl. Microbiol. *88*, 633–640.

Lim, D.V., Simpson, J.M., Kearns, E.A., and Kramer, M.F. (2005). Current and developing technologies for monitoring agents of bioterrorism and biowarfare. Clin. Microbiol. Rev. *18*, 583–607.

Lindberg A., Skarin, H., Knutsson, R., Blomqvist, G., and Båverud, V. (2010). Real-time PCR for *Clostridium botulinum* type C neurotoxin (BoNTC) gene, also covering a chimeric C/D sequence – application on outbreaks of botulism in poultry. Vet. Microbiol. *146(1–2)*:118–23.

Lindstrom, M., Keto, R., Markkula, A., Nevas, M., Hielm, S., and Korkeala, H. (2001). Multiplex PCR assay for detection and identification of *Clostridium botulinum* types A, B, E, and F in food and fecal material. Appl. Environ. Microbiol. *67*, 5694–5699.

Loiez, C., Herwegh, S., Wallet, F., Armand, S., Guinet, F., and Courcol, R.J. (2003). Detection of *Yersinia pestis* in sputum by real-time PCR. J. Clin. Microbiol. *41*, 4873–4875.

McAvin, J.C., McConathy, M.A., Rohrer, A.J., Huff, W.B., Barnes, W.J., and Lohman, K.L. (2003). A real-time fluorescence polymerase chain reaction assay for the identification of *Yersinia pestis* using a field-deployable thermocycler. Mil. Med. *168*, 852–855.

McAvin, J.C., Morton, M.M., Roudabush, R.M., Atchley, D.H., and Hickman, J.R. (2004). Identification of *Francisella tularensis* using real-time fluorescence polymerase chain reaction. Mil. Med. *169*, 330–333.

McBride, M.T., Masquelier, D., Hindson, B.J., Makarewicz, A.J., Brown, S., Burris, K., Metz, T., Langlois, R.G., Tsang, K.W., Bryan, R., *et al.* (2003). Autonomous detection of aerosolized *Bacillus anthracis* and *Yersinia pestis*. Anal. Chem. *75*, 5293–5299.

Mackay, I.M. (2004). Real-time PCR in the microbiology laboratory. Clin. Microbiol. Infect. *10*, 190–212.

McKillip, J.L., and Drake, M. (2004). Real-time nucleic acid-based detection methods for pathogenic bacteria in food. J.Food Prot. *67*, 823–832.

Melo, A.C., Almeida, A.M., and Leal, N.C. (2003). Retrospective study of a plague outbreak by multiplex-PCR. Lett. Appl. Microbiol. 37, 361–364.

Millar, B.C., Xu, J., and Moore, J.E. (2002). Risk assessment models and contamination management: implications for broad-range ribosomal DNA PCR as a diagnostic tool in medical bacteriology. J. Clin. Microbiol. *40*, 1575–1580.

Miller, D.N. (2001). Evaluation of gel filtration resins for the removal of PCR-inhibitory substances from soils and sediments. J. Microbiol. Methods *44*, 49–58.

Miller, S.C., LiPuma, J.J., and Parke, J.L. (2002). Culture-based and non-growth-dependent detection of the *Burkholderia cepacia* complex in soil environments. Appl. Environ. Microbiol. *68*, 3750–3758.

Mitchell, P.S., Germer, J.J., and Patel, R. (2004). Molecular Microbiology: Diagnostic Principles and Practice (ASM Press, Washington, DC).

Newby, D.T., Hadfield, T.L., and Roberto, F.F. (2003). Real-time PCR detection of *Brucella abortus*: a comparative study of SYBR green I, 5′-exonuclease, and hybridization probe assays. Appl. Environ. Microbiol. *69*, 4753–4759.

Nguyen, L.T., Gillespie, B.E., Nam, H.M., Murinda, S.E., and Oliver, S.P. (2004). Detection of *Escherichia coli* O157:H7 and *Listeria monocytogenes* in beef products by real-time polymerase chain reaction. Foodborne. Pathog. Dis. *1*, 231–240.

Nguyen T.Q., Clark, N.; 2006 N.Y.C Anthrax Working Group, Karpati, A., Goldberg, A., Paykin, A., Tucker, A., Baker, A., Almiroudis, A., Fine, A., *et al.* (2010). Public health and environmental response to the first case of naturally acquired inhalational anthrax in the United States in 30 years: infection of a new york city resident who worked with dried animal hides. J. Public Health Manag. Pract. *16*, 189–200.

Nitsche, A., Ellerbrok, H., and Pauli, G. (2004). Detection of orthopoxvirus DNA by real-time PCR and identification of variola virus DNA by melting analysis. J. Clin. Microbiol. *42*, 1207–1213.

Panning, M., Asper, M., Kramme, S., Schmitz, H., and Drosten, C. (2004). Rapid detection and differentiation of human pathogenic orthopox viruses by a fluorescence resonance energy transfer real-time PCR assay. Clin. Chem. *50*, 702–708.

Pedley, S., and Pond, K. (2003). Emerging issues in water and infectious disease (World Health Organization). Available at: http://www.who.int/water_sanitation_health/emerging/emerging.pdf

Perdue, M.L. (2003). Molecular diagnostics in an insecure world. Avian Dis. *47*, 1063–1068.

Perelle, S., Dilasser, F., Malorny, B., Grout, J., Hoorfar, J., and Fach, P. (2004). Comparison of PCR-ELISA and LightCycler real-time PCR assays for detecting *Salmonella* spp. in milk and meat samples. Mol. Cell Probes *18*, 409–420.

Persing, D.H. (1991). Polymerase chain reaction: trenches to benches. J. Clin. Microbiol. *29*, 1281–1285.

Ramisse, V., Patra, G., Garrigue, H., Guesdon, J.L., and Mock, M. (1996). Identification and characterization of *Bacillus anthracis* by multiplex PCR analysis of sequences on plasmids pXO1 and pXO2 and chromosomal DNA. FEMS Microbiol. Lett. *145*, 9–16.

Ryu, C., Lee, K., Yoo, C., Seong, W.K., and Oh, H.B. (2003). Sensitive and rapid quantitative detection of anthrax spores isolated from soil samples by real-time PCR. Microbiol. Immunol. *47*, 693–699.

Sachdeva A., Defibaugh-Chávez S.L., Day, J.B., Zink, D., and Sharma, S.K. (2010). Detection and confirmation of Clostridium botulinum in water used for cooling at a plant producing low-acid canned foods. Appl Environ Microbiol. *76*, 7653–7657.

Salles, J.F., De Souza, F.A., and van Elsas, J.D. (2002). Molecular method to assess the diversity of *Burkholderia* species in environmental samples. Appl. Environ. Microbiol. *68*, 1595–1603.

Schaad, N.W., Frederick, R.D., Shaw, J., Schneider, W.L., Hickson, R., Petrillo, M.D., and Luster, D.G. (2003). Advances in molecular-based diagnostics in meeting crop biosecurity and phytosanitary issues. Annu. Rev. Phytopathol. *41*, 305–324.

Shepard, J.E., Egan, C., and Cirino, N.M. (2006). Water Biosurveillance Systems (Elsevier Science/Academic Press, San Diego, CA).

Sjostedt, A., Eriksson, U., Berglund, L., and Tarnvik, A. (1997). Detection of *Francisella tularensis* in ulcers of patients with tularemia by PCR. J. Clin. Microbiol. *35*, 1045–1048.

Sutlovic, D., Definis, G.M., Andelinovic, S., Gugic, D., and Primorac, D. (2005). Taq polymerase reverses inhibition of quantitative real time polymerase chain reaction by humic acid. Croat. Med. J. *46*, 556–562.

Synder, J.W., Jungkind, D., and Persing, D.H. (2004). Molecular Microbiology: Diagnostic Principles and Practice (ASM Press, Washington, DC).

Szabo, E.A., Pemberton, J.M., Gibson, A.M., Eyles, M.J., and Desmarchelier, P.M. (1994). Polymerase chain reaction for detection of Clostridium botulinum types A, B and E in food, soil and infant faeces. J. Appl. Bacteriol. *76*, 539–545.

Templeton, K.E., Scheltinga, S.A., Beersma, M.F., Kroes, A.C., and Claas, E.C. (2004). Rapid and sensitive method using multiplex real-time PCR for diagnosis of infections by influenza a and influenza B viruses, respiratory syncytial virus, and parainfluenza viruses *1*, 2, 3, and 4. J. Clin. Microbiol. *42*, 1564–1569.

Teshale, E.H., Painter, J., Burr, G.A., Mead, P., Wright, S.V., Cseh, L.F., Zabrocki, R., Collins, R., Kelley, K.A., Hadler, J.L., *et al.* (2002). Environmental sampling for spores of *Bacillus anthracis*. Emerg. Infect. Dis. *8*, 1083–1087.

Tomaso, H., Reisinger, E.C., Al, D.S., Frangoulidis, D., Rakin, A., Landt, O., and Neubauer, H. (2003). Rapid detection of *Yersinia pestis* with multiplex real-time PCR assays using fluorescent hybridisation probes. FEMS Immunol. Med. Microbiol. *38*, 117–126.

Tong, H.I., and Lu, Y. (2011). Effective detection of human adenovirus in hawaiian waters using enhanced PCR methods. Virol. J. *8*, 57.

Tong, Y., Tang, W., Kim, H.J., Pan, X., Ranalli, T., and Kong, H. (2008). Development of isothermal TaqMan assays for detection of biothreat organisms. Biotechniques *45*, 543–547.

Tsai, Y.L., and Olson, B.H. (1992). Rapid method for separation of bacterial DNA from humic substances in sediments for polymerase chain reaction. Appl. Environ. Microbiol. *58*, 2292–2295.

van De Rijke, F., Zijlmans, H., Li, S., Vail, T., Raap, A.K., Niedbala, R.S., and Tanke, H.J. (2001). Up-converting phosphor reporters for nucleic acid microarrays. Nat. Biotechnol. *19*, 273–276.

Versage, J.L., Severin, D.D., Chu, M.C., and Petersen, J.M. (2003). Development of a multitarget real-time TaqMan PCR assay for enhanced detection of *Francisella tularensis* in complex specimens. J. Clin. Microbiol. *41*, 5492–5499.

Volkmann, H., Schwartz, T., Bischoff, P., Kirchen, S., and Obst, U. (2004). Detection of clinically relevant antibiotic-resistance genes in municipal wastewater using real-time PCR (TaqMan). J. Microbiol. Methods *56*, 277–286.

Wannamaker, B., Denio, L., Dodson, W.E., Dreifuss, F., Crosby, C., Santilli, N., Duffner, P., Ryan-Dudeck, P., Conboy, C., Ellis, E., *et al.* (1989). Immunochromatographic measurement of phenobarbital in whole blood with a non-instrumented assay. Neurology *39*, 1215–1218.

Whelen, A.C., and Persing, D.H. (1996). The role of nucleic acid amplification and detection in the clinical microbiology laboratory. Annu. Rev. Microbiol. *50*, 349–373.

Wilson, I.G. (1997). Inhibition and facilitation of nucleic acid amplification. Appl. Environ. Microbiol. *63*, 3741–3751.

Yamamoto, Y. (2002). PCR in diagnosis of infection: detection of bacteria in cerebrospinal fluids. Clin. Diagn. Lab. Immunol. *9*, 508–514.

Yoon, S.Y., Chung, G.T., Kang, D.H., Ryu, C., Yoo, C.K., and Seong, W.K. (2005). Application of real-time PCR for quantitative detection of *Clostridium botulinum* type A toxin gene in food. Microbiol. Immunol. *49*, 505–511.

Veterinary Applications of Real-time PCR for Detection and Diagnosis of Infectious Agents

5

Alan McNally

Abstract

The detection and diagnosis of veterinary infectious diseases is an area in which the potential of Real-time PCR has been best demonstrated. In particular, real-time PCR has been successfully applied as a front-line tool in the diagnostic algorithm for notifiable veterinary viral pathogens such as avian influenza, foot-and-mouth disease, bluetongue virus, as well as rabies and Newcastle disease virus. The rapidly transmissible nature of these agents necessitates near real-time detection and diagnosis in suspected infected animals to allow implementation of control procedures. This chapter will highlight the importance of real-time PCR in facilitating this rapid diagnosis, and the effect such rapid detection has had on containing and controlling veterinary infectious disease outbreaks.

Introduction

The advent of real-time PCR technology promised great advances in the field of detection and diagnostics of infectious agents, yet with the odd exception of a handful of large commercial successes in the human diagnostics market, this impact has been felt greatest in the field of veterinary microbiology. In particular, the detection of notifiable veterinary viral pathogens such as avian influenza, foot-and-mouth, Newcastle disease, bluetongue virus and rabies have been greatly impacted by the introduction of real-time PCR based detection assays, and represent possibly the greatest success stories in terms of the use of Real-time PCR for *in vitro* diagnostics (Hoffmann *et al.*, 2009). This chapter will review the impact of real-time PCR based detection on surveillance and transmission of veterinary viral pathogens, as well as the increasing impact on detection of bacterial infectious agents.

Avian influenza (AI)

Influenza viruses are segmented, ss-RNA orthomyxoviruses, and are divided into Influenza A, B, or C. Only type A Influenza viruses are known to cause infections in avian species. Influenza A viruses are further divided into subtypes based on the antigenic properties of the haemagglutinin (HA) and neuraminidase (NA) surface glycoproteins. To date 15 unique HA subtypes (H1–H15) and 9 NA subtypes (N1–N9) have been identified, with each virus having one HA and one NA protein, potentially in any combination (Alexander, 2007). Influenza viruses infecting domestic fowl are further classified based on their virulence in the host. Highly pathogenic avian influenza (HPAI) viruses cause a systemic disease with rapid onset of death and a mortality rate approaching 100%. Low pathogenic avian influenza viruses (LPAI) cause a localized infection with little or no disease, generally associated with secondary pathogens. To date, all HPAI isolates have been H5 or H7 subtypes, although not all H5 or H7 viruses cause HPAI (Alexander, 2007).

Highly pathogenic avian influenza (HPAI) is a disease of poultry with a high flock mortality rate. A number of HPAI outbreaks in poultry caused by viruses of the H5N1 subtype had been reported intermittently in China and Hong Kong since the mid 1990s, but in December 2003 and January 2004 eight countries in South-East and East Asia reported outbreaks (Sims *et al.*, 2005). By late July 2005, H5N1 had spread beyond its apparent original foci in East Asia to the Russian Federation and adjacent parts of Kazakhstan, with infection reported in wild birds without apparent associated outbreaks in poultry. By October 2005 the virus had spread to Europe, and was reported in Turkey, Romania and Croatia, and by December in Ukraine. By March 2006 the virus had spread further into western Europe, and by May 2006 a total of 18 countries had reported the presence of HPAI H5N1 virus, mostly in wild birds, with fewer incursions into domestic poultry (Brown *et al.*, 2007).

Current theories suggest that the spread of the virus, both from Asia and within Europe may be facilitated by the carriage of HPAI H5N1 by wild birds. This hypothesis is supported by the fact that the viruses isolated from the European outbreaks are almost identical to those recovered from 6000 dead migratory birds at Qinghai lake nature reserve and the surrounding area in central China in April 2005, which is purported to be devoid of poultry (Liu *et al.*, 2005).

Development of avian influenza diagnostics

All suspected cases of both low- and high-pathogenic avian influenza must be reported through an official reference laboratory, with confirmatory diagnosis only possible after characterization of virus isolated from inoculated embryonated egg (Anonymous). This process can take as long as 10 days from receipt of sample in the case of an old sample containing a low viral titre. As H5N1 spread rapidly from it's original foci it became clear that a test with a much more rapid turnaround was required for efficient, detection-based control strategies to be implemented. Two key published studies provided the platform for development of robust real-time PCR assays for use in routine diagnosis. The first was the development of a High-pathogenic specific assay for H5 and H7 viruses developed by the David Suarez group (Spackman *et al.*, 2002). The second was the first report of the standard use of Real-time PCR for avian influenza diagnosis during the 2003 Dutch outbreak of H7N1 (Elbers *et al.*, 2004; Fouchier *et al.*, 2004). These two landmark reports opened the way for the use of robust, reliable molecular testing for avian influenza, and the development of a validated H5 (Slomka *et al.*, 2007a) and H7 real-time RT-PCR assay through EU approved blind ring trials, for routine veterinary diagnostics (Slomka *et al.*, 2007b, 2009). Moreover the Eurasian outbreak of H5N1 presented an ideal opportunity for the validation of these assays in an outbreak environment (Slomka *et al.*, 2010).

Routine avian influenza testing

Following the successful development and utilization of reverse real-time PCR (RRT-PCR) tests for avian influenza, the use of molecular RRT-PCR assays in avian influenza reference laboratory detection and diagnosis is authorized by the World Organisation for Animal Health (OIE) and the Food and Agriculture Organization (FAO) (Breed *et al.*, 2010; Slomka *et al.*, 2010). Oropharyngeal or cloacal swabs, or infected tissues are processed using commercial RNA extraction protocols to isolate viral RNA. In larger reference laboratories this is often an automated process using extraction robotics. Once isolated, an Influenza A M gene specific RRT-PCR should be performed to determine the presence of an influenza A virus. Once performed, an HPAI H5 or H7 specific RRT-PCR should then be performed on the same RNA sample to determine the virotype present. Often the three tests are run simultaneously in parallel to speed up the process, giving a full diagnostic profile within 1 hour of having extracted RNA from sample (Slomka *et al.*, 2010). Laboratories may use their own in-house assays assuming they meet validation criteria set down by OIE or FAO standards, though increasingly most laboratories will use recommended primers and protocols developed collaboratively by reference laboratories (a good example is provided by the EU, OIE, FAO avian influenza reference laboratory at the Veterinary Laboratories Agency in the UK (http://www.defra.gov.uk/vla/science/sci_ai_reflab_prot.htm).

The benefits of molecular testing for avian influenza are enormous. Whilst it is still a requirement for confirmatory diagnosis to isolate infectious viral particles from a sample using embryonated eggs, all appropriate control procedures can now be implemented immediately following a RRT-PCR result, which has become the model for how all molecular real-time PCR tests are incorporated into the veterinary diagnostic algorithm (Fig. 5.1). This allows restriction of movement procedures and surveillance zones to be set up as quickly as 24 hours after suspected cases, limiting the opportunity for spread of infection. It is widely believed that such rapid detection and response times have limited the incursion of H5N1 virus into Western Europe. The other great benefit is the large number of tests which can be processed. Using RRT-PCR it is possible to perform batch processing of samples, from robotic automated extraction of RNA from samples through to PCR set up and results analysis. This allows the laboratory to analyse hundreds of samples per hour, depending on their PCR and sample extraction capabilities. Where this has most impact is in the EU wild bird surveillance programme, which screens the European wild bird population for HPAI H5 and H7 viruses (Breed *et al.*, 2010). Owing to the high-throughput nature of testing it is possible to analyse thousands of samples per week in EU influenza reference laboratories, greatly enhancing Europe's surveillance capacity for avian influenza, and assisting in the efforts to prevent H5N1 becoming an endemic virus in Western Europe.

Foot-and-mouth disease virus (FMDv)

Foot-and-mouth disease is a highly contagious disease of cloven-hoofed animals caused by the foot-and-mouth disease virus, FMDv, an *Aphthovirus* of the Picornaviridae. Disease in animals is characterized by the development of infectious vesicles on the tongue, dental pad and snout, as well as in the interdigital space of the hoof, and is accompanied by fever. The lesions lead to lameness, loss of appetite, excess salivation and depression in the animal with symptoms more severe in cattle and pigs than in sheep. The disease has an incubation period of anywhere between one and fourteen days depending on the infectious dose of virus the animal has been exposed to, with farm-to-farm incubation periods of 4–14 days (Belsham, 1993).

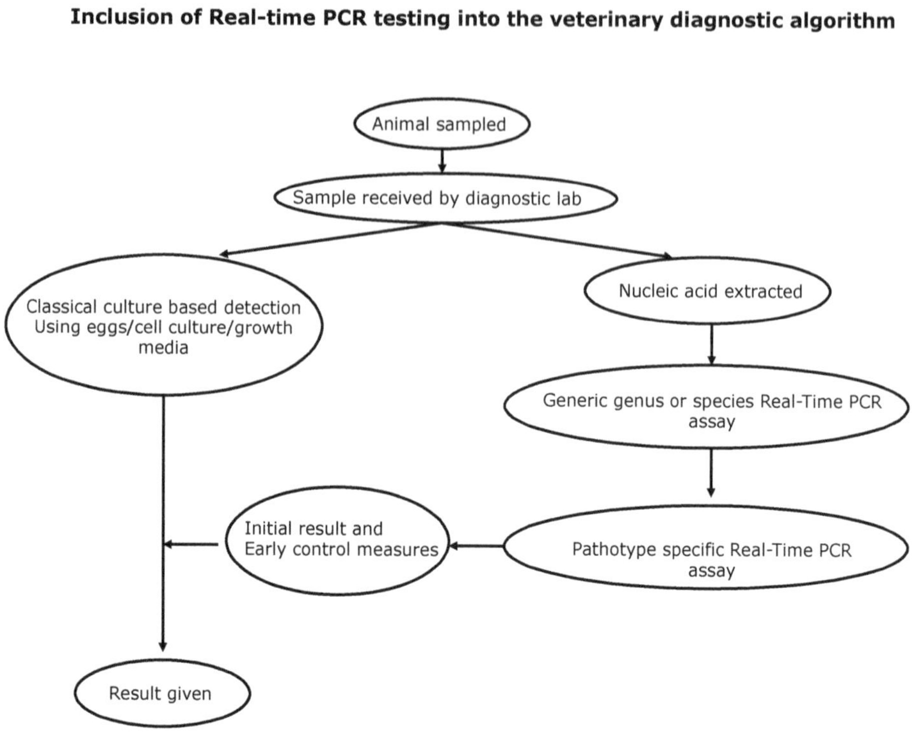

Figure 5.1 Flow diagram of the way in which real-time PCR is testing is incorporated into reference laboratory approved diagnostic algorithms for veterinary infectious diseases.

The virus is highly infectious and can be spread by direct contact between animals, by mechanical transmission via fomites, people, vehicles etc., and by airborne transmission. Contaminated animal products contain high titres of infectious viral particles and can also act as a route of transmission via ingestion, particularly in pigs. Infection of animals leads to devastating livestock production losses and prevention of animal and animal product trade. The 2001 FMD epidemic in the UK was the largest recorded outbreak of FMD alongside the 1967/68 UK epidemic, resulting in 2030 outbreaks of infection and a £3.1 billion loss to the UK agricultural industry (Thompson *et al.*, 2002).

Confirmatory diagnosis of FMD in livestock requires a positive serology test by ELISA or Complement fixation test, and/or isolation of infectious viral particles from samples using primary thyroid or kidney cells, BHK21 cell lines, or from infected mice, as prescribed by the OIE diagnostic manual (Anonymous). This results in an extended timeline for FMD confirmatory diagnosis of anywhere between 24 hours and 5 days. Eradication of the virus is through slaughter of infected animals and close contact animals in conjunction with controls on animal movements. Vaccines are available but offer only short lived immunity, and more importantly are virus-type specific, which creates a problem in that there are seven distinct serotypes of FMDv circulating. Vaccination is also avoided in non-endemic countries as vaccinated animals can revert to a carrier state if exposed to new virus infections without showing disease symptoms.

Benefits of molecular testing in controlling FMD outbreaks

In August 2007 an outbreak of FMD was confirmed on a farm in Surrey, resulting in a FMD outbreak lasting into September affecting a total of eight farms in the neighbouring counties of Surrey and Berkshire (Ryan *et al.*, 2008). The duration and effect of the 2007 episode is in stark contrast to the 2001 UK epidemic of 2030 outbreaks, and is attributable to the first ever use of RRT-PCR in diagnostic support during an FMD epidemic. All samples received by the FMD world reference laboratory (Institute of Animal Health, Pirbright, Surrey, UK) were subjected to automated viral RNA extraction alongside conventional virology and/or serology testing (Reid *et al.*, 2009). The RRT-PCR strategy used was a dual target approach using primers designed to amplify highly conserved regions of the virus genome allowing pan-serotype detection of FMDv. Specifically the reference laboratory used a combination of an assay to detect the 5′ untranslated region (5′ UTR) of the virus, and the RNA polymerase (3D) region of the RNA genome. Developmental research on these assays showed that when used in concert they provided near 100% sensitivity for detection of FMDv across a diverse selection of field isolates (King *et al.*, 2006). As the outbreak grew, isolated viruses were sequenced to confirm the presence of a single circulating strain, and from this information bioinformatics analysis was performed to confirm that the 3D RRT-PCR alone would be sufficient to detect FMDv in samples. This fine tuning of the diagnostic algorithm reduced cost and time of sample analysis, making diagnosis in the outbreak situation more streamlined and effective (Reid *et al.*, 2009).

The 2007 FMDv outbreak is an excellent example of the true power of RT-PCR-based molecular diagnostics during infectious disease outbreaks. The speed with which confirmatory diagnosis could be achieved led to a more real-time approach to transmission tracking of the outbreak, as well as ruling in or out infected premises more efficiently. The automated nature of testing also allowed a vast number of samples to be tested quickly, meaning a more thorough investigation of spread of the virus in the crucial early stages of the outbreak was possible. The reference laboratory and subsequent independent reports into the outbreak both stated the impact that rapid molecular testing had on containing the spread and impact of the outbreak was significant, and that molecular testing be employed in all suspected FMD outbreaks (Ryan *et al.*, 2008).

Use of real-time PCR for detection of other veterinary viral pathogens of note

Rabies

The success of real-time PCR in surveillance and detection of veterinary viral pathogens is not restricted to FMD and AI. RRT-PCR has been utilized in the routine surveillance of rabies for several years now, notably in the United Kingdom through the OIE rabies reference laboratory. Currently RRT-PCR is not accepted by the OIE for rabies detection (Anonymous); however, the reference laboratory does have an accredited assay which it uses for confirmatory diagnoses, and more importantly in large scale epidemiological surveys (Hoffmann *et al.*, 2010). In particular RRT-PCR has played a significant role in national bat surveillance programmes, such as the UK bat surveillance programme set up in 2003. Daubenton's and Serotine bats in Scotland and England are routinely captured, ringed and sampled for rabies. During the process mouth swabs are taken of the bats and the saliva is processed for viral RNA extraction. A RRT-PCR is then performed using primers targeting the nucleoprotein gene to determine the presence of bat lyssavirus, the predominant rabies virus in Western Europe (Hoffmann *et al.*, 2009). Recently a RRT-PCR assay has been developed using bioinformatics analysis of all available EBLV (European Bat Lyssa Viruses) and rabies virus sequences, using a dual assay approach as with AI and FMD. The twin PCR gave a 100% detection rate in clinical samples, and has been suggested for use in routine analysis in rabies reference laboratories (Hoffmann *et al.*, 2010).

Newcastle disease virus

Newcastle disease virus (NDV) is an avian paramyxovirus which causes a devastating disease of domesticated and wild birds, with 27 of the 50 orders of birds susceptible to infection. NDV strains can be classified as highly virulent (velogenic), intermediate (mesogenic) or non-virulent (lentogenic) (Alexander, 2000). The classification is based on the results of mean death times in chicken eggs during virus characterization. Velogenic/mesogenic and lentogenic strains differ in the consensus sequence of their F protein cleavage site [^{112}(R/K)RQ(R/K)RF^{117} for velogenic strains and ^{112}(G/E)(K/R)Q(G/E)RL^{117} for lentogenic strains]. The difference in protein cleavage sites results in differences in substrate specificity of host cellular proteases (Kawahara *et al.*, 1992). Lentogenic virus F proteins can only be cleaved by trypsin like enzymes, commonly found in the respiratory and intestinal tracts, whereas virulent strain F proteins can be cleaved by a host of proteases found in a diverse range of tissues. Subsequently, infection with a virulent strain results in fatal systemic infection. For this reason the F proteins cleavage site is postulated as the primary virulence determinant of NDV. Virus is most consistently isolated from faeces and cloacal swabs of infected birds, as well as tracheal or oropharyngeal swabs, and is also commonly isolated from the lung, spleen, liver, and brain of animals which die from infection.

As with AI, diagnosis of NDV is only possible by isolation of infectious virus particles from inoculated embryonated eggs, and the use of molecular testing is not approved by the OIE diagnostic manual (Anonymous). However as with rabies, NDV RRT-PCR developed by the OIE reference laboratory can be used for confirmatory diagnosis and front-line testing/surveillance. The RRT-PCR developed targets the F gene, allowing differential diagnosis of velogenic isolates, and has been successfully utilized in several NDV outbreaks (Fuller *et al.*, 2009). As with FMD and AI, it is widely believed that the utilization of RRT-PCR for NDV detection has had a dramatically positive effect on impact of NDV outbreaks, which have been largely confined to their foci of origin when molecular testing has been used in the front line of the diagnostic algorithm.

Bluetongue virus

Bluetongue is a disease of sheep and cattle characterized by prolonged fever, which shows more severe symptoms in sheep. The virus targets endothelial cells leading to capillary weakening and leakage causing reddening and swelling of the lips, mouth nasal lining and eyelids. This is usually accompanied by nasal discharge and frothing of the mouth. Muscle degeneration and lameness can then occur rapidly, and death will follow if untreated, particularly in sheep. Bluetongue virus is a vector-borne disease and is spread by biting midges; however, the virus can migrate over vast distances when midges are carried by prevailing winds. The virus is a Reoviridae from the *Orbivirus* genus, of which there are

24 classified serotypes, and is commonly found in warmer climates of Africa and south-east Asia, with incursions into southern Europe in the late twentieth and early twenty-first century creating a significant bluetongue epidemic (MacLachlan, 1994).

In 2006, northern Europe reported a first ever incursion of bluetongue with outbreaks in Belgium, France, Luxembourg, Germany and the Netherlands, with the virus re-emerging in the spring and summer of 2007. In September 2007 prevailing strong winds carried infected midges from Belgium to the UK, and outbreaks were reported in the southern coastal county of Suffolk. The virus causing the European outbreak was serotype 8, and genome sequencing showed that the virus had originated in sub-Saharan Africa (Batten *et al.*, 2009). The 2006–2007 incursion of bluetongue to northern Europe created another opportunity to demonstrate the power of Real-time PCR based surveillance and detection in controlling veterinary viral diseases. As with other infections, prior to the outbreak molecular testing was not accepted as an official diagnostic for bluetongue by the OIE. However, during the 2006 incursion a real-time PCR assay was developed using two sets of primers and probes to detect the presence of all known bluetongue virus variants. The target is the highly conserved genome segment 1, encoding viral polymerase, and was shown to provide detection with 100% specificity (Shaw *et al.*, 2007).

The assay was utilized by all national reference laboratories in northern Europe during the outbreak, including the UK, where it was crucially used to rule out infection in some 2255 animals imported from northern Europe during the outbreak (Shaw *et al.*, 2007). The use of an available tool that can quickly confirm infection status and prevent unnecessary destruction of valuable livestock is as key a use of RRT-PCR in surveillance and detection as finding infection foci.

Use of real-time PCR for surveillance of veterinary bacterial pathogens

Perhaps due to the nature of the diseases caused, and by which they spread, there has been much more published success on the use of RT-PCR in surveillance of veterinary viral pathogens (Hoffmann *et al.*, 2009) than bacterial pathogens, primarily driven by the need for such rapid determination of infection status to control spread of airborne viral infections. However the use of real-time molecular detection for bacterial infectious agents does have benefits, and there are a myriad of publications reporting development of in-house RT-PCR assays for surveillance of notifiable bacterial pathogens of veterinary livestock such as *Salmonella* and *Brucella*, as well as zoonotic agents such as *Campylobacter*. One draw back of using molecular detection methods for bacteria is in finding target amplicons which are truly unique to the organism to be detected, and as modern sequence technologies result in an explosion in bacterial genome sequences, it is becoming clearer that what was once thought to be a sequence motif unique to a particular pathogenic species or subspecies may actually be present in a myriad of semi-related species and genera of bacteria (Pallen and Wren, 2007). On the positive side of this explosion in genomic technology, it is now, or in the near future will be, easily possible to design single nucleotide polymorphism (SNP) based real-time assays which can both detect and subtype bacteria. This is essential for bacterial detection, where the presence of pathogenic species and subspecies within larger, more complex, genera creates the main problem in the use of molecular detection methods in front-line diagnoses.

Salmonella

Salmonella is a genus of great importance to both veterinary livestock and human health, and can range from veterinary pathogens to zoonotic human pathogens. In humans the most commonly isolated subspecies are *Salmonella enterica* Enteriditis, *S.* Typhimurium, and *S.* Newport, which are also commonly found in both diseased and asymptomatic cattle, pigs and poultry, though particularly pigs and poultry (Andrews-Polymenis *et al.*, 2010). Another particular threat to the health of poultry is the avian pathogenic subspecies *S.* Gallinarum (Chappell *et al.*, 2009). As a result the EU zoonoses regulations require by law that all EU member states operate a National Control Programme (NCP) to detect and control *Salmonella* of public health concern in chicken breeding flocks, laying hens, broilers, turkeys, pig herds for slaughter and breeding pigs. Currently, EU guidelines require isolation of organism by culture for a confirmed positive sample; however, many laboratories now use RT-PCR as a front-line screen, similar to the diagnostic and detection algorithm used for viruses.

In the case of *Salmonella* most laboratories will target a region which gives an unambiguous result for the organism. In the case of most bacteria this is most easily done with 16S rDNA targets, which can distinguish any species with appropriately designed primers (Sontakke *et al.*, 2009). Whilst this is fine for in house testing, 16S based PCR assays face a problem if the assay designers wish to commercialize their assay in that the use of 16S sequences as a molecular detection tool are protected by patent. There are also problems with weak primer hybridization against 16S regions of closely related species and genera, or subspecies not of public health importance. As such many *Salmonella* assays employ primers designed to amplify the *invA* gene. This gene is found in all pathogenic members of the *S. enterica* species such as *S.* Enteriditis, *S.* Typhimurium, and *S.* Gallinarum, but not in non-pathogenic members of the genus such as *S. bongori* (Rahn *et al.*, 1992). There are also numerous publications highlighting the use of multiplex assays employing primers directed against regions unique to the particular subspecies (Pugliese *et al.*, 2011), as well as assays additionally identifying antimicrobial resistance and virulence gene profiles, primarily through multiplex PCR (Aarts *et al.*, 2010) and PCR-based SNP typing or melt analysis (Song *et al.*, 2010). However in laboratories employing high-throughput screening of veterinary samples, most screening is performed with a front-line general assay such as *invA* or 16S, and any positives are subjected to culture analysis and characterization by classical typing methods.

Brucella

Brucellosis is a highly infectious disease of cattle, sheep, pigs, and goats caused by pathogenic species of *Brucella*. The organism infects the reproductive organs, particularly the uterus of pregnant animals, leading to late gestation abortion, upon which there is massive excretion of organism into the environment, leading to substantial environmental contamination and subsequent high level spread of infection among herds. Additionally the organism is zoonotic and infection of humans easily occurs through consumption of dairy produce or occupational exposure, resulting in a flu-like illness. As such, brucellosis is a notifiable veterinary disease in the UK, and there is a national screening programme for *Brucella* spp. Like *Salmonella*, there are pathogenic species of note, such as *B. melitensis*, *B. abortus*, *B. suis*, *B. ovis*, and *B. canis*, with laboratories often wishing to determine the species present in a sample (Seleem *et al.*, 2010).

As with other bacterial pathogens, many laboratories will use a front-line test such as 16S or, in the case of *Brucella*, primers such as those targeted against the *bcsp31* gene found in all species and biovars of the organism (Mayer-Scholl *et al.*, 2010). This allows an initial screen against samples for positives, which can be further investigated in more detail. Also similarly to *Salmonella*, there are numerous publications highlighting the use of more advanced real-time PCR based assays such as amplification and SNP typing to allow identification between species and biovars (Gopaul *et al.*, 2008). Whilst these assays provide fantastic scope for fast accurate differential detection of the pathogen, most laboratories screening large numbers of samples will opt for the cheaper option of a generic first pass screen.

Campylobacter

Campylobacter jejuni and *Campylobacter coli* are the most common causes of food-borne bacterial intestinal disease in the developed world, with infection primarily caused by the ingestion of organisms from raw or undercooked poultry. The organisms are members of the intestinal microbiota of poultry and often cattle and pigs, and as such enter the food chain in large numbers (Cody *et al.*, 2010). In addition, *C. venerealis* and *C. fetus* are associated with abortion and infertility in cattle and sheep (Sauerwein *et al.*, 1993). Despite the significant public health risk posed by *Campylobacter* there are no international or indeed national surveillance and monitoring programmes, as improvement of biosecurity to prevent incursion of the organism into commercial poultry house facilities is considered by policy makers as the most appropriate method to reduce cases of human intestinal disease due to the organism.

Despite this there are many programmes to determine the true level of *Campylobacter* in livestock, and an increasing shift within the research community to move towards real-time surveillance of flocks within poultry houses, as little success has been achieved in limiting incursions of the organism into commercial flocks. The ability to perform real-time monitoring of poultry on commercial premises would allow farmers to identify when *Campylobacter* enters a house, meaning the farmer can ensure this is the last house dealt with during routine work, ensuring the organism is not spread to subsequent houses. In addition the farmer could identify poultry from houses with a *Campylobacter* positive status, ensuring they can be transported and slaughtered either lastly or in isolation, minimizing spread to other poultry carcasses and reducing bacterial numbers entering the food chain.

In order for such testing to be feasible there must be robust rapid detection assays in place, and most work on *Campylobacter* RT-PCR based detection assays have focused on 16S targeting (Randall *et al.*, 2010), or targeting of the *flaA* flagella gene which is conserved across the hyperthermophilic members of the genus which cause disease (Oyofo *et al.*, 1992). The other benefit of targeting *fla* is that this can also be utilized in a SNP base detection assay to epidemiologically type the organisms, which would be of necessity in the example given for controlling movement of *Campylobacter* within a commercial poultry facility (Ridley *et al.*, 2008).

Concluding remarks

The advent of real-time PCR has led to a significant and extremely beneficial step-change in the way viral pathogen surveillance and detection is undertaken, particularly in the surveillance and detection of notifiable veterinary viral infections. The simplistic nature of viral genomes means that identifying a unique sequence motif to allow differential detection of the virus of interest is a relatively easy process. The highly mutagenic nature of viruses does pose a problem in creating an assay that will detect all known variants of the virus of interest whilst differentiating viral subtypes, but this is easily subverted by use of multiplexes and/or degenerate nucleotide primers. The success with which RRT-PCR has been incorporated into the approved diagnostic algorithm for detection of pathogens such as avian influenza, and the resulting effect on control and surveillance of the pathogen provides the greatest example of the success of real-time PCR based infectious disease diagnosis.

Whilst RT-PCR is just as applicable to detection of veterinary bacterial and zoonotic pathogens, there are numerous factors which have limited the success of such approaches. The primary one is the relative ease and inexpensive nature of culturing bacteria from samples, which is cheaper, less labour intensive, takes around 24 hours for most bacteria, and requires less skills training than RT-PCR testing, in direct contrast to viral pathogens where virus culture is costly, time consuming and can take days or weeks. The other significant problem for the wide spread adaptation of RT-PCR for bacterial detection and surveillance is the genetic nature of bacterial genera, species and subspecies. Very often there is a need to subtype bacteria down to a level beyond the auspices of simple detection of a handful of genes in order to distinguish pathogen from environmental or commensal. Additionally, as our knowledge of the microbial pan-genome grows exponentially through the use of next generation sequencing projects it is becoming clear that the genetic diversity amongst species and genera is less defined than once believed, and that many species and genera share genes once considered unique to certain pathogens (Pallen and Wren, 2007).

References

Anonymous. OIE Manual of Diagnostic Tests and Vaccines for Terrestrial Animals 2010. Available at: http://www.oie.int/manual-of-diagnostic-tests-and-vaccines-for-terrestrial-animals/.

Aarts, H.J., Vos, P., Larsson, J.T., van Hoek, A.H., Huehn, S., Weijers, T., Gronlund, H.A., and Malorny, B. (2010). A multiplex ligation detection assay for the characterization of *Salmonella enterica* strains. Int. Food, J. Microbiol. *145(Suppl. 1)*, S68–S78.

Alexander, D.J. (2007). An overview of the epidemiology of avian influenza. Vaccine *25*, 5637–5644.

Alexander, D.J. (2000). Newcastle disease and other avian paramyxoviruses. Rev. Sci. Tech. *19*, 443–462.

Andrews-Polymenis, H.L., Baumler, A.J., McCormick, B.A., and Fang, F.C. (2010). Taming the elephant: *Salmonella* biology, pathogenesis, and prevention. Infect. Immun. *78*, 2356–2369.

Batten, C.A., Sanders, A.J., Bachanek-Bankowska, K., Bin-Tarif, A., and Oura, C.A. (2009). Bluetongue virus: European Community proficiency test (2007). to evaluate ELISA and RT-PCR detection methods with special reference to pooling of samples. Vet. Microbiol. *135*, 380–383.

Belsham, G.J. (1993). Distinctive features of foot-and-mouth disease virus, a member of the picornavirus family; aspects of virus protein synthesis, protein processing and structure. Prog. Biophys. Mol. Biol. *60*, 241–260.

Breed, A.C., Harris, K., Hesterberg, U., Gould, G., Londt, B.Z., Brown, I.H., and Cook, A.J. (2010). Surveillance for avian influenza in wild birds in the European Union in 2007. Avian Dis. *54*, 399–404.

Brown, I.H., Pittman, M., Irza, V., and Laddomada, A. (2007). Experiences in control of avian influenza in Europe, the Russian Federation and the Middle East. Dev. Biol. (Basel) *130*, 33–38.

Chappell, L., Kaiser, P., Barrow, P., Jones, M.A., Johnston, C., and Wigley, P. (2009). The immunobiology of avian systemic salmonellosis. Vet. Immunol. Immunopathol. *128*, 53–59.

Cody, A.J., Colles, F.M., Sheppard, S.K., and Maiden, M.C. (2010). Where does *Campylobacter* come from? A molecular odyssey. Adv. Exp. Med. Biol. *659*, 47–56.

Elbers, A.R., Fabri, T.H., de Vries, T.S., de Wit, J.J., Pijpers, A., and Koch, G. (2004). The highly pathogenic avian influenza A (H7N7) virus epidemic in The Netherlands in 2003 – lessons learned from the first five outbreaks. Avian Dis. *48*, 691–705.

Fouchier, R.A., Schneeberger, P.M., Rozendaal, F.W., Broekman, J.M., Kemink, S.A., Munster, V., Kuiken, T., Rimmelzwaan, G.F., Schutten, M., Van Doornum, G.J., *et al.* (2004). Avian influenza A virus (H7N7) associated with human conjunctivitis and a fatal case of acute respiratory distress syndrome. Proc. Natl. Acad. Sci. U.S.A. *101*, 1356–1361.

Fuller, C.M., Collins, M.S., and Alexander, D.J. (2009). Development of a real-time reverse-transcription PCR for the detection and simultaneous pathotyping of Newcastle disease virus isolates using a novel probe. Arch. Virol. *154*, 929–937.

Gopaul, K.K., Koylass, M.S., Smith, C.J., and Whatmore, A.M. (2008). Rapid identification of *Brucella* isolates to the species level by real time PCR based single nucleotide polymorphism (SNP) analysis. BMC Microbiol. *8*, 86.

Hoffmann, B., Beer, M., Reid, S.M., Mertens, P., Oura, C.A., van Rijn, P.A., Slomka, M.J., Banks, J., Brown, I.H., Alexander, D.J., *et al.* (2009). A review of RT-PCR technologies used in veterinary virology and disease control: sensitive and specific diagnosis of five livestock diseases notifiable to the World Organisation for Animal Health. Vet. Microbiol. *139*, 1–23.

Hoffmann, B., Freuling, C.M., Wakeley, P.R., Rasmussen, T.B., Leech, S., Fooks, A.R., Beer, M., and Muller, T. (2010). Improved safety for molecular diagnosis of classical rabies viruses by use of a TaqMan real-time reverse transcription-PCR 'double check' strategy. J. Clin. Microbiol. *48*, 3970–3978.

Kawahara, N., Yang, X.Z., Sakaguchi, T., Kiyotani, K., Nagai, Y., and Yoshida, T. (1992). Distribution and substrate specificity of intracellular proteolytic processing enzyme(s) for paramyxovirus fusion glycoproteins. Gen, J. Virol. *73(Pt 3)*, 583–590.

King, D.P., Ferris, N.P., Shaw, A.E., Reid, S.M., Hutchings, G.H., Giuffre, A.C., Robida, J.M., Callahan, J.D., Nelson, W.M., and Beckham, T.R. (2006). Detection of foot-and-mouth disease virus: comparative diagnostic sensitivity of two independent real-time reverse transcription-polymerase chain reaction assays. J. Vet. Diagn. Invest. *18*, 93–97.

Liu, J., Xiao, H., Lei, F., Zhu, Q., Qin, K., Zhang, X.W., Zhang, X.L., Zhao, D., Wang, G., Feng, Y., *et al.* (2005). Highly pathogenic H5N1 influenza virus infection in migratory birds. Science *309*, 1206.

MacLachlan, N.J. (1994). The pathogenesis and immunology of bluetongue virus infection of ruminants. Comp. Immunol. Microbiol. Infect. Dis. *17*, 197–206.

Mayer-Scholl, A., Draeger, A., Gollner, C., Scholz, H.C., and Nockler, K. (2010). Advancement of a multiplex PCR for the differentiation of all currently described *Brucella* species. J. Microbiol. Methods *80*, 112–114.

Oyofo, B.A., Thornton, S.A., Burr, D.H., Trust, T.J., Pavlovskis, O.R., and Guerry, P. (1992). Specific detection of *Campylobacter jejuni* and *Campylobacter coli* by using polymerase chain reaction. J. Clin. Microbiol. *30*, 2613–2619.

Pallen, M.J., and Wren, B.W. (2007). Bacterial pathogenomics. Nature *449*, 835–42.

Pugliese, N., Circella, E., Pazzani, C., Pupillo, A., and Camarda, A. (2011). Validation of a seminested PCR approach for rapid detection of *Salmonella enterica* subsp. enterica serovar Gallinarum. J. Microbiol. Methods *85*, 22–27.

Rahn, K., De Grandis, S.A., Clarke, R.C., McEwen, S.A., Galan, J.E., Ginocchio, C., Curtiss, R., 3rd, and Gyles, C.L. (1992). Amplification of an invA gene sequence of *Salmonella* typhimurium by polymerase chain reaction as a specific method of detection of Salmonella. Mol. Cell. Probes *6*, 271–279.

Randall, L., Lemma, F., Rodgers, J., Vidal, A., and Clifton-Hadley, F. (2010). Development and evaluation of internal amplification controls for use in a real-time duplex PCR assay for detection of *Campylobacter coli* and *Campylobacter jejuni*. J. Med. Microbiol. *59*, 172–178.

Reid, S.M., Ebert, K., Bachanek-Bankowska, K., Batten, C., Sanders, A., Wright, C., Shaw, A.E., Ryan, E.D., Hutchings, G.H., Ferris, N.P., *et al.* (2009). Performance of real-time reverse transcription polymerase chain reaction for the detection of foot-and-mouth disease virus during field outbreaks in the United Kingdom in 2007. J. Vet. Diagn. Invest. *21*, 321–330.

Ridley, A.M., Allen, V.M., Sharma, M., Harris, J.A., and Newell, D.G. (2008). Real-time PCR approach for detection of environmental sources of *Campylobacter* strains colonizing broiler flocks. Appl. Environ. Microbiol. *74*, 2492–2504.

Ryan, E., Gloster, J., Reid, S.M., Li, Y., Ferris, N.P., Waters, R., Juleff, N., Charleston, B., Bankowski, B., Gubbins, S., *et al.* (2008). Clinical and laboratory investigations of the outbreaks of foot-and-mouth disease in southern England in 2007. Vet. Rec. *163*, 139–147.

Sauerwein, R.W., Bisseling, J., and Horrevorts, A.M. (1993). Septic abortion associated with *Campylobacter fetus* subspecies fetus infection: case report and review of the literature. Infection *21*, 331–333.

Seleem, M.N., Boyle, S.M., and Sriranganathan, N. (2010). Brucellosis: a re-emerging zoonosis. Vet. Microbiol. *140*, 392–398.

Shaw, A.E., Monaghan, P., Alpar, H.O., Anthony, S., Darpel, K.E., Batten, C.A., Guercio, A., Alimena, G., Vitale, M., Bankowska, K., *et al.* (2007). Development and initial evaluation of a real-time RT-PCR assay to detect bluetongue virus genome segment 1. J. Virol. Methods *145*, 115–126.

Sims, L.D., Domenech, J., Benigno, C., Kahn, S., Kamata, A., Lubroth, J., Martin, V., and Roeder, P. (2005). Origin and evolution of highly pathogenic H5N1 avian influenza in Asia. Vet. Rec. *157*, 159–164.

Slomka, M.J., Pavlidis, T., Banks, J., Shell, W., McNally, A., Essen, S., and Brown, I.H. (2007a). Validated H5 Eurasian real-time reverse transcriptase-polymerase chain reaction and its application in H5N1 outbreaks in 2005–2006. Avian Diseases *51*, 373–3777.

Slomka, M.J., Coward, V.J., Banks, J., Londt, B.Z., Brown, I.H., Voermans, J., Koch, G., Handberg, K.J., Jorgensen, P.H., Cherbonnel-Pansart, M., *et al.* (2007b). Identification of sensitive and specific avian influenza polymerase chain reaction methods through blind ring trials organized in the European Union. Avian Dis. *51*, 227–234.

Slomka, M.J., Pavlidis, T., Coward, V.J., Voermans, J., Koch, G., Hanna, A., Banks, J., and Brown, I.H. (2009). Validated RealTime reverse transcriptase PCR methods for the diagnosis and pathotyping of Eurasian H7 avian influenza viruses. Influenza Other Respi. Viruses 3, 151–164.

Slomka, M.J., Irvine, R.M., Pavlidis, T., Banks, J., and Brown, I.H. (2010). Role of real-time RT-PCR platform technology in the diagnosis and management of notifiable avian influenza outbreaks: experiences in Great Britain. Avian Dis. *54*, 591–596.

Song, Y., Roumagnac, P., Weill, F.X., Wain, J., Dolecek, C., Mazzoni, C.J., Holt, K.E., and Achtman, M. (2010). A multiplex single nucleotide polymorphism typing assay for detecting mutations that result in decreased fluoroquinolone susceptibility in *Salmonella enterica* serovars Typhi and Paratyphi A. J. Antimicrob. Chemother. *65*, 1631–1641.

Sontakke, S., Cadenas, M.B., Maggi, R.G., Diniz, P.P., and Breitschwerdt, E.B. (2009). Use of broad range16S rDNA PCR in clinical microbiology. J. Microbiol. Methods *76*, 217–225.

Spackman, E., Senne, D.A., Myers, T.J., Bulaga, L.L., Garber, L.P., Perdue, M.L., Lohman, K., Daum, L.T., and Suarez, D.L. (2002). Development of a Real-Time reverse transcriptase PCR assay for type A Influenza virus of the avian H5 and H7 haemaglutinin subtypes. J. Clin. Microbiol. *40*, 3256–3260.

Thompson, D., Muriel, P., Russell, D., Osborne, P., Bromley, A., Rowland, M., Creigh-Tyte, S., and Brown, C. (2002). Economic costs of the foot and mouth disease outbreak in the United Kingdom in 2001. Rev. Sci. Tech. *21*, 675–687.

Applications in Clinical Microbiology

6

Andrew D. Sails

Abstract

The introduction of real-time PCR technology to diagnostic clinical microbiology laboratories has led to significant improvements in the diagnosis of infectious disease. It has been particularly useful to detect slow-growing or difficult to grow infectious agents therefore much of its initial impact was in diagnostic virology. However, in more recent years real-time PCR-based methods have been introduced in diagnostic bacteriology, mycology and parasitology and there are few areas of clinical microbiology which remain unaffected by real-time PCR methodologies. One area where it has had great impact is its use for quantitation of viral pathogens. The ability to monitor the PCR reaction in real-time allows accurate quantitation of target sequence over at least six orders of magnitude. In addition, the closed-tube format removes the need for post-amplification manipulation of the PCR products also reducing the likelihood of amplicon carryover to subsequent reactions reducing the risk of false-positives. The inherent sensitivity of real-time PCR means that contamination between samples and from previously amplified product can lead to false-positive results. Therefore diagnostic labs utilizing real-time PCR methods have to strictly adhere to good laboratory practice to reduce the likelihood of cross-contamination. In addition individual laboratories must ensure quality of diagnostic testing by participating in external quality assurance schemes.

Introduction

The first description of the application of PCR in microbiology was for the detection of HIV in established infected cell lines and in cells cultured from infected individuals (Kwok *et al.*, 1986). Since this initial paper there has been an explosion of interest in utilizing PCR in clinical microbiology. Initially methods where based on conventional PCR and this mainly limited their use to reference laboratories and academic research. However the introduction of real-time PCR methods had a profound effect on the uptake of PCR into routine diagnostics in clinical microbiology. Since the launch of the first real-time PCR instruments in the late 1990s their use has increased dramatically and such instrumentation has now become commonplace in larger diagnostic microbiology laboratories. This has led to significant changes in the way we diagnose infectious disease within the laboratory and has also indirectly improved our understanding of the disease process for many infections. In this chapter an overview is provided of some of the areas real-time PCR methods that have had a significant impact in clinical microbiology and the diagnosis of infectious disease. Parasitology applications are not included and the focus is on applications in bacteriology and virology. This review is not intended to be exhaustive and does not attempt to describe every real-time PCR study which has been published. Prediction of where the technology may be going and what further benefits this will bring to the diagnosis of infectious disease is included.

Why use real-time PCR, what are the advantages?

The first PCR methods to be described for clinical microbiology utilized gel electrophoresis for the detection of PCR amplification products. Although these assays proved useful, their specificity and sensitivity was compromised by this cumbersome end-point detection method. Specificity of detection could be improved by incorporating a solid phase hybridization such as Southern blotting; however, this was labour intensive and time-consuming, requiring further manipulation of the PCR product. Detection of PCR products by solid phase hybridization also limited the numbers of samples that could be processed, and the methods used were difficult to standardize between laboratories. The overall time taken to produce a result from a PCR assay could be two or three days and the test required a significant level of technical skill limiting the use of PCR to specialized laboratories only. The introduction of enzyme-linked hybridization probe formats (PCR-ELISA) for the detection of amplification products did improve the detection process however they still required manipulation of the amplification products following PCR. Manipulation of the amplified product increases the risk of contaminating subsequent PCR reactions leading to false-positives by amplicon carryover. PCR-ELISA methods did allow the first semi-accurate quantitation of gene target within samples (quantitative PCR or QPCR) however the range and accuracy of quantitation was very limited.

The introduction of real-time instrumentation and chemistry for PCR has revolutionized molecular diagnostic detection methods in clinical microbiology. These closed tube systems reduce the risk of amplicon carryover because the samples are not opened following thermal cycling. Many of these new platforms process samples more rapidly than conventional block-based thermal cyclers making pathogen testing much more rapid. In addition, the ability to monitor the reaction in real-time provides results immediately after cycling and facilitates quantitation of the original target sequence over many orders of magnitude. This has facilitated the development of assays for viral genome quantitation which have become increasingly useful in the diagnosis and management of viral disease. This is discussed more fully in the section 'viral genome quantitation by real-time PCR'.

Real-time PCR instruments measure fluorescence over a wide range of wavelengths and there are a multitude of fluorescent dyes now available which facilitates the multiplexing of assays. Therefore assays can be multiplexed to detect a range of pathogens within the same tube. This has proven to be particularly useful for diagnosing conditions such as respiratory infections where the same symptoms can be caused by a range of pathogens.

Real-time PCR instrumentation

The first two commercial real-time PCR instruments were introduced in the mid-1990s; however there is now a wide choice of instrumentation available with new equipment becoming available each year. Improvements in technology have led to more rapid thermocycling times, increased multiplexing capacity, higher throughput and more user friendly software. Some of the more commonly used instruments are the TaqMan 7500 system, the Rotorgene, the Light Cycler and the Smart Cycler.

Detection of viral pathogens by real-time PCR

Real-time PCR methods have proven to be useful tools in the diagnosis and management of a wide range of viral diseases (Niesters, 2002). Table 6.1 lists some of the real-time PCR assays that have been described which target a wide range of viral disease agents. The speed and specificity, plus their ability to directly quantify the target organism without post-PCR manipulations have made them the method of choice for a wide range of applications. Some of the areas in which real-time PCR methods have been applied in clinical virology are reviewed here.

Table 6.1 Real-time PCR methods for the detection of pathogenic viruses

Organism	Detection/ quantitation	Sensitivity	Gene(s) targeted	Reference
Adenovirus	Q	1.5×10^1 to 1.5×10^8 copies/PCR	Hexon gene	Heim *et al.* (2003)
Human astrovirus	D	0.0052 IU	ORF1a	Royuela *et al.* (2006)
Human bocavirus	D	10 copies	NS1 and NP-1 genes	Lu *et al.* (2006)
Human bocavirus	D	Not reported	NP1 gene	Kleines *et al.* (2007)
BK virus	Q	10 copies/PCR	T-ag gene	McNees *et al.* (2005)
BK virus	Q	10 copies/ml	LT-ag gene	Si-Mohamed *et al.* (2006)
Human cytomegalovirus (HCMV)	Q	10^1 to 10^7 copies/ PCR	major immediate-early gene	Nitsche *et al.* (2000)
HCMV	Q	20 to 10^7 copies/ PCR	US17 gene	Machida *et al.* (2000)
HCMV	Q	10 to 10^6 copies/ PCR	UL83 region	Gault *et al.* (2001)
HCMV	Q	10 to 10^4 copies/ PCR	UL83 gene	Griscelli *et al.* (2001)
HCMV	Q	2×10^3 to 5×10^8 CMV DNA copies/ml blood	glycoprotein B	Kearns *et al.* (2001)b
HCMV	Q	10^2 to 10^6 copies/ PCR	DNA polymerase gene	Sanchez *et al.* (2002)
HCMV	Q	500 to 50,000 CMV DNA copies/ml plasma	US17 gene	Stocher and Berg (2002)
HCMV	Q	250 copies/ml plasma	UL123 gene	Leruez-Ville *et al.* (2003)
HCMV	Q	Not reported	immediate-early antigen gene	Li *et al.* (2003)

Table 6.1 (Continued)

Organism	Detection/ quantitation	Sensitivity	Gene(s) targeted	Reference
Coronavirus (SARS associated)	D	Not reported	N gene	Hui *et al.* (2004)
Dengue virus	Q	10 to 10^7 PFU/ml	conserved core regions	Shu *et al.* (2003)
Dengue virus	D	0.1 to 1.1 PFU detection limit	N5S, capsid (C), UTR	Callahan *et al.* (2001)
Ebola, Marburg, Lassa, Crimean-Congo haemorrhagic fever, Rift Valley fever, dengue, yellow fever virus	Q	8.6 to 16 RNA copies/PCR	L, GPC, NP, G2, 5′ non-coding region, 3′ non-coding region	Drosten *et al.* (2002)
Epstein–Barr virus (EBV)	Q	2 to 10^7 copies EBV DNA/PCR	BALF5 gene	Kimura *et al.* (1999)
EBV	Q	100 to 10^7 copies/ml plasma or serum	BNRF1 p143 gene	Niesters *et al.* (2000)
EBV	Q	10 copies/PCR	EBNA1 gene	Stevens *et al.* (2002)
EBV	Q	10 to 10^9 copies/ PCR	BZLF1 gene	Patel *et al.* (2003)
Enterovirus	D	11.8 enterovirus GE/ PCR	5′ non-coding region	Verstrepen *et al.* (2001)
Enterovirus	D	50 enterovirus GE/ PCR	5′ non-coding region	Monpoeho *et al.* (2002)
Enterovirus	D	Not reported	5′ non-coding region	Nijhuis *et al.* (2002)
Enterovirus	D	0.1 $TCID_{50}$	5′ non-coding region	Rabenau *et al.* (2002)
Enterovirus	D	510 copies/ml CSF	5′ non-coding region	Lai *et al.* (2003)
Enterovirus	Q	<10 copies/PCR	5′ non-coding region	Mohamed *et al.* (2004)
Hepatitis B virus (HBV)	Q	10^1 to 10^8 DNA copies/PCR	S and X genes	Abe *et al.* (1999)
HBV	Q	373 to 10^8 copies/ PCR	Pre S gene	Pas *et al.* (2000)
Hepatitis C virus (HCV)	Q	10^3 to 10^7 copies/ PCR	5′ non-coding region	Martell *et al.* (1999)
HCV	Q	64 to 4,180,000 IU/ ml serum	5′ non-coding region	Kleiber *et al.* (2000)
HCV	Q	10^3 to 10^7 RNA copies/PCR	not reported	Yang *et al.* (2002)
Herpes simplex virus (HSV)	D	Not reported	DNA polymerase	Espy *et al.* (2000b)
HSV	D	12.5 copies/PCR	DNA polymerase	Kessler *et al.* (2000)

Organism	Detection/ quantitation	Sensitivity	Gene(s) targeted	Reference
HSV, varicella zoster virus (VZV), CMV	D	580 copies/ml (HSV-1), 430 copies (HSV-2)	gD, gene, gG gene, DNA polymerase, gene 38	van Doornum *et al.* (2003)
HSV	Q	<10 to 10^8 copies/ PCR	glycoprotein gene	Ryncarz *et al.* (1999)
HSV	D	335 copies/ml cervical lavage fluid	DNA polymerase	Aryee *et al*. (2005)
HSV, VZV	D	<10 plasmid copies/ PCR	gD, gene, gG gene, DNA polymerase	Weidmann *et al.* (2003)
HSV (genital herpes infection)	D	Not reported	glycoprotein B gene	Scoular *et al.* (2002)
HSV	D	Not reported	gB region	Namvar *et al.* (2005)
Human Herpesvirus 6 (HHV-6)	Q	10 GE/PCR	U22 gene	Collot *et al.* (2002)
HHV-6	Q	10 GE/PCR	U65-U66 genes	Gautheret-Dejean *et al.* (2002)
HHV-6	Q	1 GE/PCR (1 to 10^6 GE/PCR)	U67 gene	Locatelli *et al.* (2000)
HHV-6	Q	10 GE/PCR	Large tegument protein gene	Kearns *et al.* (2001a)
HHV-6, human herpesvirus 7 (HHV-7)	Q	10 GE/PCR	DNA polymerase gene	Safrontez *et al.* (2003)
Human HHV-7	Q	14 GE/PCR	U100 gene	Fernandez *et al*. (2002)
Human herpesvirus (HHV-8)	Q	10 GE/PCR	Major capsid protein gene	Broccolo *et al.* (2002)
HHV-8	Q	5–50 GE/PCR	K5, ORF25, ORF37, ORF47, ORF56	Stamey *et al.* (2001)
Human metapneumovirus (hMPV)	D	Not reported	N gene	Mackay *et al.* (2003)
hMPV	D	100 copies/PCR	N, M, F, P, and L genes	Cote *et al.* (2003)
hMPV	Q	10 copies/PCR	fusion protein gene	Kuypers *et al.* (2005)
hMPV	D	5–10 copies/PCR	Nucleoprotein gene	Maertzdorf *et al.* (2004)
hMPV	D	Not reported	Nucleocapsid gene	Sumino *et al.* (2005)
hMPV	D	Not reported	L gene	Scheltinga *et al.* (2005)

Table 6.1 (Continued)

Organism	Detection/ quantitation	Sensitivity	Gene(s) targeted	Reference
Influenza viruses A and B	Q	13 copies/PCR (A) and 11 copies/PCR (B)	matrix protein (A) and haemagglutinin (B)	van Elden *et al.* (2001)
Influenza viruses A and B	D	50 copies/PCR	Matrix protein and fusion gene	Boivin *et al.* (2004)
Influenza viruses A and B	D	Not reported	Matrix protein	Smith *et al.* (2003)
Influenza A virus	D	120–350 copies/PCR	Matrix gene	Stone *et al.* (2004)
Influenza A (H5)	Q	Not stated	Haemagglutinin	Ng *et al.* (2005)
Influenza A (H5N1)	D	10^2–10^3 copies/µl	H5, N1 and M genes	Payungporn *et al.* (2006)
JK virus	Q	10 copies/PCR	T-ag gene	McNees *et al.* (2005)
Norovirus	Q	10 copies/PCR	ORF 1-ORF 2	Kageyama *et al.* (2003)
Norovirus	Q	5 copies/PCR	Polymerase, capA	Pang *et al.* (2004)
Paraechovirus	D	75 copies per reaction	5′ untranslated region	Benschop *et al.* (2008)
Paraechovirus	D	500 copies/mL	5′ untranslated region	Renaud *et al.* (2011)
Respiratory syncytial virus (RSV)	Q	2.3×10^{-3} to 2.3×10^{2} PFU	N gene	Hu *et al.* (2003)
RSV	Q	Not reported	N gene	Gueudin *et al.* (2003)
RSV	Q	10 copies/PCR	matrix protein gene/polymerase	Kuypers *et al.* (2004)
Human rhinovirus	D	Not reported	5′ non-coding region	Scheltinga *et al.* (2005)
Smallpox virus	D	5 to 10 GE/PCR	haemagglutinin	Espy *et al.* (2002)
Smallpox virus	D	25 GE	haemagglutinin	Sofi Ibrahim *et al.* (2003)
SV-40 virus	Q	10 copies/PCR	T-ag gene	McNees *et al.* (2005)
VZV	D	Not reported	gene 28 and gene 29	Espy *et al.* (2000)a
West Nile virus (WNV)	D	<1 PFU virus/PCR	novel sequence	Lanciotti *et al.* (2000)
WU and KI virus	D	10 copies per reaction	Regulatory gene region and large T antigen	Bialasiewicz *et al.* (2007).
WU and KI virus	D	5 copies per reaction	Capsid protein VP1	Lindau *et al.* (2009)

UTR, untranslated 3′ region; TCID50, 50% tissue culture infective dose; ORF, open reading frame.

Detection of viral respiratory pathogens by real-time PCR

Viruses that commonly cause respiratory disease include influenza virus, parainfluenza virus, respiratory syncytial virus, adenovirus, human metapneumovirus, coronavirus (NL63 and OC43), and rhinovirus. The past ten years has seen the emergence of corona virus-associated severe acute respiratory syndrome (SARS), influenza A H5N1 (avian influenza) and pandemic influenza H1N1 (swine variant) which has highlighted the need for rapid and specific diagnostic tests to differentiate between patients suffering from more common respiratory pathogens and those infected with these highly pathogenic viruses. In addition rapid recognition of influenza infection facilitates timely initiation of antiviral therapy using new antiviral compounds such as neuraminidase inhibitors. Prior to the introduction of real-time PCR methods detection of respiratory viruses was based on virus culture, which is slow and relatively insensitive, ELISA or immunofluorescence (IF) which also has limited sensitivity and specificity. The first real-time PCR assays described for the detection of respiratory viruses targeted individual pathogens such as the influenza virus assay described by van Elden *et al.* (2001). This multiplex assay simultaneously detected influenza virus A and B using primers designed to target the matrix protein of influenza A virus and the haemagglutinin gene segment of influenza B virus. Application of the assays to the detection of influenza virus in 98 clinical samples from patients demonstrated that the PCR assay was more sensitive than the conventional methods. Real-time assays for other respiratory viruses were also described including respiratory syncytial virus (RSV). RSV is the most common aetiological agent of lower respiratory tract disease in children and the virus has been classified into two subgroups, A and B based on antigenic and genetic variations in the structural proteins. Hu *et al.* (2003) described a real-time PCR assay for the detection, subgrouping and quantitation of RSV, which utilized two pairs of primers targeting the conserved nucleocapsid (N) gene specific for detection and subgrouping of the A and B virus subtypes. The assays were applied to the detection of RSV in children with respiratory disease with the results being compared to culture and IF with the new assay being demonstrated to be more sensitive than culture and IF for the detection of RSV. Human metapneumovirus (hMPV) is a member of the *Paramyxoviridae* which was first described in 2001 by van den Hoogen *et al.* (2001). The clinical symptoms of hMPV infection are similar to those caused by human RSV infection, ranging from upper respiratory tract disease to severe bronchiolitis and pneumonia. Similar to other respiratory viruses these proved difficult to grow in culture and initially there were no IF tests available for their detection. A number of real-time assays were developed for the detection of hMPV in clinical samples (Mackay *et al.* 2003; Cote *et al.* 2003) which targeted several genes including phosphoprotein (P), and polymerase (L) genes. These assays were demonstrated to be sensitive and specific for the detection of the virus in clinical samples from symptomatic patients. However, more importantly the rapid availability of a sensitive and specific test for their detection facilitated the demonstration of their role in causing respiratory disease (Stockton *et al.*, 2003).

Multiplex detection of viral respiratory pathogens

Since the description of these assays a multitude of additional real-time PCR assays has been described for a wide range of individual viral respiratory pathogens (Table 6.1). However, their use was often limited to research based applications or studies and they were rarely adopted as the first line test in diagnostic laboratories. This was partly due to the perceived difficulties associated with PCR method adoption with many regarding such techniques as being technically too difficult. In addition, because a wide range of viruses could be associated with similar symptoms in patients many laboratories continued to use culture and IF methods. However the development of multiplex screening assays which could detect a wide range of respiratory pathogens in a single screen have revolutionized the diagnosis of viral respiratory disease. The first multiplex assay was described by Templeton *et al.* (2004) which targeted influenza A and B viruses, respiratory syncytial virus, and parainfluenza viruses 1, 2, 3 and 4. The assay was a two-tube multiplex which used molecular beacon probes to discriminate the pathogens. Application of the assay to 358 respiratory samples demonstrated that using real-time PCR increases the sensitivity for respiratory viral diagnosis significantly. In addition, because the results could be obtained in a timely manner (within 6 hours) unlike conventional laboratory approaches, the authors concluded that application of such multiplex screening assays could improve patient management and infection control. There have been several additional multiplex screening assays described since this initial publication. The development of these assays has effectively relegated viral culture to specialist reference laboratories and most clinical microbiology laboratories now routinely screen respiratory samples for a wide range of viral pathogens including influenza A and B, RSV, hMPV, parainfluenza virus (1–4), human rhinovirus, adenovirus using multiplex real-time PCR methods. This 'syndromic' approach to real-time PCR screening has been adopted in other areas of clinical microbiology which will be described later in this chapter.

Detection of emerging viral respiratory pathogens

Since 2000, more than 10 newly emerging respiratory viruses have been described including avian influenza viruses (H5N1, H7N7, and H7N3), pandemic H1N1 (swine flu), human metapneumovirus (hMPV), severe acute respiratory syndrome (SARS) CoV, human CoVs (i.e. HCoV NL63 and HKU1), human bocavirus, WU virus and KI virus. The emergence of pandemic influenza A H1N1 (swine variant) in 2009 highlighted the need for laboratories to be able to respond quickly to a growing public health emergency. Prior familiarity with real-time PCR methods and existing infrastructure and instrumentation facilitated the rapid development and more importantly the dissemination of validated real-time PCR methods not just within nations but across the globe as a whole. Within weeks of the virus being recognized laboratories all over the world were able to specifically and sensitively detect this influenza variant facilitating the introduction of directed public health controls. Prior to the introduction of molecular diagnostic methods such rapid test development and adoption would not have been possible therefore public health interventions would not have been aided by directed laboratory testing.

Detection of herpes viruses by real-time PCR

The human herpes virus family includes the clinically important pathogen herpes simplex viruses 1 and 2 (HSV-1 and HSV-2), varicella-zoster virus (VZV), and the human herpes viruses 6, 7 and 8 (HHV-6, HHV-7, HHV-8). HSV causes a wide spectrum of clinical disease with a variety of anatomical sites being infected including the skin, genital tract, lips, oral cavity, eyes, and the CNS. Disseminated HSV infection is a serious condition that can occur in immunocompromised patients and in neonatal infection acquired by transmission of the virus through the infected birth canal of the mother. Although CNS infection by HSV can be fatal, effective therapeutic management of infections is possible. However, rapid diagnosis is required to enable timely administration of antiviral therapy.

Prior to the introduction of molecular diagnostic methods laboratories were reliant on conventional methods for the detection of HSV such as cell culture, which was the gold standard for all specimens except CSF samples from patients with CNS infections. However, such methods were very slow with turnaround times far too long to influence patient management. PCR has become the new gold standard for HSV detection in human samples. The first real-time PCR assay for the detection of HSV was described in 1999 by Ryncarz *et al.*, with the assay having a linear range of detection of < 10 to 10^8 copies of HSV DNA per reaction. Application to 335 genital tract specimens and 380 CSF samples demonstrated that the new assay was as sensitive and specific as conventional tissue culture. However the real-time PCR was much more rapid than conventional methods with results being available on the same day as the sample was taken.

In the study of Scoular *et al.* (2002) a real-time PCR assay targeting the glycoprotein B gene was compared to viral culture for the diagnosis of genital herpes in a genitourinary medicine (GUM) clinic. One hundred and nine patients had a positive result for HSV by either culture or PCR, with 88 patients being both culture and PCR positive and twenty-one patients positive only by PCR. Overall the use of PCR increased the sensitivity of detection by 24% leading the authors to advocate the use of real-time PCR as an efficient, sensitive and rapid test to aid the diagnosis of genital herpes infection. Data from such studies has led to the adoption of HSV PCR as the primary diagnostic method for HSV genital infections in many laboratories.

Laboratory differentiation of herpes simplex virus (HSV) and *Treponema pallidum* (syphilis) infections is important as they can be clinically indistinguishable. This has led to the development of multiplex HSV and syphilis assays which help to streamline the diagnosis of genitourinary infection. Scott *et al.* (2010) described the development and evaluation of a multiplexed assay which was applied to 692 samples, with 139 positive for HSV-1, 136 for HSV-2, 15 for syphilis, one for both syphilis and HSV-1, and 401 were negative. The PCR test was demonstrated to be more sensitive than either dark ground microscopy or serological testing for the diagnosis of primary syphilis. The authors concluded that multiplex PCR testing for HSV and syphilis on ulcer swabs improved detection rates and reduced both turnaround time and cost, ultimately lowering the onwards transmission risk of primary syphilis and the development of sequelae.

VZV, the causative agent of chickenpox and shingles can cause severe CNS or respiratory tract disease in immunocompromised patients. CNS infections can be fatal; however, rapid diagnosis allows timely initiation of antiviral therapy. Weidmann *et al.* (2003) described the development and evaluation of a real-time multiplex assays for the detection of HSV-1, HSV-2, and VZV in clinical samples. Application of the new assays to 106 clinical samples including CSF, skin swabs, biopsies, vitreous body samples, blood and BAL (bronchoalveolar lavage) samples demonstrated that the assay was as sensitive and specific as assays used previously in the author's laboratory. Similarly, van Doornum *et al.* (2003) described a set of assays for the detection of HSV-1, HSV-2, VZV, and human cytomegalovirus (HCMV). Seal herpes virus type 1 (PhHV-1), was added to each clinical sample prior to extraction and testing as a control DNA to monitor DNA extraction and subsequent DNA amplification. Each of the viral targets was detected in separate assays because multiplexing the primers reduced the sensitivity of detection. The PhHV-1 was added at a low concentration to each clinical sample and was then extracted and detected to provide an internal control for inter-assay precision and reproducibility. The use of such internal controls is discussed further later in this chapter.

Detection of enterovirus infections by real-time PCR

Enteroviruses (EV) are common causes of CNS infections including viral meningitis and encephalitis and it has been estimated that they are responsible for 70–80% of the cases of viral meningitis. Prior to the introduction of PCR-based methods the gold standard for the diagnosis of enteroviral infections of the CNS was growth of the virus in cell culture. However, similar to other diagnostic tests reliant on cell culture, this was slow requiring several days to produce results. Furthermore, some EV types were difficult to propagate in cell culture. The introduction of PCR-based diagnostic tests for the rapid and reliable diagnosis of EV in CNS infections has facilitated more informed patient management with reduced use of unnecessary procedures and irrelevant treatment. Verstrepen *et al.* (2001) developed a 5′ nuclease real-time PCR assay utilizing a single-tube reverse transcriptase and amplification format. The assay, which targeted the 5′ non-coding region was applied to the detection of EV in 70 CSF specimens from patients with suspected viral meningitis. Nineteen of these samples were positive in the PCR assay demonstrating the increased sensitivity over culture which gave 17 positives. Enterovirus real-time PCR methods are now common place in many diagnostic laboratories where they have replaced cell culture and other less rapid and less sensitive tests as the primary diagnostic test for viral CNS infections.

Viral genome quantitation by real-time PCR

Viral genome quantitation by molecular diagnostic methods has become very useful in the diagnosis and management of viral disease. For example viral load (VL) determined by real-time PCR testing is a useful diagnostic marker in human cytomegalovirus (CMV) infection. It has also proven to be a useful prognostic marker for assessing disease progression and as a therapeutic marker for monitoring the efficacy of antiviral chemotherapy in patients with chronic HIV, HBV or HCV infection. The role of VL testing by PCR in clinical virology has been previously reviewed by Niesters (2001), Berger and Preiser (2002) and Mackay *et al.* (2002). The first quantitative PCR (QPCR) methods were described prior to the introduction of real-time PCR and used end-point detection utilizing internal or external controls of known concentrations. These were amplified in parallel with the samples of interest. Following PCR cycling the unknown test sample and control values were compared and a quantitative value assigned to the test samples. Because the final amount of accumulated PCR product is very susceptible to minor variations in reagents and sample matrices, the accuracy of QPCR methods based on end-point detection was poor. The introduction of real-time PCR methods provided more accurate quantification because reactions are quantified by the cycle at which amplification is detected rather than the amount of PCR product accumulated after a fixed number of cycles.

Quantitative detection of human CMV virus by real-time PCR

Quantitation of CMV by real-time QPCR has become a very useful diagnostic test for the management of immunocompromised patients following solid organ or bone marrow transplantation (BMT) and in other immunosuppressed patients. Various CMV genes have been targeted in these assays including the major immediate-early gene locus (Nitsche *et al.*, 2000), the US17 gene (Machida *et al.*, 2000), and the glycoprotein B gene (Kearns *et al.*, 2001b). The application of real-time QPCR assays to patients at risk of CMV disease has demonstrated that monitoring of viral loads assists the assessment of post-transplantation CMV disease risk for patients with higher viral loads (i.e. those carrying higher risks). In addition, an increasing viral load on antiviral therapy was found to be an indication to continue full-dose antiviral therapy.

Quantitative detection of EBV virus by real-time PCR

Epstein–Barr virus (EBV) infects more than 90% of the population worldwide and in immunocompetent individuals the virus establishes a life-long asymptomatic infection. In immunocompromised patients, active EBV infection is a strong risk factor for the development of post-transplant lymphoproliferative disease (PTLD), AIDS-related lymphoma, and X-linked proliferative syndrome. The monitoring of EBV load in transplant patients has been shown to be a useful tool in the diagnosis and management of PTLD. A number of real-time QPCR assays for EBV VL testing have been described which target the viral DNA polymerase gene (Kimura *et al.*, 1999) and the BNRF1 p143 gene encoding the non-glycosylated membrane protein (Niesters *et al.*, 2000).

The detection of latent EBV in immunosuppressed patients without symptoms of EBV disease is in keeping with previous reports suggesting the presence of EBV DNA alone does not always indicate active EBV disease. Although real-time QPCR assays have proven useful in the diagnosis and management of patients with active EBV disease (Fig. 6.1), inter-laboratory standardization of these methods has not yet been achieved. Future prospective studies investigating the EBV VL in immunosuppressed patients using these assays will

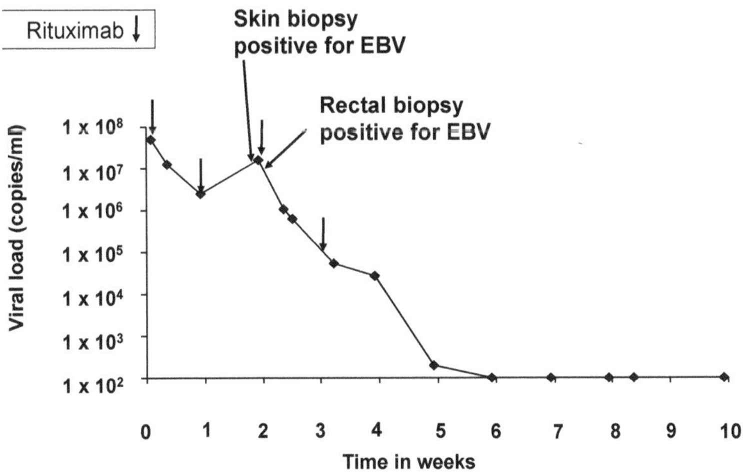

Figure 6.1 Monitoring Epstein–Barr virus (EBV) viral load in a paediatric severe combined immunodeficiency patient. EBV viral load was measured in EDTA blood specimens from the patient using a TaqMan real-time PCR assay targeting the BNRF1 gene. The patient was treated with four doses of Rituximab over a four week period. This demonstrates the usefulness of viral load testing to monitor response to therapy.

help to establish clinically relevant cut-off values, and preferred sample types and extraction methods.

Quantitative detection of adenovirus by real-time PCR

There are six species of human adenovirus (HAdV) with 51 types associated with a variety of diseases affecting all organ systems and HAdV can cause serious life-threatening infections in immunocompromised hosts such as haematopoietic stem cell transplant patients. These HAdV infections are frequently associated with viraemia with one or more organs being involved producing symptoms such as pneumonia, meningitis, hepatitis, rash, diarrhoea, or cystitis. Prior to the development of real-time PCR the gold standard for the detection of HAdV was virus isolation however that was slow and required up to three weeks for cytopathic effects to develop. Quantitation of HAdV DNA in blood samples by PCR provides more informative data for predicting the risk of disseminated infection in immunocompromised patients. Heim *et al.* (2003) described a real-time PCR assay for the sensitive detection and quantitation of all 51 HAdV types in blood samples from immunocompromised patients. The assay targeted a conserved region of the hexon gene and demonstrated a sensitivity of approximately 15 copies per reaction with a linear range of quantitation of 1.5×10^1 to 1.5×10^8 per reaction. Application of the assay to 234 clinical samples including blood, serum, eye swabs, and faeces demonstrated the assay to be more sensitive than a previously developed conventional in-house PCR assay. The HAdV VL were greater in the paediatric patients than in adult patients or controls. A small number of immunocompromised children had very high VL (up to 1.1×10^{10} copies/ml) which was associated with symptoms of disseminated disease. HAdV viraemia was detected in 4 of 27 paediatric and 8 of 93 adult stem cell transplant recipients but only in 5 of 306 healthy controls. Real-time quantitative assays are now considered a useful tool in the management of immunocompromised patients at risk of HAdV infection and may provide simplified and more rapid diagnosis of disseminated disease. Although there is no specific antiviral therapy for disseminated HAdV infection, cidofovir and ribavirin have been used to treat severe infections in patients. In addition there are new investigational compounds as well as novel immunotherapeutic strategies emerging (Lenaerts *et al.*, 2008). This QPCR assay may therefore prove useful for monitoring HAdV VL in patients during clinical trials of these new treatment options for disseminated infection.

Quantitative detection of BK virus by real-time PCR

BK virus (BCV) is a non-enveloped, double stranded DNA virus which affects immunocompromised hosts. It is particularly associated with nephropathy (BKVN) in renal transplant patients receiving immunosuppression therapy. It is thought to affect up to 8% of kidney transplant patients and the syndrome is characterized by persistent graft dysfunction and frequent graft loss. BKVN is difficult to diagnose because the disease can occur without clinical signs or symptoms. The gold standard for diagnosing BKVN is renal transplant biopsy to demonstrate viral cytopathic changes in the renal tubular epithelium. However, measurement of a less invasive surrogate marker, BK DNA in urine and blood by quantitative real-time PCR is used to make decisions regarding clinical intervention (Bechert *et al.*, 2010). Serial monitoring of BK viral load is useful for monitoring the course of BKV infection and several studies have described real-time PCR assays for this purpose. Randhawa *et al.* (2004) described a quantitative real-time PCR assay which targeted the capsid protein-1 gene. In the study the authors applied the assay to urine, plasma and kidney biopsy samples in three clinical settings: (1) patients with asymptomatic BKVN; (2) patients with active BKVN; and (3) patients with resolved BKVN. The study demonstrated that active BKVN was associated with levels of BK viraemia greater than 5×10^3 copies/ml and with BK viruria greater than 1×10^7 copies/ml in all cases. Interestingly, resolution of nephropathy also led to BK viraemia becoming undetectable and decreased viruria levels. The viral loads in patients with asymptomatic BKVN were generally lower however in some cases the levels were similar to those more typical of BKVN. The authors concluded that serial measurement of viral loads by real-time quantitative PCR is a useful tool in monitoring BKVN however the results of the assay should be interpreted in conjunction with the clinical symptoms and biopsy findings. In 2005 an expert panel recommended the use of either urine cytology or nucleic-acid based testing to screen renal-transplant recipients for 2 years post transplant if they presented with evidence of graft dysfunction and accompanying renal biopsy (Hirsch *et al.*, 2005). The guidelines included quantitative cut-offs for BK viral loads such that urine DNA loads $> 10^7$ copies/ml or plasma DNA loads $> 10^4$ copies/ml, persisting for more than 3 weeks were indicative of presumptive BKVN and renal biopsy should be considered. However, several assays have described the use of different sets of primers and probes and standards limiting the viability of comparing viral loads measured in different institutions by different assays. This led Hoffman and colleagues (2008) to compare seven real-time PCR assays for the quantitative detection of BKV in a panel of 251 urine samples. The authors demonstrated that there was significant disagreement between the viral loads detected by individual assays which were attributable to primer and probe design or the reference standard used. Primer and probe mismatches due to subtype-associated polymorphisms, particularly among subtype III and IV isolates were a significant source of error. To overcome these mismatch issues the authors used available subtype DNA sequence data plus data on subtype distribution to design a new assay to reliably measure all subtypes encountered in their study population. The new assay demonstrated significantly improved performance across all subtypes and the authors concluded that the use of standardized reference standards and the new assay would facilitate the comparison of viral loads measured by laboratories.

Quantitative detection of hepatitis viruses by real-time PCR

Approximately 300 million people worldwide are chronically infected with hepatitis B virus (HBV). HBV VL testing is useful to assess treatment response in patients undergoing antiviral therapy and VL can also be used as a marker for the emergence of resistant viral strains in patients undergoing antiviral treatment. Conventional QPCR assays were developed for the quantitation of HBV in patients including several commercially produced tests. Although these commercially produced assays have proven to be useful many have a limited linear range of quantitation. The introduction of real-time PCR has facilitated more accurate quantitation of HBV and several in-house assays have been described. The assay of Abe *et al.* (1999) targets the HBV surface gene (S gene) and the X gene with a sensitivity of ten copies/PCR and a linear range of quantitation from 10^1 to 10^8 DNA copies per PCR. The assay described by Pas *et al.* (2000) targets the pre-S gene of HBV and the dynamic range of detection was demonstrated to be 373 to 10^{10} genome copies per ml. Application of the assay to the monitoring of four patients during antiviral treatment demonstrated the results of the new assay correlated closely with those from two commercial assays.

Hepatitis C (HCV) is the most common cause of non-A and non-B viral hepatitis and most infection is asymptomatic; however, in approximately 70–80% of patients, the infection becomes chronic. Chronic HCV infection can lead to a range of clinical outcomes from mild non-progressive liver damage, to severe chronic hepatitis that can progress to cirrhosis, end-stage liver disease, and hepatocellular carcinoma. Chronic infection is characterized by persistent viraemia, which is the standard diagnostic marker for chronic infection. HCV VL testing has only limited value for determining disease prognosis but is a valuable tool for monitoring the success of antiviral therapy. HCV VL testing is also useful for estimating infectivity and determining the risk of mother-to-infant transmission. As for HBV, several in-house assays have been described for quantitative detection of HCV including the assays of Martell *et al.* (1999) and Kleiber *et al.* (2000). Commercial diagnostic companies have developed integrated instrument and real-time assay systems for the accurate quantitation of HBV, HCV and HIV. These test platforms incorporate extraction, amplification and detection and have very high levels of reproducibility, accuracy with relatively minor user input. This has led to many institutions adopting the commercial systems for the detection and quantitation of these viruses in preference to in-house tests. Some of these systems will be described further in the section on 'Automation of real-time PCR workflow in clinical diagnostics'.

Detection of bacterial pathogens by real-time PCR

Detection of bacterial meningitis and sepsis by real-time PCR

Meningitis can be a devastating disease that affects the central nervous system (CNS) causing significant morbidity and mortality. *Neisseria meningitidis, Streptococcus pneumoniae* and *Haemophilus influenzae* are responsible for the majority of bacterial meningitis infections. Prior to the introduction of PCR-based methods diagnosis of bacterial meningitis was reliant on the culture of the causative organism from cerebrospinal fluid (CSF). However, culture methods are slow and may take up to three days to produce a result and if the clinical sample is taken after the initiation of antimicrobial therapy the likelihood of recovering the causative organism by culture is reduced. The first real-time PCR assay for the detection of *N. meningitidis* in samples including whole blood, CSF, plasma, and serum samples targeted the meningococcal capsular transfer gene (*ctrA*), and two assays targeting the sialyltransferase gene (*siaD*) for serogroup B and C determination (Guiver *et al.* 2000) (Table 6.2). The *ctrA* assay was demonstrated to be specific for *N. meningitidis* and the authors adopted the *ctrA* assay as their primary screening test followed by serogroup determination using the *siaD* B and C assays. Implementation of the assay resulted in a 56% increase in the number of laboratory confirmed cases of meningococcal disease compared to culture only confirmed cases. A real-time multiplex assay for the simultaneous detection of *N. meningitidis, H. influenzae,* and *S. pneumoniae* in suspected cases of meningitis and septicaemia was developed by Corless *et al.* (2001). The assay targeted the *ctrA* gene of *N. meningitidis,* the capsulation (*bexA*) gene of *H. influenzae* and the pneumolysin gene (*ply*) of *S. pneumoniae.* This multiplex PCR was applied to 4113 culture-negative clinical samples and cases of meningococcal, *H. influenzae* and pneumococcal disease that had not previously been confirmed by culture were identified. Application of this multiplex PCR in a routine screening service provides improved non-culture diagnosis and case ascertainment of bacterial meningitis and septicaemia.

Detection of bacterial respiratory pathogens by real-time PCR

Bordetella pertussis causes whooping cough, a respiratory disease occurring mainly in adolescents. Prior to the advent of PCR-based methods laboratory diagnosis was reliant on isolation of the organism by culture. Although culture is highly specific its sensitivity is poor and dependent on the quality of specimen provided. In addition, isolation may take from 3 to 12 days because the organism is slow growing and fastidious. Rapid diagnosis facilitates the administration of appropriate treatment and prophylaxis of contacts. Because of these issues real-time PCR-based diagnostic testing has become the method of choice for the diagnosis of whooping cough, particularly in children.

Most PCR tests for *Bordetella* are based on detection of insertion sequences which are usually present in multiple copies per genome, increasing the sensitivity of the tests. The insertion sequence IS481 is the most well described and validated target for *B. pertussis* detection and there are more than 50 copies of IS481 per *B. pertussis* genome (Kösters *et al.*, 2001; Reischl *et al.*, 2001). However, IS481 is also present in *B. holmesii* at 8–10 copies per genome and is occasionally found in human isolates of *B. bronchiseptica*. Although the IS481 based assays do cross-react with other closely related species of the genus the consensus is that the sensitivity of using the IS481 as a target outweighs this limitation. Other IS elements and single-copy genes have been developed as targets in real-time PCR assays and a recent review has been published by *Loeffelholz* (2012) which outlines the different diagnostic algorithms and target genes available for the diagnosis of whooping cough.

Legionnaires' disease is a community acquired, or occasionally nosocomial, pulmonary infection caused by *Legionella pneumophila*, which can be severe and life-threatening. Infection occurs in the elderly and immunocompromised and is acquired through inhalation of *L. pneumophila* from contaminated environmental sources such as the water systems of large buildings such as hospitals. The gold standard for the diagnosis of *Legionella* infection is laboratory culture of the organism from clinical samples with diagnosis being made from a variety of clinical specimen types. However, isolation of *Legionella* from patient samples can take up to two weeks and rapid diagnosis of infection, in immunocompromised patients where the fatality rate can be as high as 50%, is important because the infection often responds to antimicrobial therapy.

Rapid tests such as urine antigen detection tests and direct fluorescent antibody assays (DFA) have been developed. However, DFA tests can cross-react with other species leading to false-positive results and it also has a low sensitivity. Antigen testing of urine has a relatively high sensitivity (85%) but can detect only a limited range of pathogenic *Legionella* species. Several real-time PCR-based assays for the detection of *Legionella* species in clinical samples have been described based on the detection of target regions in the 16S rRNA, 5S rRNA or the *mip* gene Hayden *et al.* (2001). Many of these assays have been applied to the detection of *Legionella* in BAL and open lung biopsy samples and the results compared to those of culture and DFA. Often these studies have used relatively small sets of samples so it has been difficult to validate the use of real-time PCR in routine clinical diagnostics and the use of such methods has therefore remained largely restricted to specialist and reference laboratories. This is due in part, to the lack of robust data regarding their performance

Table 6.2 Real-time PCR methods for the detection of bacterial pathogens

Target organism	Detection/ quantitation	Sensitivity	Gene(s) targeted	Reference
Bartonella species	D	Not reported	*ribC*	Zeaiter *et al.* (2003)
Bordetella pertussis, *B. holmesii*	D	Not reported	IS481	Reischl *et al.* (2001)
B. pertussis, *B. parapertussis*	D	0.1 CFU and 10 CFU/PCR	IS481, IS1001	Kösters *et al.* (2001)
B. pertussis, *B. parapertussis*	D	0.75 CFU/PCR	IS481, IS1001, *ptg*	Sloan *et al.* (2002)
B. pertussis, *B. parapertussis*	D	0.1 CFU and 10 CFU/PCR	IS481, IS1001	Kösters *et al.* (2002)
B. pertussis, *B. parapertussis*	D	1 CFU and 5 CFU/PCR	IS481, IS1001	Cloud *et al.* (2003)
Borrelia burgdorferi	Q	1–3 spirochetes/ PCR	Flagellin	Pahl *et al.* (1999)
B. burgdorferi	Q	Not reported	*rec*A	Morrison *et al.* (1999)
Borrelia garnii, *Borrelia afzelii*, *B. burgdorferi*	D	Not reported	*rec*A	Pietila *et al.* (2000)
Borrelia species	Q	1 to 10 spirochetes/PCR	*osp*A	Rauter *et al.* (2002)
B. burgdorferi	Q	1 to 10 spirochetes/PCR	Flagellin	Piesman *et al.* (2001)
B. burgdorferi	Q	1 to 10 spirochetes/PCR	Flagellin	Zeidner *et al.* (2001)
B. burgdorferi	D	Not reported	*recA*, p66	Mommert *et al.* (2001)
Campylobacter species	D	Not reported	16S rRNA	Logan *et al.* (2001)
Campylobacter jejuni	Q	12 GE	Novel ORF	Sails *et al.* (2003)
C. jejuni	Q	1 CFU/PCR	Novel sequence	Nogva *et al.* (2000)
Chlamydophila pneumoniae	D	0.001 IFU/PCR	*ompA*	Tondella *et al.* (2002)
C. pneumoniae	Q	Not reported	*ompA*	Kuoppa *et al.* (2002)
C. pneumoniae	Q	10^{-6} IFU/PCR	*ompA*	Apfalter *et al.* (2003)
C. pneumoniae	D	0.02 IFU/PCR	16S rRNA	Reischl *et al.* (2003)
C. pneumoniae	D	10 gene copies	*Pst1* gene fragment	Welti *et al.* (2003)
Chlamydia trachomatis	Q	1–3 copies/PCR	Cryptic plasmid, MOMP	Jalal *et al.* (2006)
C. trachomatis	Q	25–50 copies/ PCR	Cryptic plasmid, *omp*1 (L-2 specific)	Halse *et al.* (2006)
Clostridium difficile	D	10 gene copies	*tcdA*, *tcdB*	Bélanger *et al.* (2003)
Escherichia coli (VTEC)	D	1 CFU/PCR	stx_1, stx_2	Bellin *et al.* (2001)

Target organism	Detection/ quantitation	Sensitivity	Gene(s) targeted	Reference
E. coli (VTEC)	D	10 gene copies	*stx_1*, *stx_2*	Bélanger *et al.* (2002)
E. coli (VTEC)	D	5.8–580 CFU/ beef, 1.2–1200CFU/ faeces	*stx_1*, *stx_2*, *eae*	Sharma *et al.* (1999)
E. coli (VTEC)	D	Not reported	*stx_1*, *stx_2*, *eae*, *hly*A	Reischl *et al.* (2002a)
E. coli (VTEC)	D	Not reported	*stx_1*, *stx_2*, *eae*	Nielson and Anderson (2003)
E. coli (VTEC)	D	10^4 CFU/g faeces	*stx_1*, *stx_2*, *eae*	Sharma and Dean-Nystrom (2003)
Haemophilus influenzae	D	Not reported	*bexA*	Corless *et al.* (2001)
H. influenzae	Q	1 CFU/PCR	Region 2 capsulation locus	Marty *et al.* (2004)
Helicobacter pylori	Q	10 gene copies	*ureC*	He *et al.* (2002)
Legionella pnuemophila	D	2.5 CFU/PCR	*mip*	Ballard *et al.* (2000)
Legionella species, *L. pnuemophila*	D	<10 CFU/PCR	5S rRNA, *mip*	Hayden *et al.* (2001)
Legionella species, *L. pnuemophila*	D	1fg DNA/PCR	16S rRNA, *mip*	Wellinghausen *et al.* (2001)
Legionella species, *L. pnuemophila*	D	3 GE/PCR	16S rRNA	Reischl *et al.* (2002b)
Legionella species	D	2 CFU/PCR	16S rRNA	Rantakokko-Jalava and Jalava (2001)
L. pnuemophila	D	10 CFU/PCR	*mip*	Wilson *et al.* (2003)
L. pnuemophila	D	10 GE/PCR	*mip*	Welti *et al.* (2003)
Moraxella catarrhalis	D	1 CFU/PCR	*copB*	Greiner *et al.* (2003)
Mycobacterium tuberculosis	Q	Not reported	IS6110	Desjardin *et al.* (1998)
M. tuberculosis complex	D	800 gene copies/ PCR	ITS	Kraus *et al.* (2001)
M. tuberculosis complex	D	Not reported	ITS	Miller *et al.* (2002)
M. tuberculosis complex	D	Not reported	16S rRNA gene	Shrestha *et al.* 2003
M. tuberculosis complex	D	50 genomes/ PCR	16S rRNA gene	Burggraf *et al.* (2005)
Mycoplasma genitalium	Q	10 gene copies/ PCR	16S rRNA	Deguchi *et al.* (2002)
M. genitalium	Q	10 gene copies/ PCR	16S rRNA	Yoshida *et al.* (2002)
Mycoplasma pneumoniae	D	10 gene copies	P1 gene	Welti *et al.* (2003)

Table 6.2 (Continued)

Target organism	Detection/ quantitation	Sensitivity	Gene(s) targeted	Reference
M. pneumoniae	D	20–200 cells/ PCR	P1 gene	Ursi *et al.* (2003)
M. pneumoniae	D	Not reported	P1 gene	Hardegger *et al.* (2000)
Neisseria gonorrhoeae	D	Not reported	*cppB* gene	Whiley *et al.* (2002)
N. gonorrhoeae	D	Not reported	*porA* psuedogene	Whiley *et al.* (2004)
N. gonorrhoeae	D	Not reported	*cppB*, 16S rRNA	Boel *et al.* (2005)
N. gonorrhoeae	D	1 GE/PCR	*opa*	Geraats-Peters *et al.* (2005)
Neisseria meningitidis	D	Not reported	*ctrA*	Corless *et al.* (2001)
N. meningitidis	D	Not reported	16S rRNA, *sacC*, *siaD*, *porA*	Molling *et al.* (2002)
Staphylococcus aureus	D	2 CFU/PCR	*nucA*	Palomares *et al.* (2003)
CNS	D	Not reported	16S rRNA	Edwards *et al.* (2001a)
S. aureus, MRSA	D	25 GE/PCR	Novel sequence, *mecA*	Reischl *et al.* (2000)
S. aureus, MRSA	D	Not reported	Sa442, *mecA*	Tan *et al.* (2001)
MRSA	D	Not reported	Sa442, *mecA*	Grisold *et al.* (2002)
MRSA	D	Not reported	*nucA*, *mecA*	Fang *et al.* (2003)
MRSA	D	25 CFU/sample	SCC*mec*, *orfX*	Huletsky *et al.* (2004)
Group A *streptococci*	D	Not reported	not stated	Uhl *et al.* (2003)
Streptococcus pneumoniae	Q	1 CFU/PCR	*ply*	Greiner *et al.* (2001)
S. pneumoniae	D	4 GE/PCR	*lytA*	McAvin *et al.* (2001)
S. pneumoniae	D	Not reported	*ply*	Corless *et al.* (2001)
Tropheryma whipplei	Q	Not reported	16S 23S rRNA ITS, *rpoB*	Fenollar *et al.* (2002)
Yersinia enterocolitica	D	Not reported	*bipA* and novel target sequence	Aarts *et al.* (2001)

CNS, coagulase-negative staphylococcus; IFU, inclusion-forming unit; MRSA, methicillin-resistant *S. aureus*.

in clinical diagnostics. In a recent study by Mentasti *et al.* (2012) a real-time PCR assay targeting the *mip* gene was applied to the detection of *Legionella* in 161 respiratory samples. It was demonstrated that PCR was significantly more sensitive than culture for samples taken after ≤2 days of hospitalization (94.7% vs. 79.6%), with the difference being even more marked for samples taken between 3 and 14 days (79.3% vs. 47.8%). Overall, the real-time PCR assay was approximately 30% more sensitive than culture although the authors cautioned that further prospective clinical validation studies were required.

Real-time PCR assays have been described for a number of other respiratory pathogens including *Streptococcus pneumoniae* (Greiner *et al.*, 2001). *Mycoplasma pneumoniae* (Hardegger *et al.*, 2000; Welti *et al.*, 2003; Ursi *et al.*, 2003), *Haemophilus influenzae* (Marty *et al.*, 2004), and *Moraxella catarrhalis* Greiner *et al.*, 2003).

Detection of antibiotic resistance, antiviral susceptibility and toxin genes by real-time PCR

Real-time PCR methods have been applied to the detection of a wide range of antimicrobial resistance determinants and toxin genes. Examples are presented in Table 6.3. A single base mutation within the peptidyltransferase-encoding region of the 23S rRNA gene in *Helicobacter pylori* is associated with clarithromycin resistance. The three mutations observed are responsible for the replacement of adenine residues at positions 2143 and 2144 with guanine (A2143G and A2144G) or cytosine (A2143C). Several assays targeting the 23S rRNA gene have been described for the discrimination of clarithromycin sensitive and clarithromycin-resistant strains (Gibson *et al.*, 1999; Matsumura *et al.*, 2001). In the study of Matsumura *et al.* the authors demonstrated that they could identify clarithromycin resistant strains directly in gastric biopsy samples. Similarly, in the study of Oleastro *et al.* (2003) a real-time PCR assay was designed to target the three point mutations (A2142C, A2142G, and A2143G) to determine clarithromycin susceptibility directly in gastric biopsy sample from patients with *H. pylori* infection. The assay was also shown to be able to detect resistant strains among sensitive wild-type strains in mixed populations of *H. pylori*. Rapid detection of resistance genotypes of *H. pylori* directly in gastric biopsy specimens may prove useful in determining appropriate treatment regimens for the successful eradication of *H. pylori* from the patient.

Assays for the detection of ciprofloxacin resistance have been described for *Campylobacter jejuni* (Wilson *et al.*, 2000), *Salmonella enterica* serotype *Typhimurium* DT104 (MR DT104) (Walker *et al.*, 2001), and *Yersinia pestis* (Lindler *et al.*, 2001). Resistance to quinolones such as ciprofloxacin is chromosomally mediated with mutations within the quinolone resistance-determining region (QRDR) of the DNA gyrase gene (*gyrA*) playing a major role in gram-negative bacteria. An assay for the discrimination of ciprofloxacin-sensitive and ciprofloxacin-resistant strains of *C. jejuni* was described by Wilson *et al.* (2000). The assay targeted the C to T transition in codon 86 in the QRDR of *gyrA* and was able to rapidly and reliably differentiate between of ciprofloxacin-sensitive and ciprofloxacin-resistant strains of *C. jejuni*.

Lindler *et al.* (2001) demonstrated that ciprofloxacin resistance in *Y. pestis* is due to a single nucleotide mutation in the QRDR of *gyrA*. The authors developed a real-time PCR assay to differentiate between all four point observed mutations within the *gyrA* gene that resulted in ciprofloxacin resistance. The assay successfully discriminated between sensitive and resistant strains of *Y. pestis* on the basis of probe melting curve analysis following PCR amplification. Several other real-time PCR assays have been described for the detection of

Table 6.3 Real-time PCR methods for the detection of antibiotic susceptibility, antibiotic resistance, antiviral susceptibility and toxin genes in clinical microbiology

Organism	Function of assay	Gene(s) targeted	Reference
Bordetella pertussis	Pertussis toxin gene variants	ptxS1A, ptxS1B, ptxS1D, ptxS1E	Makinen *et al.* (2002)
Campylobacter jejuni	Ciprofloxacin resistance	*gyrA*	Wilson *et al.* (2000)
Corynebacterium diphtheriae	Toxin gene detection	A and B subunit toxin genes	Mothershed *et al.* (2002)
Coxiella burnetti	Antibiotic susceptibility testing	*com1* gene	Brennan and Samuel (2003)
Enterococcus faecalis, *Enterococcus faecium*	Oxazolidinione resistance	G2576U rRNA mutation	Woodford *et al.* (2002)
Helicobacter pylori	Clarithromycin resistance	23S rRNA	Oleastro *et al.* (2003)
H. pylori	Clarithromycin resistance	23S rRNA	Matsumura *et al.* (2001)
H. pylori	Clarithromycin resistance	23S rRNA	Gibson *et al.* (1999)
H. pylori	Tetracycline resistance	16S rDNA	Lawson *et al.* (2005)
Mycobacterium tuberculosis	Rifampicin and isoniazid resistance	*rpoB* and *katG*	Garcia de Viedma *et al.* (2002)
M. tuberculosis	Rifampicin and isoniazid resistance	*rpoB* and *inhA*	Torres *et al.* (2000)
M. tuberculosis	Isoniazid resistance	*katG*	van Doorn *et al.* (2003)
M. tuberculosis	Rifampicin resistance	*rpoB*	Edwards *et al.* (2001b)
M. tuberculosis	Rifampicin and isoniazid resistance	*rpoB, katG and inhA*	Ruiz *et al.* (2004)
M. tuberculosis	Rifampicin, isoniazid and ethambutol resistance	*rpoB, katG and embB*	Wada *et al.* (2004)
M. tuberculosis	Rifampicin and isoniazid resistance	*rpoB* and *katG*	Marín *et al.* (2004)
M. tuberculosis	Rifampicin resistance	*rpoB*	Kocagoz *et al.* (2005)
Salmonella enterica serotype *Typhimurium* DT104	Ciprofloxacin resistance	*gyrA*	Walker *et al.* (2001)
Streptococcus pneumoniae	Penicillin susceptibility	*ptp2b* gene	Kearns *et al.* (2002a)
Staphylococcus species	Methicillin resistance	*mecA*	Killgore *et al.* (2000)
Staphylococcus aureus	Fluoroquinolone resistance	*grlA*	Lapierre *et al.* (2003)
S. aureus	Panton-Valentine leukocidin gene	*mecA*, *nuc*, PVL	McDonald *et al.* (2005)
S. aureus	Panton-Valentine leukocidin gene	PVL	Deurenberg *et al.* (2004)
Vancomycin resistant enterococcus (VRE)	Vancomycin resistance	*vanA* and *vanB*	Palladino *et al.* (2003a)
VRE	Vancomycin resistance	*vanA* and *vanB*	Palladino *et al.* (2003b)
Yersinia pestis	Ciprofloxacin resistance	*gyrA*	Lindler *et al.* (2001)

antibiotic resistance genes within a number of microbial pathogens. These include assays for the detection of the methicillin resistance gene, *mecA* the most frequent cause of methicillin resistance in staphylococci (Killgore *et al.*, 2000). The *vanA* and *vanB* genes of vancomycin-resistant enterococci were targeted in the assays of Palladino *et al.* (2003a,b).

Real-time PCR assays have also been developed for the rapid detection of bacterial toxin genes. Laboratory diagnosis of diphtheria is reliant on isolation of the causative organism (*Corynebacterium diphtheriae*) from the patient and demonstration of the toxigenic potential of the isolate using the Elek test. Unfortunately, the viable organisms that are present are not always detectable. This deficiency has led to the development of PCR-based detection tests, which are often the only laboratory test available for the diagnosis of diphtheria in the absence of culturable organisms. A real-time PCR assay for the detection and differentiation of the A and B subunits of the toxin gene of *C. diphtheriae* was developed by Mothershed *et al.* (2002). Application of the assay to 36 clinical specimens from patients with confirmed clinical diphtheria demonstrated that the assay detected one or both of the subunits of the toxin gene in 34 of the 36 specimens compared to only nine specimens being found to be positive in the conventional PCR. This assay provides a more sensitive and rapid alternative to conventional PCR for the detection of toxigenic strains of this important pathogen.

Practical aspects of using real-time PCR in clinical microbiology

Sample processing and nucleic acid extraction

Prior to detection by real-time assays the nucleic acid in the clinical sample must be extracted in order for it to be detectable in the assay. For some specimens such as CSF this may be as simple as boiling the specimen however for more complex sample matrices such as pulmonary secretions more complex extraction procedures are necessary. Many types of clinical samples such as faeces and whole blood contain substances inhibitory to the PCR process and therefore the extraction method must also remove these to prevent inhibition of the PCR. Originally manual extraction methods were the only methods available to clinical laboratories however the development of automated or semi-automated platforms have helped to standardize sample extraction methods. However, before new extraction instruments are integrated into the laboratory workflow they require validation to ensure they are suitable for the wide range of sample types found in clinical microbiology, these include swabs, fluids, pulmonary secretions, tissues, CSF, urine, faeces, blood, and pus. The optimum type of sample for a particular assay and a particular pathogen also needs to be established and standardized. Clinical samples may contain very low numbers of pathogenic organisms against a background flora of many other microbes. Generic nucleic acid extraction methods produce nucleic acid extracts which contain a small amount of nucleic acid from the target pathogen against a background of nucleic acid from the rest of the mixed flora in the sample. A small volume of this sample is then tested in the PCR assay (usually between 1 and 10 μl) and therefore the LOD is always compromised when there are only small numbers of target organism in the original sample. Pathogen-specific sample preparation methods which concentrate the target organisms before extraction or concentrate the target nucleic acid during extraction help to increase the sensitivity of detection of assays. This factor becomes increasingly important if the volume of the PCR reaction is further reduced through the development of miniature real-time PCR systems based on microfluidics and microchips (Liao *et al.*, 2005).

Automation of real-time PCR workflow in clinical diagnostics

The development of software interfaces for analytical instruments such as serological analysers which allow bidirectional communication between the laboratory information management systems (LIMS) have reduced the need for the use of paper in the laboratory. Such interfaces allow sample test lists to be generated and imported into analysers and following processing the results are uploaded directly into the LIMS. This eliminates transcription errors and removes the need to generate paper worksheets. The introduction of automated sample extraction instruments has increased throughput and reduced the amount of operator intervention required for real-time PCR diagnostics as discussed earlier. However, such systems still require the samples to be transferred from the extraction instrument into the individual PCR reactions which have to be set up by the laboratory technician. This lack of LIMS connectivity and integration of workflow has been addressed by several instrument systems which have become available over the last few years. These instrument systems integrate parts of the workflow and also provide LIMS connectivity via a bidirectional interface for uploading the results. There are several systems currently available which integrate sample extraction and PCR setup into a single instrument. These include the Cobas 4800 system, the Cobas AmpliPrep, the Cobas TaqMan system (Roche Diagnostics Ltd), the m2000 system (Abbott Diagnostics) and the Qiasymphony RGQ (Qiagen Ltd). The current systems still require some user intervention such as the transfer of the prepared PCR reactions to the real-time PCR instrument however three of the systems (Cobas 4800 system/Cobas AmpliPrep-TaqMan/m2000 system) include a results algorithm which eliminates the need for manual curve analysis. These three systems are intended primarily for performing commercial IVD tests such as quantitative HIV, hepatitis B and C virus and/or qualitative tests for human papillomavirus and *C. trachomatis* and *N. gonnorrhoeae*. Although they do provide workflow solutions for a limited number of target organisms they cannot address the lack of integration of technology for in-house tests a subject discussed further below (see section on future developments).

Instruments which integrate sample processing and extraction, plus PCR setup and automated result analysis have been developed recently. The GeneXpert System (Cepheid) is a closed, self-contained, fully integrated real-time instrument which is scalable from 1, 2, 4 or 16 individual test modules. The system requires minimal user intervention other than to add the sample to the individual test cartridge and then load this onto the instrument. The modular design which features independently operated PCR modules facilitates on-demand molecular testing using the supplier's pathogen detection kits. There are currently a range of kits available but the portfolio does not include all of the targets currently covered by either other commercial kits or in-house assays. In addition, this instrument is not an 'open' platform so users cannot run their own in-house assays and the purchase price of the instruments is significant which may also limit their use to larger institutions.

The BD MAX instrument (Becton Dickinson) is a fully integrated instrument for the purification, amplification and detection of nucleic acids. The instrument utilizes an extraction cartridge which contains all of the extraction reagents and required pipette tips. It can also be loaded with real-time PCR reagents either from an IVD kit from the supplier or the users own in-house assays. Up to 24 samples can be loaded onto the system per run with the PCR reaction being carried out in a micro fluidic cartridge which is loaded automatically by the instrument and up to five channels can be monitored. The instrument can be utilized in several different modes including *in vitro* diagnostic mode (IVD) using IVD kits from the supplier; however, there is currently only a very limited menu of tests available. In 'open' mode the DNA extraction and amplification reagents are provided lyophilized and primers and probes are added to the extraction cartridge prior to processing. In 'research' mode the operator can run assays utilizing user-defined enzyme and primer and probe combinations therefore facilitating the transfer of previously developed in house assays directly onto the system. The total processing time for 24 samples is approximately two and a half hours depending on the assay type.

Appropriate controls for real-time PCR in clinical microbiology

The application of real-time methods to aid the diagnosis and management of infectious disease is one of the most rapidly growing fields in PCR diagnostics with an increasing number of assays being reported in the literature every month. Each assay run needs to include appropriate controls to demonstrate that the assay is performing correctly (see Chapter 2). The most basic control format may be positive control material in the form of nucleic acid extracted from a culture of the target organism. This only acts as a control to demonstrate that the amplification and detection process is working adequately. This type of control is now being supplemented or replaced by a more sophisticated approach in which an extraction control virus or bacteria (depending on the assay target) is 'spiked' into the clinical sample prior to testing. In our laboratory we use MS2 and T4 bacteriophages as our internal amplification controls which are 'spiked' into clinical samples prior to nucleic acid extraction. Total nucleic acid is then extracted from the sample with the MS2 RNA or T4 DNA also being co-extracted with that from any pathogens present. A separate set of primers and a hydrolysis probe specific for MS2 and/or T4 phage are included in the real-time PCR assay so they can be detected. This approach controls for the whole of the assay process including the extraction and detection steps and most closely resembles the pathogen detection process. This approach was recently evaluated by Ninove *et al.* (2011) using a series of 8950 clinical samples (representing 36 distinct specimen types). The study authors concluded that the use of T4 and MS2 phages as ICs proved to be cost-effective and flexible and was a valuable strategy for enhancing the quality of molecular diagnosis using in-house designed assays.

Working standards or 'run controls' to control for inter-assay variability are essential. These consist of either previously extracted nucleic acid from the target organism or an aliquot of simulated sample containing inactivated virus or bacteria. The samples are prepared by diluting the material to give a crossing point or Ct value of between 32 and 35 cycles. Inter-assay variation results in the Ct value of these samples changing between each individual assay run. To determine if assay performance is changing over time the Ct value of the run controls over time is monitored. This can help establish if there is a change in sensitivity of the assay perhaps due to a change in reagent batch. Inclusion of appropriate negative controls is also required to monitor for cross-contamination and reagent contamination. To monitor for possible cross-contamination between samples it is essential to include a negative control (often viral transport media or PCR-grade water) which is extracted within the same run as the patient samples. The inclusion of a 'reagent blank' composed of all of the reaction components except nucleic acid template is also helpful.

Reducing the likelihood of cross-contamination and the risk of false-positives

The inherent sensitivity of real-time PCR means that contamination between samples and from previously amplified product can lead to false-positive results. Good laboratory practice such as the wearing of gloves and working within either PCR cabinets or laminar flow hoods reduces the likelihood of contamination. Incorporation of uracil DNA glycosylase and UTP in PCR master mixes can also reduce the likelihood of amplicon carryover to subsequent reactions. Although the risk of contamination with amplified product from previous real-time assay runs should be minimal assuming the tubes are not opened following amplification, measures to reduce the likelihood of this occurring are very important. However, if amplified reactions are opened post cycling the risk of false-positives in subsequent assay runs is greatly increased. If such manipulation is necessary, for example when performing antiviral resistance assays, it should preferably be carried out in a separate laboratory away from where other reactions are set up. Good laboratory practice, including the use of separate rooms for master mix preparation, nucleic acid extraction, template addition and cycling and any post-PCR manipulation reduces the likelihood of cross-contamination. Each room should have dedicated equipment including pipettes and other small laboratory equipment. Additional procedures can include using paper disposable gowns in the PCR clean areas, which can then be thrown away at the end of each working day.

Following good laboratory practice also reduces the likelihood of specimen to specimen contamination. The use of automated extraction instruments and PCR set-up robots reduces the possibility of manual pipetting errors. The likelihood of specimen to specimen contamination is increased if samples containing high numbers of target organisms (such as a sample with a high viral load) are processed in the same run as samples which are likely to contain very low or negative titres of the same target. An example of this would be the processing of a genital swab for herpes simplex virus which may have a very high viral titre in the same run as a set of cerebrospinal fluid specimens from patients with possible viral meningitis where the viral titre may be very low. To reduce the possibility of specimen to specimen contamination between such samples it is advisable to run them on different instruments.

External quality assurance for real-time PCR in clinical microbiology

Quality assurance and proficiency testing has a fundamental role in the implementation of molecular diagnostic testing for the diagnosis of infectious disease. Quality Control for Molecular Diagnostics is an independent international external quality assessment (EQA) and proficiency testing (PT) organization (www.qcmd.org) which distributes proficiency panels of simulated clinical samples containing a wide range of viral and bacterial pathogens for molecular diagnostic assays. More than 2000 participants in over 100 countries regularly participate in the programme, which is endorsed by the European Society for Clinical Virology and the European Society for Microbiology and Infectious Disease. The aim of the EQA programmes is to help monitor and improve quality within the clinical laboratory by assessing the ability of the participating laboratory to detect pathogens within the routine clinical setting. The laboratory is sent a set of samples that resemble clinical specimens and is then required to test these samples using standard procedures and report the results back to QCMD. A full EQA report plus an individual performance report is sent back to the participating laboratory to allow them to assess the performance of their own assay in comparison with other participating laboratories. In the UK the National External Quality Assessment Service (NEQAS) and the Clinical Virology Network (CVN) also distribute EQA panels for a more limited set of pathogens. Laboratories providing molecular diagnostic testing should participate in these types of scheme to ensure quality of testing.

Future developments

This chapter has illustrated some of the areas in clinical microbiology in which the introduction of real-time methods have made a dramatic impact. Although much has been achieved in the last 15 years several areas still require further development. For in-house testing most clinical diagnostic laboratories use a range of extraction platforms, PCR set-up robots and real-time PCR instruments often from different manufacturers. Some instruments have limited interfacing capability however many cannot directly communicate with instruments from other manufacturers thus limiting the overall connectivity of the real-time PCR workflows. This is a problem which needs to be addressed by the instrument manufacturers who need to agree a consensus data format to allow direct interfacing of instrumentation. In an era of wi-fi and Bluetooth connectivity where mobile phones and tablet computers can communicate effortlessly it seems unlikely that different manufacturers' laboratory instrumentation will remain unable to communicate directly.

In the last few years we have seen the development of self-contained, fully integrated real-time PCR instruments. In future years further automation and miniaturization of instrumentation and the development of wider panels of assay targets is likely to occur. However, the development of new commercial assays takes a significant period of time from concept to becoming available to the customer. Therefore, commercial manufacturers of assays will not be able to react quickly to the emergence or re-emergence of a new species or strain of pathogen. This means that testing laboratories still need to retain expertise in developing and running 'in-house' methodologies to enable a quick response to emerging threats.

The introduction of real-time PCR methods in clinical microbiology has improved the detection of infectious disease agents and led to improvements in patient management and care. However, the longer term future of real-time PCR in clinical microbiology may not be as sure. Other emerging technologies such as next-generation sequencing (NGS) will supplement and may even completely replace real-time PCR and other molecular methodologies in the future. As the price of NGS instrumentation falls and the cost of applying these tools to diagnostic clinical microbiology becomes more affordable we may replace our real-time assays targeting limited numbers of genes with whole genome based methodologies.

References

Aarts, H.J., Joosten, R.G., Henkens, M.H., Stegeman, H., and van Hoek, A.H. (2001). Rapid duplex PCR assay for the detection of pathogenic *Yersinia enterocolitica* strains. J. Microbiol. Methods. *47*, 209–217.

Abe, A., Inoue, K., Tanaka, T., Kato, J., Kajiyama, N., Kawaguchi, R., Tanaka, S., Yoshiba, M., and Kohara, M. (1999). Quantitation of Hepatitis B virus genomic DNA by real-time detection PCR. J. Clin. Microbiol. 37, 2899–2903.

Apfalter, P., Barousch, W., Nehr, M., Makristathis, A., Willinger, B., Rotter, M., and Hirschl, A.M. (2003). Comparison of a new quantitative ompA-based real-time PCR TaqMan assay for detection of *Chlamydia pneumoniae* DNA in respiratory specimens with four conventional PCR assays. J. Clin. Microbiol. *41*, 592–600.

Aryee, E.A., Bailey, R.L., Natividad-Sancho, A., Kaye, S., and Holland, M.J. (2005). Detection, quantification and genotyping of Herpes Simplex Virus in cervicovaginal secretions by real-time PCR: a cross sectional survey. Virol. J. *2*, 61.

Ballard, A.L., Fry, N.K., Chan, L., Surman, S.B., Lee, J.V., Harrison, T.G., and Towner, K.J. (2000). Detection of *Legionella pneumophila* using a real-time PCR hybridisation assay. J. Clin. Microbiol. *38*, 4215–4218.

Bechert, C.J., Schnadig, V.J., Payne,. D.A., and Dong, J. (2010). Monitoring of BK viral load in renal allograft recipients by real-time PCR assays. Am. J. Clin. Pathol. *133*, 242–250.

Bélanger, S.D., Boissinot, M., Ménard, C., Picard, F.J., and Bergeron, M.G. (2002). Rapid detection of shiga toxin-producing bacteria in faeces by multiplex PCR with molecular beacons on the SmartCycler. J. Clin. Microbiol. *40*, 1436–1440.

Bélanger, S.D., Boissinot, M., Clairoux, N., Picard, F.J., and Bergeron, M.G. (2003). Rapid detection of *Clostridium difficile* in faeces by real-time PCR. J. Clin. Microbiol. *41*, 730–734.

Bellin, T., Pulz, M., Matussek, A., Hempen, H.G., and Gunzer, F. (2001). Rapid detection of enterohaemorrhagic *Escherichia coli* by real-time PCR with fluorescent hybridisation probes. J. Clin. Microbiol. *39*, 370–374.

Benschop, K., Molenkamp, R., van der Ham, A., Wolthers, K., and Beld, M. (2008). Rapid detection of human parechoviruses in clinical samples by real-time PCR. J. Clin. Virol. *41*, 69–74.

Berger, A., and Preiser, W. (2002). Viral genome quantification as a tool for improving patient management: the example of HIV, HBV, HCV and CMV. J. Antimicrob. Chemother. *49*, 713–721.

Bialasiewicz, S., Whiley, D.M., Lambert, S.B., Gould, A., Nissen, M.D., and Sloots, T.P. (2007). Development and evaluation of real-time PCR assays for the detection of the newly identified KI and WU polyomaviruses. J. Clin. Virol. *40*, 9–14.

Boel, C.H., van Herk, C.M., Berretty, P.J., Onland, G.H., and van den Brule, A.J. (2005). Evaluation of conventional and real-time PCR assays using two targets for confirmation of results of the COBAS AMPLICOR *Chlamydia trachomatis/Neisseria gonorrhoeae* test for detection of *Neisseria gonorrhoeae* in clinical samples. J. Clin. Microbiol. *43*, 2231–2235.

Boivin, G., Cote, S., Dery, P., De Serres, G., and Bergeron, M.G. (2004). Multiplex real-time PCR assay for detection of influenza and human respiratory syncytial viruses. J. Clin. Microbiol. *42*, 45–51.

Brennan, R.E., and Samuel, J.E. (2003). Evaluation of *Coxiella burnetii* antibiotic susceptibilities by real-time PCR assay. J. Clin. Microbiol. *41*, 1869–1874.

Broccolo, F., Locatelli, G., Sarmati, L., Piergiovanni, S., Veglia, F., Andreoni, M., Buttò, S., Ensoli, B., Lusso, P., and Malnati, M.S. (2002). Calibrated real-time PCR assay for quantitation of human Herpesvirus 8 DNA in biological fluids. J. Clin. Microbiol. *40*, 4652–4658.

Burggraf, S., Reischl, U., Malik, N., Bollwein, M., Naumann, L., and Olgemöller, B. (2005a). Comparison of an internally controlled, large-volume LightCycler assay for detection of *Mycobacterium tuberculosis* in clinical samples with the COBAS AMPLICOR Assay. J. Clin. Microbiol. *43*, 1564–1569.

Burggraf, S., and Olgemoller, B. (2005b). Straightforward procedure for internal control of real-time reverse transcription amplification assays. Clin. Chem. *51*, 1508–1510.

Callahan, J.D., Wu, S.J., Dion-Schultz, A., Mangold, B.E., Peruski, L.F., Watts, D.M., Porter, K.R., Murphy, G.R., Suharyono, W., King, C.C., *et al.* (2001). Development and evaluation of serotype- and group-specific fluorogenic reverse transcriptase PCR (TaqMan) assays for dengue virus. J. Clin. Microbiol. *39*, 4119–4124.

Cloud, J.L., Hymas, W.C., Turlak, A., Croft, A., Reischl, U., Daly, J.A., and Carroll, K.C. (2003). Description of a multiplex *Bordetella pertussis* and *Bordetella parapertussis* LightCycler PCR assay with inhibition control. Diagn. Microbiol. Infect. Dis. *46*, 189–195.

Collot, S., Petit, B., Bordessoule, D., Alain, S., Touati, M., Denis, F., and Ranger-Rogez, S. (2002). Real-Time PCR for quantification of human Herpesvirus 6 DNA from lymph nodes and saliva. J. Clin. Microbiol. *40*, 2445–2451.

Corless, C.E., Guiver, M., Borrow, R., Edwards-Jones, V., Fox, A.J., and Kaczmarski, E.B. (2001). Simultaneous detection of *Neisseria meningitidis, Haemophilus influenzae,* and *Streptococcus pneumoniae* in suspected cases of meningitis and septicaemia using real-time PCR. J. Clin. Microbiol. *39*, 1553–1558.

Cote, S., Abed, Y., and Boivin, G. (2003). Comparative evaluation of real-time PCR assays for detection of the human Metapneumovirus. J. Clin. Microbiol. *41*, 3631–3635.

Deguchi, T., Yoshida, T., Yokoi, S., Ito, M., Tamaki, M., Ishiko, H., and Maeda, S. (2002). Longitudinal quantitative detection by real-time PCR of *Mycoplasma genitalium* in first-pass urine of men with recurrent non-gonococcal urethritis. J. Clin. Microbiol. *40*, 3854–3856.

Desjardin, L.E., Chen, Y., Perkins, M.D., Teixeira, L., Cave, M.D., and Eisenach, K.D. (1998). Comparison of the ABI 7700 system (TaqMan) and competitive PCR for quantification of IS6110 DNA in sputum during treatment of tuberculosis. J. Clin. Microbiol. *36*, 1964–1968.

Deurenberg, R.H., Vink, C., Driessen, C., Bes, M., London, N., Etienne, J., and Stobberingh, E.E. (2004). Rapid detection of Panton-Valentine leukocidin from clinical isolates of *Staphylococcus aureus* strains by real-time PCR. FEMS. Microbiol. Lett. *240*, 225–228.

van Doorn, H.R., Claas, E.C., Templeton, K.E., van der Zanden, A.G., te Koppele Vije, A., de Jong, M.D., Dankert, J., and Kuijper, E.J. (2003). Detection of a point mutation associated with high-level isoniazid resistance in *Mycobacterium tuberculosis* by using real-time PCR technology with 3′-minor groove binder-DNA probes. J. Clin. Microbiol. *41*, 4630–4635.

van Doornum, G.J.J., Schouls, L.M., Pijl, A., Cairo, I., Buimer, M., and Bruisten, S. (2001). Comparison between the LCx probe system and the COBAS AMPLICOR system for detection of *Chlamydia trachomatis* and *Neisseria gonorrhoeae* infections in patients attending a clinic for treatment of sexually transmitted diseases in Amsterdam, the Netherlands. J. Clin. Microbiol. *39*, 829–835.

van Doornum, G.J., Guldemeester, J., Osterhaus, A.D., and Niesters, H.G. (2003). Diagnosing herpesvirus infections by real-time amplification and rapid culture. J. Clin. Microbiol. *41*, 576–580.

Drosten, C., Göttig, S., Schilling, S., Asper, M., Panning, M., Schmitz, H., and Günther, S. (2002). Rapid detection and quantification of RNA of Ebola and Marburg viruses, Lassa virus, Crimean-Congo Haemorrhagic Fever virus, Rift Valley Fever virus, Dengue virus, and Yellow Fever virus by real-time reverse transcription-PCR. J. Clin. Microbiol. *40*, 2323–2330.

Edwards, K.J., Kaufmann, M.E., and Saunders, N.A. (2001a). Rapid and accurate identification of coagulase-negative Staphylococci by real-time PCR. J. Clin. Microbiol. *39*, 3047–3051.

Edwards, K.J., Metherel, L.A., Yates, M., and Saunders, N.A. (2001b). Detection of *rpoB* mutations in *Mycobacterium tuberculosis* by biprobe analysis. J. Clin. Microbiol. *39*, 3350–3352.

van Elden, L.J.R., Nijhuis, M.., Schipper, P., Schuurman, R., and van Loon, A.M. (2001). Simultaneous detection of Influenza Viruses A and B using real-time quantitative PCR. J. Clin. Microbiol. *39*, 196–200.

Espy, M.J., Teo, R., Ross, T.K., Svien, K.A., Wold, A.D., Uhl, J.R., and Smith, T.F. (2000a). Diagnosis of Varicella-Zoster virus infections in the clinical laboratory by LightCycler PCR. J. Clin. Microbiol. *38*, 3187–3189.

Espy, M.J., Uhl, J.R., Mitchell, P.S., Thorvilson, J.N., Svien, K.A., Wold, A.D., and Smith, T.F. (2000b). Diagnosis of Herpes Simplex virus infections in the clinical laboratory by LightCycler PCR. J. Clin. Microbiol. *38*, 795–799.

Espy, M.J., Cockerill, F.R., Meyer, R.F., Bowen, M.D., Poland, G.A., Hadfield, T.L., and Smith, T.F. (2002). Detection of Smallpox virus DNA by LightCycler PCR. J. Clin. Microbiol. *40*, 1985–1988.

Espy, M.J., Uhl, J.R., Sloan, L.M., Buckwalter, S.P., Jones, M.F., Vetter, E.A., Yao, J.D.C., Wengenack, N.L., Rosenblatt, J.E., Cockerill, I.I.I F.R., *et al.* (2006). Real-time PCR in clinical microbiology: applications for routine laboratory testing. Clin. Microbiol. Rev. *19*, 165–256.

Fang, H., and Hedin, G. (2003). Rapid screening and identification of methicillin-resistant *Staphylococcus aureus* from clinical samples by selective-broth and real-time PCR assay. J. Clin. Microbiol. *41*, 2894–2899.

Fenollar, F., Fournier, P.E., Raoult, D., Gerolami, R., Lepidi, H., and Poyart, C. (2002). Quantitative detection of *Tropheryma whipplei* DNA by real-time PCR. J. Clin. Microbiol. *40*, 1119–1120.

Fernandez, C., Boutolleau, D., Manichanh, C., Mangeney, N., Agut, H., and Gautheret-Dejean, A. (2002). Quantitation of HHV-7 genome by real-time polymerase chain reaction assay using MGB probe technology. J. Virol. Methods *106*, 11–16.

Garcia de Viedma, D., del Sol Diaz Infantes, M., Lasala, F., Chaves, F., Alcala, L., and Bouza, E. (2002). New real-time PCR able to detect in a single tube multiple rifampin resistance mutations and high-level isoniazid resistance mutations in *Mycobacterium tuberculosis*. J. Clin. Microbiol. *40*, 988–995.

Gault, E., Michel, Y., Dehée, A., Belabani, C., Nicolas, J.-C., and Garbarg-Chenon, A. (2001). Quantification of human Cytomegalovirus DNA by real-time PCR. J. Clin. Microbiol. *39*, 772–725.

Gautheret-Dejean, A., Manichanh, C., Thien-Ah-Koon, F., Fillet, A.M., Mangeney, N., Vidaud, M., Dhedin, N., Vernant, J.P., and Agut, H. (2002). Development of a real-time polymerase chain reaction assay for the diagnosis of human herpesvirus-6 infection and application to bone marrow transplant patients. J. Virol. Methods *100*, 27–35.

Geraats-Peters, C.W., Brouwers, M., Schneeberger, P.M., van der Zanden, A.G., Bruisten, S.M., Weers-Pothoff, G., Boel, C.H., van den Brule, A.J., Harmsen, H.G., and Hermans, M.H. (2005). Specific and sensitive detection of *Neisseria gonorrhoeae* in clinical specimens by real-time PCR. J. Clin. Microbiol. *43*, 5653–5659.

Geretti, A.M., and Brown, D.W. (2005). National survey of diagnostic services for genital herpes. Sex. Transm. Infect. *81*, 316–317.

Gibson, J.R., Saunders, N.A., Burke, B., and Owen, R.J. (1999). Novel method for rapid determination of clarithromycin sensitivity in *Helicobacter pylori*. J. Clin. Microbiol. *37*, 3746–3748.

Greiner, O., Day, P.J., Bosshard, P.P., Imeri, F., Altwegg, M., and Nadal, D. (2001). Quantitative detection of *Streptococcus pneumoniae* in nasopharyngeal secretions by real-time PCR. J. Clin. Microbiol. *39*, 3129–3134.

Greiner, O., Day, P.J., Altwegg, M., and Nadal, D. (2003). Quantitative detection of *Moraxella catarrhalis* in nasopharyngeal secretions by real-time PCR. J. Clin. Microbiol. *41*, 1386–1390.

Griscelli, F., Barrois, M., Chauvin, S., Lastere, S., Bellet, D., and Bourhis, J.-H. (2001). Quantification of human Cytomegalovirus DNA in bone marrow transplant recipients by real-time PCR. J. Clin. Microbiol. *39*, 4362–4369.

Grisold, A.J., Leitner, E., Mühlbauer, G., Marth, E., and Kessler, H.H. (2002). Detection of methicillin-resistant *Staphylococcus aureus* and simultaneous confirmation by automated nucleic acid extraction and Real-Time PCR. J. Clin. Microbiol. *40*, 2392–2397.

Gueudin, M., Vabret, A., Petitjean, J., Gouarin, S., Brouard, J., and Freymuth, F. (2003). Quantitation of respiratory syncytial virus RNA in nasal aspirates of children by real-time RT-PCR assay. J. Virol. Methods. *109*, 39–45.

Guiver, M., Borrow, R., Marsh, J., Gray, S.J., Kaczmarski, E.B., Howells, D., Boseley, P., and Fox, A.J. (2000). Evaluation of the applied biosystems automated Taqman polymerase chain reaction system for the detection of meningococcal DNA. FEMS Immunol. Med. Microbiol. *28*, 173–179.

Hackett, S.J., Guiver, M., Marsh, J., Sillis, J.A., Thomson, A.P., Kaczmarski, E.B., and Hart, C.A. (2002). Meningococcal bacterial DNA load at presentation correlates with disease severity. Arch. Dis. Child. *86*, 44–46.

Halse, T.A., Musser, K.A., and Limberger, R.J. (2006). A multiplexed real-time PCR assay for rapid detection of *Chlamydia trachomatis* and identification of serovar L-2, the major cause of Lymphogranuloma venereum in New York. Mol. Cell. Probes *20*, 290–297.

Hardegger, D., Nadal, D., Bossart, W., Altwegg, M., and Dutly, F. (2000). Rapid detection of *Mycoplasma pneumoniae* in clinical samples by real-time PCR. J. Microbiol. Methods *41*, 45–51.

Hardick, J., Yang, S., Lin, S., Duncan, D., and Gaydos, C. (2003). Use of the Roche LightCycler instrument in a real-time PCR for *Trichomonas vaginalis* in urine samples from females and males. J. Clin. Microbiol. *41*, 5619–5622.

Hayden, R.T., Uhl, J.R., Qian, X., Hopkins, M.K., Aubry, M.C., Limper, A.H., Lloyd, R.V., and Cockerill, F.R.,(2001). Direct detection of *Legionella* species from bronchoalveolar lavage and open lung biopsy specimens: comparison of LightCycler PCR, in situ hybridisation, direct fluorescence antigen detection and culture. J. Clin. Microbiol. *39*, 2618–2626.

He, Q., Wang, J.-P., Osato, M., and Lachman, L.B. (2002). Real-time quantitative PCR for detection of *Helicobacter pylori*. J. Clin. Microbiol. *40*, 3720–3728.

Heim, A., Ebnet, C., Harste, G., and Pring-Akerblom, P. (2003). Rapid and quantitative detection of human adenovirus DNA by real-time PCR. J. Med. Virol. *70*, 228–239.

Hirsch, H.H., Brennan, D.C., Drachenberg, C.B., Ginevri, F., Gordon, J., Limaye, A.P., Mihatsch, M.J., Nickeleit, V., Ramos, E., Randhawa, P., *et al.* (2005). Polyomavirus-associated nephropathy in renal transplantation: interdisciplinary analyses and recommendations. Transplantation. *79*, 1277–1286.

Hoffman, N.G., Cook, L., Atienza, E.E., Limaye, A.P., and Jerome, K.R. (2008). Marked variability of BK virus load measurement using quantitative real-time PCR among commonly used assays. J. Clin. Microbiol. *46*, 2671–2680.

van den Hoogen, B.G., de Jong, J.C., Groen, J., Kuiken, T., de Groot, R., Fouchier, R.A., and Osterhaus, A.D. (2001). A newly discovered human pneumovirus isolated from young children with respiratory tract disease. Nat Med. *7*, 719–724.

van den Hoogen, B.G., Herfst, S., Sprong, L., Cane, P.A., Forleo-Neto, E., de Swart, R.L., Osterhaus, A.D., and Fouchier, R.A. (2004). Antigenic and genetic variability of human metapneumoviruses. Emerg. Infect. Dis. *10*, 658–666.

Hu, A., Colella, M., Tam, J.S., Rappaport, R., and Cheng, S.M. (2003). Simultaneous detection, subgrouping and quantitation of Respiratory Syncytial Virus A and B by real-time PCR. J. Clin. Microbiol. *41*, 149–154.

Hui, R.K., Zeng, F., Chan, C.M., Yuen, K.Y., Peiris, J.S., and Leung, F.C. (2004). Reverse transcriptase PCR diagnostic assay for the coronavirus associated with severe acute respiratory syndrome. J. Clin. Microbiol. *42*, 1994–1999.

Huletsky, A., Giroux, R., Rossbach, V., Gagnon, M., Vaillancourt, M., Bernier, M., Gagnon, F., Truchon, K., Bastien, M., Picard, F.J., *et al.* (2004). New real-time PCR assay for rapid detection of methicillin-resistant *Staphylococcus aureus* directly from specimens containing a mixture of staphylococci. J. Clin. Microbiol. *42*, 1875–1884.

Jalal, H., Stephen, H., Curran, M.D., Burton, J., Bradley, M., and Carne, C. (2006). Development and validation of a rotor-gene real-time PCR assay for detection, identification, and quantification of *Chlamydia trachomatis* in a single reaction. J. Clin. Microbiol. *44*, 206–213.

Kageyama, T., Kojima, S., Shinohara, M., Uchida, K., Fukushi, S., Hoshino, F.B., Takeda, N., and Katayama, K. (2003). Broadly reactive and highly sensitive assay for norwalk-like viruses based on real-time quantitative reverse transcription-PCR. J. Clin. Microbiol. *41*, 1548–1557.

Kearns, A.M., Turner, A.J., Taylor, C.E., George, P.W., Freeman, R., and Gennery, A.R. (2001a). LightCycler-based quantitative PCR for rapid detection of human herpesvirus 6 DNA in clinical material. J. Clin. Microbiol. *39*, 3020–3021.

Kearns, A.M., Guiver, M., James, V., and King, J. (2001b). Development of a real-time quantitative PCR for the detection of human cytomegalovirus. J. Virol. Methods *95*, 121–131.

Kearns, A.M., Draper, B., Wipat, W., Turner, A.J., Wheeler, J., Freeman, R., Harwood, J., Gould, F.K., and Dark, J.H. (2001c). LightCycler-based quantitative PCR for detection of cytomegalovirus in blood, urine, and respiratory samples. J. Clin. Microbiol. *39*, 2364–2365.

Kearns, A.M., Graham, C., Burdess, D., Heatherington, J., and Freeman, R. (2002a). Rapid real-time PCR for determination of penicillin susceptibility in pneumococcal meningitis, including culture-negative cases. J. Clin. Microbiol. *40*, 682–684.

Kearns, A.M., Turner, A.J., Eltringham, G.J., and Freeman, R. (2002b). Rapid detection and quantification of CMV DNA in urine using LightCycler-based real-time PCR. J. Clin. Virol. *24*, 131–134.

Kessler, H.H., Mühlbauer, G., Rinner, B., Stelzl, E., Berger, A., Dörr, H.-.W., Santner, B., Marth, E., and Rabenau, H. (2000). Detection of Herpes Simplex Virus DNA by real-time PCR. J. Clin. Microbiol. *38*, 2638–2642.

Killgore, G.E., Holloway, B., and Tenover, F.C. (2000). A 5′ nuclease PCR (TaqMan) high-throughput assay for detection of the *mecA* gene in staphylococci. J. Clin. Microbiol. *38*, 2516–2519.

Kimura, H., Morita, M., Yabuta, Y., Kuzushima, K., Kato, K., Kojima, S., Matsuyama, T., and Morishima, T. (1999). Quantitative analysis of Epstein–Barr virus load by using a real-time PCR assay. J. Clin. Microbiol. *37*, 132–136.

Kleiber, J., Walter, T., Haberhausen, G., Tsang, S., Babiel, R., and Rosenstraus, M. (2000). Performance characteristics of a quantitative, homogeneous TaqMan RT-PCR test for HCV RNA. J. Mol. Diagn. *2*, 158–166.

Kleines, M., Scheithauer, S., Rackowitz, A., Ritter, K., and Häusler, M. (2007). High prevalence of human bocavirus detected in young children with severe acute lower respiratory tract disease by use of a standard PCR protocol and a novel real-time PCR protocol. J. Clin. Microbiol. *45*, 1032–1034.

Kocagoz, T., Saribas, Z., and Alp, A. (2005). Rapid determination of rifampin resistance in clinical isolates of *Mycobacterium tuberculosis* by real-time PCR. J. Clin. Microbiol. *43*, 6015–6019.

Koenig, M.G., Kosha, S.L., Doty, B.L., and Heath, D.G. (2004). Direct comparison of the BD ProbeTec ET system with in-house LightCycler PCR assays for detection of *Chlamydia trachomatis* and *Neisseria gonorrhoeae* from clinical specimens. J. Clin. Microbiol. *42*, 5751–5756.

Kösters, K., Riffelmann, M., Wirsing von Konig, C.H. (2001). Evaluation of a real-time PCR assay for detection of *Bordetella pertussis* and *B. parapertussis* in clinical samples. J. Med. Microbiol. *50*, 436–440.

Kösters, K., Reischl, U., Schmetz, J., Riffelmann, M., and Wirsing von König, C.H. (2002). Real-time LightCycler PCR for detection and discrimination of *Bordetella pertussis* and *Bordetella parapertussis*. J. Clin. Microbiol. *40*, 1719–1722.

Kraus, G., Cleary, T., Miller, N., Seivright, R., Young, A.K., Spruill, G., and Hnatyszyn, H.J. (2001). Rapid and specific detection of the *Mycobacterium tuberculosis* complex using fluorogenic probes and real-time PCR. Mol. Cell. Probes *15*, 375–383.

Kuoppa, Y., Boman, J., Scott, L., Kumlin, U., Eriksson, I., and Allard, A. (2002). Quantitative detection of respiratory *Chlamydia pneumoniae* infection by real-time PCR. J. Clin. Microbiol. *40*, 2273–2274.

Kuypers, J., Wright, N., and Morrow, R. (2004). Evaluation of quantitative and type-specific real-time RT-PCR assays for detection of respiratory syncytial virus in respiratory specimens from children. J. Clin. Virol. *31*, 123–129.

Kuypers, J., Wright, N., Corey, L., and Morrow, R.J. (2005). Detection and quantification of human metapneumovirus in paediatric specimens by real-time RT-PCR. Clin. Virol. *33*, 299–305.

Kwok, S., Mack, D.H., Mullis, K.B., Poiesz, B., Ehrlich, G., Blair, D., Friedman-Kien, A., and Sninsky, J.J. (1987). Identification of human immunodeficiency virus sequences by using *in vitro* enzymatic amplification and oligomer cleavage detection. J Virol. *61*, 1690–1694.

Lai, K.K., Cook, L., Wendt, S., Corey, L., and Jerome, K.R. (2003). Evaluation of real-time PCR versus PCR with liquid-phase hybridisation for detection of Enterovirus RNA in cerebrospinal fluid. J. Clin. Microbiol. *41*, 3133–3141.

Lanciotti, R.S., Kerst, A.J., Nasci, R.S., Godsey, M.S., Mitchell, C.J., Savage, H.M., Komar, N., Panella, N.A., Allen, B.C., Volpe, K.E., *et al.* (2000). Rapid detection of west nile virus from human clinical specimens, field-collected mosquitoes, and avian samples by a TaqMan reverse transcriptase-PCR assay. J. Clin. Microbiol. *38*, 4066–4071.

Lapierre, P., Huletsky, A., Fortin, V., Picard, F.J., Roy, P.H., Ouellette, M., and Bergeron, M.J. (2003). Real-time PCR assay for detection of fluoroquinolone resistance associated with *grlA* mutations in *Staphylococcus aureus*. J. Clin. Microbiol. *41*, 3246–3251.

Lawson, A.J., Elviss, N.C., and Owen, R.J. (2005). Real-time PCR detection and frequency of 16S rDNA mutations associated with resistance and reduced susceptibility to tetracycline in *Helicobacter pylori* from England and Wales. J. Antimicrob. Chemother. *56*, 282–286.

Lenaerts, L., De Clercq, E., and Naesens, L. (2008). Clinical features and treatment of adenovirus infections. Rev. Med. Virol. *18*, 357–374.

Leruez-Ville, M., Ouachée, M., Delarue, R., Sauget, A.-S., Blanche, S., Buzyn, A., and Rouzioux, C. (2003). Monitoring Cytomegalovirus infection in adult and paediatric bone marrow transplant recipients by a real-time PCR assay performed with blood plasma. J. Clin. Microbiol. *41*, 2040–2046.

Li, H., Dummer, J.S., Estes, W.R., Meng, S., Wright, P.F., and Tang, Y.W. (2003). Measurement of human Cytomegalovirus loads by quantitative real-time PCR for monitoring clinical intervention in transplant recipients. J. Clin. Microbiol. *41*, 187–191.

Liao, C.S., Lee, G.B., Liu, H.S., Hsieh, T.M., and Luo, C.H. (2005). Miniature RT-PCR system for diagnosis of RNA-based viruses. Nucleic Acids Res. *33*, e156.

Lindau, C., Tiveljung-Lindell, A., Goh, S., Ramqvist, T., and Allander, T. (2009). A single-tube, real-time PCR assay for detection of the two newly characterized human KI and WU polyomaviruses. J. Clin. Virol. *44*, 24–26.

Lindler, L.E., Fan, W., and Jahan, N. (2001). Detection of ciprofloxacin-resistant *Yersinia pestis* by fluorogenic PCR using the LightCycler. J. Clin. Microbiol. *39*, 3649–3655.

Locatelli, G., Santoro, F., Veglia, F., Gobbi, A., Lusso, P., and Malnati, M.S. (2000). Real-time quantitative PCR for human herpesvirus 6 DNA. J. Clin. Microbiol. *38*, 4042–4048.

Loeffelholz, M. (2012). Towards improved accuracy of *Bordetella pertussis* nucleic acid amplification tests. J. Clin. Microbiol. *50*, 2186–2190.

Logan, J.M.J., Edwards, K.J., Saunders, N.A., and Stanley, J. (2001). Rapid identification of *Campylobacter* spp. by melting peak analysis of biprobes in real-time PCR. J. Clin. Microbiol. *39*, 2227–2232.

Lu, X., Chittaganpitch, M., Olsen, S.J., Mackay, I.M., Sloots, T.P., Fry, A.M., and Erdman, D.D. (2006). Real-time PCR assays for detection of bocavirus in human specimens. J. Clin. Microbiol. *44*, 3231–3235.

McAvin, J.C., Reilly, P.A., Roudabush, R.M., Barnes, W.J., Salmen, A., Jackson, G.W., Beninga, K.K., Astorga, A., McCleskey, F.K., Huff, W.B., *et al.* (2001). Sensitive and specific method for rapid identification of *Streptococcus pneumoniae* using real-time fluorescence PCR. J. Clin. Microbiol. *39*, 3446–3451.

McDonald, R.R., Antonishyn, N.A., Hansen, T., Snook, L.A., Nagle, E., Mulvey, M.R., Levett, P.N., and Horsman, G.B. (2005). Development of a triplex real-time PCR assay for detection of Panton-Valentine leukocidin toxin genes in clinical isolates of methicillin-resistant *Staphylococcus aureus*. J. Clin. Microbiol. *43*, 6147–6149.

McNees, A.L., White, Z.S., Zanwar, P., Vilchez, R.A., and Butel, J.S. (2005). Specific and quantitative detection of human polyomaviruses BKV, JCV, and SV40 by real time PCR. J. Clin. Virol. *34*, 52–62.

Machida, U., Kami, M., Fukui, T., Kazuyama, Y., Kinoshita, M., Tanaka, Y., Kanda, Y., Ogawa, S., Honda, H., Chiba, S., *et al.* (2000). Real-Time automated PCR for early diagnosis and monitoring of Cytomegalovirus infection after bone marrow transplantation. J. Clin. Microbiol. *38*, 2536–2542.

Mackay, I.M., Arden, K.E., and Nitsche, A. (2002). Real-time PCR in virology. Nucleic Acids Res. *30*, 1292–1305.

Mackay, I.M., Jacob, K.C., Woolhouse, D., Waller, K., Syrmis, M.W., Whiley, D.M., Siebert, D.J., Nissen, M., and Sloots, T.P. (2003). Molecular assays for detection of human metapneumovirus. J. Clin. Micro. *41*, 100–105.

Maertzdorf, J., Wang, C.K., Brown, J.B., Quinto, J.D., Chu, M., de Graaf, M., van den Hoogen, B.G., Spaete, R., Osterhaus, A.D., and Fouchier, R.A. (2004). Real-time reverse transcriptase PCR assay for detection of human metapneumoviruses from all known genetic lineages. J. Clin. Microbiol. *42*, 981–986.

Mahony, J.B., Jang, D., Chong, S., Luinstra, K., Sellors, J., Tyndall, M., and Chernesky, M. (1997). Detection of *Chlamydia trachomatis, Neisseria gonorrhoeae, Ureaplasma urealyticum*, and *Mycoplasma genitalium* in first-void urine specimens by multiplex polymerase chain reaction. Mol. Diagn. *2*, 161–168.

Makinen, J., Mertsola, J., Viljanen, M.K., Arvilommi, H., and He, Q. (2002). Rapid typing of *Bordetella pertussis* pertussis toxin gene variants by LightCycler real-time PCR and fluorescence resonance energy transfer hybridisation probe melting curve analysis. J. Clin. Microbiol. *40*, 2213–2216.

Mangold, K.A., Manson, R.U., Koay, E.S., Stephens, L., Regner, M., Thomson, R.B., Jr, Peterson, L.R., and Kaul, K.L. (2005). Real-time PCR for detection and identification of *Plasmodium* spp. J. Clin. Microbiol. *43*, 2435–2440.

Marin, M., Garcia de Viedma, D., Ruiz-Serrano, M.J., and Bouza, E. (2004). Rapid direct detection of multiple rifampin and isoniazid resistance mutations in *Mycobacterium tuberculosis* in respiratory samples by real-time PCR. Antimicrob. Agents Chemother. *48*, 4293–4300.

Martell, M., Gómez, J., Esteban, J.L., Sauleda, S., Quer, J., Cabot, B., Esteban, R., and Guardia, J. (1999). High-throughput real-time reverse transcription-PCR quantitation of Hepatitis C Virus RNA. J. Clin. Microbiol. *37*, 327–332.

Marty, A., Greiner, O., Day, P.J., Gunziger, S., Muhlemann, K., and Nadal, D. (2004). Detection of *Haemophilus influenzae* type b by real-time PCR. J. Clin. Microbiol. *42*, 3813–3815.

Matsumura, M., Hikiba, K., Ogura, G., Togo, I., Tsukuda, K., Ushikawa, Y., Shiratori, and Omata, M. (2001). Rapid detection of mutations in the 23S rRNA gene of *Helicobacter pylori* that confers resistance to clarithromycin treatment to the bacterium. J. Clin. Microbiol. *39*, 691–695.

Mengelle, C., Sandres-Saune, K., Pasquier, C., Rostaing, L., Mansuy, J.M., Marty, M., Da Silva, I., Attal, M., Massip, P., and Izopet, J. (2003). Automated extraction and quantification of human cytomegalovirus DNA in whole blood by real-time PCR assay. J. Clin. Microbiol. *41*, 3840–3845.

Mentasti, M., Fry, N.K., Afshar, B., Palepou-Foxley, C., Naik, F.C., and Harrison, T.G. (2012). Application of *Legionella pneumophila*-specific quantitative real-time PCR combined with direct amplification and sequence-based typing in the diagnosis and epidemiological investigation of Legionnaires' disease. Eur. J. Clin. Microbiol. Infect. Dis. *31*, 2017–2028.

Miller, N., Cleary, T., Kraus, G., Young, A.K., Spruill, G., and Hnatyszyn, H.J. (2002). Rapid and specific detection of *Mycobacterium tuberculosis* from acid-fast Bacillus smear-positive respiratory specimens and BacT/ALERT MP culture bottles by using fluorogenic probes and real-time PCR. J. Clin. Microbiol. *40*, 4143–4147.

Mohamed, N., Elfaitouri, A., Fohlman, J., Friman, G., and Blomberg, J. (2004). A sensitive and quantitative single-tube real-time reverse transcriptase-PCR for detection of enteroviral RNA. J. Clin. Virol. *30*, 150–156.

Molling, P., Jacobsson, S., Backman, A., and Olcen, P. (2002). Direct and rapid identification and genogrouping of meningococci and *porA* amplification by LightCycler PCR. J. Clin. Microbiol. *40*, 4531–4535.

Mommert, S., Gutzmer, R., Kapp, A., and Werfel, T. (2001). Sensitive detection of *Borrelia burgdorferi* sensu lato DNA and differentiation of *Borrelia* species by LightCycler PCR. J. Clin. Microbiol. *39*, 2663–2667.

Monpoeho, S., Coste-Burel, M., Costa-Mattioli, M., Besse, B., Chomel, J.J., Billaudel, S., and Ferre, V. (2002). Application of a real-time polymerase chain reaction with internal positive control for detection and quantification of enterovirus in cerebrospinal fluid. Eur. J. Clin. Microbiol. Infect. Dis. *21*, 532–536.

Morre, S.A., Van Valkengoed, I.G., Moes, R.M., Boeke, A.J., Meijer, C.J., and Van den Brule, A.J. (1999). Determination of *Chlamydia trachomatis* prevalence in an asymptomatic screening population: performances of the LCx and COBAS Amplicor tests with urine specimens. J. Clin. Microbiol. *37*, 3092–3096.

Morrison, T.B., Ma, Y., Weis, J.H., and Weis, J.J. (1999). Rapid and sensitive quantification of *Borrelia burgdorferi*-infected mouse tissues by continuous fluorescent monitoring of PCR. J. Clin. Microbiol. *37*, 987–992.

Mothershed, E.A., Cassiday, P.K., Pierson, K., Mayer, L.W., and Popovic, T. (2002). Development of a real-time fluorescence PCR assay for rapid detection of the diphtheria toxin gene. J. Clin. Microbiol. *40*, 4713–4719.

Namvar, L., Olofsson, S., Bergstrom, T., and Lindh, M. (2005). Detection and typing of Herpes Simplex virus (HSV) in mucocutaneous samples by TaqMan PCR targeting a gB segment homologous for HSV types 1 and 2. J. Clin. Microbiol. 43, 2058–2064.

Ng, E.K., Cheng, P.K., Ng, A.Y., Hoang, T.L., and Lim, W.W. (2005). Influenza A H5N1 detection. Emerg. Infect. Dis. *11*, 1303–1305.

Nicolas, L., Prina, E., Lang, T., and Milon, G. (2002). Real-time PCR for detection and quantitation of leishmania in mouse tissues. J. Clin. Microbiol. *40*, 1666–1669.

Nielsen, E.M., and Andersen, M.T. (2003). Detection and characterization of verocytotoxin-producing *Escherichia coli* by automated 5′ Nuclease PCR assay. J. Clin. Microbiol. *41*, 2884–2893.

Niesters, H.G. (2001). Quantitation of viral load using real-time amplification techniques. Methods *25*, 419–429.

Niesters, H.G. (2002). Clinical virology in real time. Clin, J. Virol. *25(Suppl. 3)*, S3–S12.

Niesters, H.G., van Esser, J., Fries, E., Wolthers, K.C., Cornelissen, J., and Osterhaus, A.D. (2000). Development of a real-time quantitative assay for detection of Epstein–Barr Virus. J. Clin. Microbiol. *38*, 712–715.

Ninove, L., Nougairede, A., Gazin, C., Thirion, L., Delogu, I., Zandotti, C., Charrel, R.N., and De Lamballerie, X. (2011). RNA and DNA bacteriophages as molecular diagnosis controls in clinical virology: a comprehensive study of more than 45,000 routine PCR tests. PLoS One. *6*, 16142.

Nijhuis, M., van Maarseveen, N., Schuurman, R., Verkuijlen, S., de Vos, M., Hendriksen, K., and van Loon, A.M. (2002). Rapid and sensitive routine detection of all members of the genus enterovirus in different clinical specimens by real-time PCR. J. Clin. Microbiol. *40*, 3666–3670.

Nitsche, A., Steuer, N., Schmidt, C.A., Landt, O., Ellerbrok, H., Pauli, G., and Siegert, W. (2000). Detection of human cytomegalovirus DNA by real-time quantitative PCR. J. Clin. Microbiol. *38*, 2734–2737.

Nogva, H.K., Bergh, A., Holck, A., and Rudi, K. (2000). Application of the 5′-nuclease PCR assay in evaluation and development of methods for quantitative detection of *Campylobacter jejuni*. Appl. Environ. Microbiol. *66*, 4029–4036.

Oleastro, M., Ménard, A., Santos, A., Lamouliatte, H., Monteiro, L., Barthélémy, P., and Mégraud, F. (2003). Real-time PCR assay for rapid and accurate detection of point mutations conferring resistance to clarithromycin in *Helicobacter pylori*. J. Clin. Microbiol. *41*, 397–402.

Pahl, A., Kühlbrandt, U., Brune, K., Röllinghoff, M., and Gessner, A. (1999). Quantitative detection of *Borrelia burgdorferi* by real-time PCR. J. Clin. Microbiol. *37*, 1958–1963.

Palladino S., Kay, I.D., Costa, A.M., Lambert, E.J., Flexman, J.P. (2003a). Real-time PCR for the rapid detection of *van*A and *van*B genes. Diagn. Microbiol. Infect. Dis. *45*, 81–84.

Palladino, S., Kay, I.D., Flexman, J.P., Boehm, I., Costa, A.M.G., Lambert, E.J., and Christiansen, K.J. (2003b). Rapid detection of *vanA* and *vanB* genes directly from clinical specimens and enrichment broths by real-time multiplex PCR assay. J. Clin. Microbiol. *41*, 2483–2486.

Palomares, C., Torres, M.J., Torres, A., Aznar, J., and Palomares, J.C. (2003). Rapid detection and identification of *Staphylococcus aureus* from blood culture specimens using real-time fluorescence PCR. Diagn. Microbiol. Infect. Dis. *45*, 183–189.

Pang, X., Lee, B., Chui, L., Preiksaitis, J.K., and Monroe, S.S. (2004). Evaluation and validation of real-time reverse transcription-PCR assay using the LightCycler system for detection and quantitation of norovirus. J. Clin. Microbiol. *42*, 4679–4685.

Pas, S.D., Fries, E., De Man, R.A., Osterhaus, A.D., and Niesters, H.G. (2000). Development of a quantitative real-time detection assay for Hepatitis B Virus DNA and comparison with two commercial assays. J. Clin. Microbiol. *38*, 2897–2901.

Patel, S., Zuckerman, M., and Smith, M. (2003). Real-time quantitative PCR of Epstein–Barr virus BZLF1 DNA using the LightCycler. J. Virol. Methods. *109*, 227–233.

Payungporn, S., Chutinimitkul, S., Chaisingh, A., Damrongwantanapokin, S., Buranathai, C., Amonsin, A., Theamboonlers, A., and Poovorawan, Y. (2006). Single step multiplex real-time RT-PCR for H5N1 influenza A virus detection. J. Virol. Methods. *131*, 143–147.

Pickett, M.A., Everson, J.S., Pead, P.J., and Clarke, I.N. (2005). The plasmids of *Chlamydia trachomatis* and *Chlamydophila pneumoniae* (N16): accurate determination of copy number and the paradoxical effect of plasmid-curing agents. Microbiology *151*, 893–903.

Piesman, J., Schneider, B.S., and Zeidner, N.S. (2001). Use of quantitative PCR to measure density of *Borrelia burgdorferi* in the midgut and salivary glands of feeding tick vectors. J. Clin. Microbiol. *39*, 4145–4148.

Pietila, J., He, Q., Oksi, J., and Viljanen, M.K. (2000). Rapid differentiation of *Borrelia garinii* from *Borrelia afzelii* and *Borrelia burgdorferi* Sensu Stricto by LightCycler fluorescence melting curve analysis of a PCR product of the *recA* Gene. J. Clin. Microbiol. *38*, 2756–2759.

Rabenau, H.F., Clarici, A.M., Muhlbauer, G., Berger, A., Vince, A., Muller, S., Daghofer, E., Santner, B.I., Marth, E., and Kessler, H.H. (2002). Rapid detection of enterovirus infection by automated RNA extraction and real-time fluorescence PCR. J. Clin. Virol. *25*, 155–164.

Ramaswamy, M., McDonald, C., Smith, M., Thomas, D., Maxwell, S., Tenant-Flowers, M., and Geretti, A.M. (2004). Diagnosis of genital herpes by real time PCR in routine clinical practice. Sex. Transm. Infect. *80*, 406–410.

Randhawa. P., Ho, A., Shapiro, R., Vats, A., Swalsky, P., Finkelstein, S., Uhrmacher, J., and Weck, K. (2004). Correlates of quantitative measurement of BK polyomavirus (BKV) DNA with clinical course of BKV infection in renal transplant patients. J. Clin. Microbiol. *42*, 1176–1180.

Rantakokko-Jalava, K., and Jalava, J. (2001). Development of conventional and real-time PCR assays for detection of *Legionella* DNA in respiratory specimens. J. Clin. Microbiol. *39*, 2904–2910.

Rauter, C., Oehme, R., Diterich, I., Engele, M., and Hartung, T. (2002). Distribution of clinically relevant *Borrelia* genospecies in ticks assessed by a novel, single-run, real-time PCR. J. Clin. Microbiol. *40*, 36–43.

Reischl, U., Linde, H.-J., Metz, M., Leppmeier, B., and Lehn, N. (2000). Rapid identification of methicillin-resistant *Staphylococcus aureus* and simultaneous species confirmation using real-time fluorescence PCR. J. Clin. Microbiol. *38*, 2429–2433.

Reischl, U., Lehn, N., Sanden, G.N., and Loeffelholz, M.J. (2001). Real-time PCR assay targeting IS481 of *Bordetella pertussis* and molecular basis for detecting *Bordetella holmesii*. J. Clin. Microbiol. *39*, 1963–1966.

Reischl, U., Youssef, M.T., Kilwinski, J., Lehn, N., Zhang, W.L., Karch, H., and Strockbine, N.A. (2002a). Real-time fluorescence PCR assays for detection and characterization of shiga toxin, intimin, and enterohemolysin genes from shiga toxin-producing *Escherichia coli*. J. Clin. Microbiol. *40*, 2555–2565.

Reischl, U., Linde, H.J., Lehn, N., Landt, O., Barratt, K., and Wellinghausen, N. (2002b). Direct detection and differentiation of *Legionella* spp. and *Legionella pneumophila* in clinical specimens by dual-colour real-time PCR and melting curve analysis. J. Clin. Microbiol. *40*, 3814–3817.

Reischl, U., Lehn, N., Simnacher, U., Marre, R., and Essig, A. (2003). Rapid and standardised detection of *Chlamydia pneumoniae* using LightCycler real-time fluorescence PCR. Eur. J. Clin. Microbiol. Infect. Dis. *22*, 54–57.

Renaud, C., Kuypers, J., Ficken, E., Cent, A., Corey, L., and Englund, J.A. (2011). Introduction of a novel parechovirus RT-PCR clinical test in a regional medical center. J. Clin. Virol. *51*, 50–53.

Royuela, E., Negredo, A., and Sánchez-Fauquier, A. (2006). Development of a one step real-time RT-PCR method for sensitive detection of human astrovirus. J. Virol. Methods. *133*, 14–19.

Ruiz, M., Torres, M.J., Llanos, A.C., Arroyo, A., Palomares, J.C., and Aznar, J. (2004). Direct detection of rifampin and isoniazid-Resistant *Mycobacterium tuberculosis* in auramine-rhodamine-positive sputum specimens by real-time PCR. J. Clin. Microbiol. *42*, 1585–1589.

Ryncarz, A.J., Goddard, J., Wald, A., Huang, M.L., Roizman, B., and Corey, L. (1999). Development of a high-throughput quantitative assay for detecting Herpes Simplex Virus DNA in clinical samples. J. Clin. Microbiol. *37*, 1941–1947.

Safronetz, D., Humar, A., and Tipples, G.A. (2003). Differentiation and quantitation of human herpesviruses 6A, 6B and 7 by real-time PCR. J. Virol. Methods *112*, 99–105.

Saiki, R.K., Scharf, S., Faloona, F., Mullis, K.B., Horn, G.T., Erlich, H.A., and Arnheim, N. (1985). Enzymatic amplification of beta-globin genomic sequences and restriction site analysis for diagnosis of sickle cell anemia. Science *230*, 1350–1354.

Sails, A.D., Fox, A.J., Bolton, F.J., Wareing, D.R.A., and Greenway, D.L.A. (2003). A real-time PCR assay for the detection of *Campylobacter jejuni* in foods after enrichment culture. Appl. Environ. Microbiol. *69*, 1383–1390.

Sanchez, J.L., and Storch, G.A. (2002). Multiplex, quantitative, real-time PCR assay for Cytomegalovirus and human DNA. J. Clin. Microbiol. *40*, 2381–2386.

Scheltinga, S.A., Templeton, K.E., Beersma, M.F., and Claas, E.C. (2005). Diagnosis of human metapneumovirus and rhinovirus in patients with respiratory tract infections by an internally controlled multiplex real-time RNA PCR. J. Clin. Virol. *33*, 306–311.

Scott, L.J., Gunson, R.N., Carman, W.F., and Winter, A.J. (2010). A new multiplex real-time PCR test for HSV1/2 and syphilis: an evaluation of its impact in the laboratory and clinical setting. Sex. Transm. Infect. *86*, 537–539.

Scoular, A., Gillespie, G., and Carman, W.F. (2002). Polymerase chain reaction for diagnosis of genital herpes in a genitourinary medicine clinic. Sex. Transm. Infect. *78*, 21–25.

Sharma, V.K., and Dean-Nystrom, E.A. (2003). Detection of enterohaemorrhagic *Escherichia coli* O157: H7 by using a multiplex real-time PCR assay for genes encoding intimin and Shiga toxins. Vet Microbiol. *93*, 247–260.

Sharma, V.K., Dean-Nystrom, E.A., and Casey, T.A. (1999). Semi-automated fluorogenic PCR assays (TaqMan) for rapid detection of *Escherichia coli* O157: H7 and other shiga toxigenic *E. coli*. Mol. Cell. Probes. *13*, 291–302.

Shrestha, N.K., Tuohy, M.J., Hall, G.S., Reischl, U., Gordon, S.M., and Procop, G.W. (2003). Detection and differentiation of *Mycobacterium tuberculosis* and non-tuberculous mycobacterial isolates by real-time PCR. J. Clin. Microbiol. *41*, 5121–5126.

Shu, P.Y., Chang, S.F., Kuo, Y.C., Yueh, Y.Y., Chien, L.J., Sue, C.L., Lin, T.H., and Huang, J.H. (2003). Development of group- and serotype-specific one-step SYBR Green I based real-time reverse transcription-PCR assay for Dengue Virus. J. Clin. Microbiol. *41*, 2408–2416.

Si-Mohamed, A., Goff, J.L., Desire, N., Maylin, S., Glotz, D., and Belec, L. (2006). Detection and quantitation of BK virus DNA by real-time polymerase chain reaction in the LT-ag gene in adult renal transplant recipients. J. Virol. Methods. *131*, 21–27.

Sloan, L.M., Hopkins, M.K., Mitchell, P.S., Vetter, E.A., Rosenblatt, J.E., Harmsen, W.S., Cockerill, F.R., and Patel, R. (2002). Multiplex LightCycler PCR assay for detection and differentiation of *Bordetella pertussis* and *Bordetella parapertussis* in nasopharyngeal specimens. J. Clin. Microbiol. *40*, 96–100.

Smith, A.B., Mock, V., Melear, R., Colarusso, P., and Willis, D.E. (2003). Rapid detection of influenza A and B viruses in clinical specimens by LightCycler real time RT-PCR. J. Clin. Virol. *28*, 51–58.

Sofi Ibrahim, M., Kulesh, D.A., Saleh, S.S., Damon, I.K., Esposito, J.J., Schmaljohn, A.L., and Jahrling, P.B. (2003). Real-time PCR assay to detect smallpox virus. J. Clin. Microbiol. *41*, 3835–3839.

Stamey, F.R., Patel, M.M., Holloway, B.P., and Pellett, P.E. (2001). Quantitative fluorogenic probe PCR assay for detection of human herpesvirus 8 DNA in clinical specimens. J. Clin. Microbiol. *39*, 3537–3540.

Stevens, S.J., Verkuijlen, S.A., Brule, A.J., and Middeldorp, J.M. (2002). Comparison of quantitative competitive PCR with LightCycler based PCR for measuring Epstein–Barr Virus DNA load in clinical specimens. J. Clin. Microbiol. *40*, 3986–3992.

Stocher, M., and Berg, J. (2002). Normalized quantification of human cytomegalovirus DNA by competitive real-time PCR on the LightCycler instrument. J. Clin. Micro. *40*, 4547–4553.

Stockton, J., Stephenson, I., Fleming, D., and Zambon, M. (2002). Human metapneumovirus as a cause of community-acquired respiratory illness. Emerg. Infect. Dis. *8*, 897–901.

Stone, B., Burrows, J., Schepetiuk, S., Higgins, G., Hampson, A., Shaw, R., and Kok, T. (2004). Rapid detection and simultaneous subtype differentiation of influenza A viruses by real time PCR. J. Virol. Methods. *117*, 103–112.

Sumino, K.C., Agapov, E., Pierce, R.A., Trulock, E.P., Pfeifer, J.D., Ritter, J.H., Gaudreault-Keener, M., Storch, G.A., and Holtzman, M.J. (2005). Detection of severe human metapneumovirus infection by real-time polymerase chain reaction and histopathological assessment. J. Infect. Dis. *192*, 1052–1060.

Tan, T.Y., Corden, S., Barnes, R., and Cookson, B. (2001). Rapid Identification of methicillin-resistant *Staphylococcus aureus* from positive blood cultures by real-time fluorescence PCR. J. Clin. Microbiol. *39*, 4529–4531.

Templeton, K.E., Scheltinga, S.A., van der Zee, A., Diederen, B.M., van Kruijssen, A.M., Goossens, H., Kuijper, E., and Claas, E.C. (2003). Evaluation of real-time PCR for detection of and discrimination between *Bordetella pertussis*, *Bordetella parapertussis*, and *Bordetella holmesii* for clinical diagnosis. J. Clin. Microbiol. *41*, 4121–4126.

Templeton, K.E., Scheltinga, S.A., Beersma, M.F., Kroes, A.C., and Claas, E.C. (2004). Rapid and sensitive method using multiplex real-time PCR for diagnosis of infections by influenza a and influenza B viruses, respiratory syncytial virus, and parainfluenza viruses *1*, 2, 3, and 4. J. Clin. Microbiol. *42*, 1564–1569.

Tondella, M.L.C., Talkington, D.F., Holloway, B.P., Dowell, S.F., Cowley, K., Soriano-Gabarro, M., Elkind, M.S., and Fields, B.S. (2002). Development and evaluation of real-time PCR-based fluorescence assays for detection of *Chlamydia pneumoniae*. J. Clin. Microbiol. *40*, 575–583.

Torres, M.J., Criado, A., Palomares, J.C., and Aznar, J. (2000). Use of real-time PCR and fluorimetry for rapid detection of rifampin and isoniazid resistance-associated mutations in *Mycobacterium tuberculosis*. J. Clin. Microbiol. *38*, 3194–3199.

Uhl, J.R., Adamson, S.C., Vetter, E.A., Schleck, C.D., Harmsen, W.S., Iverson, L.K., Santrach, P.J., Henry, N.K., and Cockerill, F.R. (2003). Comparison of LightCycler PCR, rapid antigen immunoassay, and culture for detection of group A streptococci from throat swabs. J. Clin. Microbiol. *41*, 242–249.

Ursi, D., Dirven, K., Loens, K., Leven, M., and Goossens, H. (2003). Detection of *Mycoplasma pneumoniae* in respiratory samples by real-time PCR using an inhibition control. J. Microbiol. Methods. *55*, 149–153.

Verstrepen, W.A., Kuhn, S., Kockx, M.M., Van De Vyvere, M.E., and Mertens, A.H. (2001). Rapid detection of enterovirus RNA in cerebrospinal fluid specimens with a novel single-tube real-time reverse transcription-PCR assay. J. Clin. Microbiol. *39*, 4093–4096.

Wada, T., Maeda, S., Tamaru, A., Imai, S., Hase, A., and Kobayashi, K. (2004). Dual-probe assay for rapid detection of drug-resistant *Mycobacterium tuberculosis* by real-time PCR. J. Clin. Microbiol. *42*, 5277–5285.

Walker, R.A., Saunders, N., Lawson, A.J., Lindsay, E.A., Dassama, M., Ward, L.R., Woodward, M.J., Davies, R.H., Liebana, E., and Threlfall, E.J. (2001). Use of a LightCycler *gyrA* mutation assay for rapid identification of mutations conferring decreased susceptibility to ciprofloxacin in multiresistant *Salmonella enterica* serotype *Typhimurium* DT104 isolates. J. Clin. Microbiol. *39*, 1443–1448.

Weidmann, M., Meyer-König, U., and Hufert, F.T. (2003). Rapid detection of herpes simplex virus and varicella-zoster virus infections by real-time PCR. J. Clin. Microbiol. *41*, 1565–1568.

Wellinghausen, N., Frost, C., and Marre, R. (2001). Detection of *Legionellae* in hospital water samples by quantitative real-time LightCycler PCR. Appl. Envir. Microbiol. *67*, 3985–3993.

Welti, M., Jaton, K., Altwegg, M., Sahli, R., Wenger, A., and Bille, J. (2003). Development of a multiplex real-time quantitative PCR assay to detect *Chlamydia pneumoniae, Legionella pneumophila* and *Mycoplasma pneumoniae* in respiratory tract secretions. Diagn. Microbiol. Infect. Dis. *45*, 85–95.

Whiley, D.M., LeCornec, G.M., Mackay, I.M., Siebert, D.J., and Sloots, T.P. (2002). A real-time PCR assay for the detection of *Neisseria gonorrhoeae* by LightCycler. Diagn. Microbiol. Infect. Dis. *42*, 85–89.

Whiley, D.M., Buda, P.J., Bayliss, J., Cover, L., Bates, J., and Sloots, T.P. (2004). A new confirmatory *Neisseria gonorrhoeae* real-time PCR assay targeting the *porA* pseudogene. Eur. J. Clin. Microbiol. Infect. Dis. *23*, 705–710.

Whiley, D.M., Buda, P.P., Freeman, K., Pattle, N.I., Bates, J., and Sloots, T.P. (2005). A real-time PCR assay for the detection of *Neisseria gonorrhoeae* in genital and extragenital specimens. Diagn. Microbiol. Infect. Dis. *52*, 1–5.

Wilson, D.L., Abner, S.R., Newman, T.C., Mansfield, L.S., and Linz, J.E. (2000). Identification of ciprofloxacin-resistant *Campylobacter jejuni* by use of a fluorogenic PCR assay. J. Clin. Microbiol. *38*, 3971–3978.

Wilson, D.A., Yen-Lieberman, B., Reischl, U., Gordon, S.M., and Procop, G.W. (2003). Detection of *Legionella pneumophila* by real-time PCR for the *mip* Gene. J. Clin. Microbiol. *41*, 3327–3330.

Woodford, N., Tysall, L., Auckland, C., Stockdale, M.W., Lawson, A.J., Walker, R.A., and Livermore, D.M. (2002). Detection of oxazolidinone-resistant *Enterococcus faecalis* and *Enterococcus faecium* strains by real-time PCR and PCR-restriction fragment length polymorphism analysis. J. Clin. Microbiol. *40*, 4298–4300.

Yang, J.H., Lai, J.P., Douglas, S.D., Metzger, D., Zhu, X.H., and Ho, W.Z. (2002). Real-time RT-PCR for quantitation of hepatitis C virus RNA. J. Virol. Methods *102*, 119–128.

Yang, S., Lin, S., Khalil, A., Gaydos, C., Nuemberger, E., Juan, G., Hardick, J., Bartlett, J.G., Auwaerter, P.G., and Rothman, R.E. (2005). Quantitative PCR assay using sputum samples for rapid diagnosis of pneumococcal pneumonia in adult emergency department patients. J. Clin. Microbiol. *43*, 3221–3226.

Yoshida, T., Deguchi, T., Ito, M., Maeda, S., Tamaki, M., and Ishiko, H. (2002a). Quantitative detection of *Mycoplasma genitalium* from first-pass urine of men with urethritis and asymptomatic men by real-time PCR. J. Clin. Microbiol. *40*, 1451–1455.

Zeaiter, Z., Fournier, P.E., Greub, G., and Raoult, D. (2003). Diagnosis of *Bartonella endocarditis* by a real-time nested PCR assay using serum. J. Clin. Microbiol. *41*, 919–925.

Zeidner, N.S., Schneider, B.S., Dolan, M.C., and Piesman, J. (2001). An analysis of spirochete load, strain, and pathology in a model of tick-transmitted Lyme borreliosis. Vector Borne Zoonotic Dis. *1*, 35–44.

The Extraction and Purification of Nucleic Acids for Analysis by PCR

7

Chaminda Salgado and Waqar Hussain

Abstract

A myriad of methods for the extraction and purification of nucleic acids prior to PCR are currently used throughout the community. While these methods have many unique and bespoke aspects, they broadly follow a sequence of lysis, isolation, washing and elution to get from a complex biological sample to purified nucleic acid that can be used in a PCR reaction. Various common methods available for each stage are described and potential sequences for particular sample types can be discerned. The potential for these methods to be automated are discussed and the process options summarized with respect to the speed of the methods, technical skill required and the resultant purity and yield that can be expected.

Introduction

In this chapter some of the more common methods for nucleic acid isolation are broadly summarized and compared. To describe all of the possible chemistries and interactions involved would warrant a dedicated book and even then many of the mechanisms and interactions would be postulated from theory as these are rarely understood well. This chapter is therefore intended to guide the reader through the common options available and to aid and inform the analyst of potential isolation routes depending on their sample type and desired target. Detailed protocols and method descriptions can be found by following the references supplied within this chapter.

Regardless of technique used all the methods undergo four basic stages. The first stage is some form of lysis, in order to gain access to the nucleic acids, within the sample. Lysis tends to be either mechanical, such as by sonication (Engler *et al.*, 1996) or bead-mill homogenization (Ho *et al.*, 2008); chemical, such as by detergents (Schüpbach *et al.*, 2006) or potassium ethyl xanthogenate (Leuko *et al.*, 2008) or biological, or a combination of the three. The choice is dependant on the sample type and the nucleic acid species under isolation. The second stage is the isolation, and forms the bulk of this review. Isolation tends to fall under two general categories of separation and capture. The isolation method tends to dictate the complexity of the subsequent steps with significant implications on the propensity of the process to automation or on the requirement for highly skilled technicians to perform the task. The choice isolation method can also entail compromises between robustness, yield, or purity and is therefore a critical process choice. The third stage are the wash steps, which depend not only on the choice of isolation used, but also the nucleic acid species desired. The wash steps can be amenable to the inclusion of a DNA or RNA digestion step to specifically select for either DNA or RNA offering potential for high levels of target purity. The forth and final stage is the elution. For compatibility with PCR there is little choice in the elution buffer as inhibition of DNA polymerases is a concern. The choice of elution buffer can play a key role in yield from the isolation method and so this step is often a compromise. Subtle changes, such as the heating of the elution buffer, can be made to improve yield and to help prevent enzymatic degradation of the nucleic acid.

This chapter will examine each of these stages and choices in turn, paying particular attention to the isolation step. A further issue to consider however is sample quality and storage. For example, for preservation of tissue samples from biopsies and to prevent degradation by proteases, DNases, and RNases, snap freezing in liquid nitrogen is the recommended technique, followed by storage at ≤−70°C. In the absence of such storage facilities, reagents such as RNA*later*©, available from Sigma, Qiagen or Life Technologies, should be used to preserve such samples at higher temperatures. Preservation by formalin fixation, as commonly used in histopathology techniques, however, is a strong inhibitor of enzymes, and should be avoided for enzyme dependent techniques such as PCR.

Sample lysis

The first issue to be tackled is gaining access to the nucleic acids within a given sample. This section explores the available options, starting from the toughest sample types, to the more delicate. Plant tissue samples, animal tissue samples and bacterial spores are among the toughest sample types from which to access the nucleic acid. As an initial step, mechanical disruption is commonly used to homogenize such sample types, followed by enzymatic lysis. The concept behind mechanical homogenization is to allow better access to the cells for subsequent lysis methods. Before entering this step of the lysis process, the analyst can take steps to help the homogenization of the sample. During sample selection, from plants or animal tissues, simply chopping the sample into smaller pieces can improve the success of the lysis process. Bead beating (Ogram *et al.*, 1987; Moré *et al.*, 1994; Yeates *et al.*, 1997; Cullen and Hirsch, 1998; Kuske *et al.*, 1998; Miller *et al.*,1999) is one of the more common mechanical disruption methods. Here beads such as, glass, ceramic, or steel beads are added to the sample, and then agitated at a high frequency in an attempt to shear the samples to a consistent homogenate. Issues to be aware of with this method are the frothing of the samples, causing difficulty in downstream processing of the samples, and heat generation, which may cause damage to the nucleic acids (particularly RNA), and result in a reduction in yield. Excessive periods of bead beating should therefore be avoided, and alternatives sought such as pulse beating, with pauses to avoid heat build up. Sonication (Porteous *et al.*, 1997) using ultrasound is another mechanical method used to break down the integrity of bacterial spores but heat generation is again a concern, as well as the generation of free radicals that are potentially damaging to nucleic acids,. Other crude methods are the use of a pestle and mortar (Volossiouk *et al.*, 1995) to physically grind snap frozen samples. A common issue to consider with all of these methods is the release of DNases and RNases from the cells which may now act upon the target molecule, reducing yield. This risk can be mitigated by homogenization at low temperatures (liquid nitrogen) or in the presence of a lysis buffer designed to inhibit or denature proteins, that will inhibit action of these enzymes.

Following homogenization which compromises the integrity of the cellular boundary the nucleic acid can be accessed by one of the following commonly used methods. Samples such as cultured cells, bacterial cells or crude virus samples, may not require homogenization and can proceed directly to enzymatic or chemical lysis. Enzymatic lysis may use proteases like Proteinase K (Gross-Bellard *et al.*, 1973) and/or lysozyme (Clewell and Helinski, 1970). Proteinase K is used to lyse the cell walls or viral capsid, but also plays a role in denaturing nucleases and proteins with a high affinity to nucleic acids. It is also a robust enzyme which remains active in the presence of several dilute detergents [i.e. sodium dodecyl sulfate (SDS)] used in cell lysis, enabling simultaneous chemical and enzymatic lysis. Lysozyme is commonly used for lysing bacterial cell walls, particularly gram negative bacterial walls, by digesting peptidoglycan in the cell walls. Chemical lysis methods in nucleic acid isolation use ionic detergents such as SDS to both denature proteins, and disrupt the lipid barrier. If chemical lysis is used alone it is commonly a strong detergent solution such as sodium hydroxide and SDS. Timing is crucial as this method not only denatures protein but also nucleic acids and, if left too long, will cause irreversible damage to the nucleic acids. As a method to isolate plasmid from genomic DNA, neutralization of the solution allows the plasmid to re-anneal but the large genomic DNA cannot, allowing for separation by centrifugation. Under centrifugal force the genomic DNA, still bound to protein, is spun down out of solution while the plasmid remains in solution. Heat denaturation by boiling or microwave (Marra *et al.*, 1999) is a method to denature the proteins encapsulating the nucleic acids, but again heat is a concern for RNA. Freeze-thaw (Erb and Wagner-Döbler, 1993) is another method for lysing cells. It is the freezing that disrupts the walls/capsid through the formation of ice crystals. Crystal formation however can also damage the nucleic acids, and reduce the yield.

Following one, or a combination of the above steps, the analyst should be left with a crude preparation of nucleic acid, proteins and lipids.

Nucleic acid binding and purification

Phenol–chloroform isolation

Phenol–chloroform isolation is a method to isolate RNA in the presence of a chaotropic denaturing solution, or DNA in the absence of this solution (Chomczynski and Sacchi, 1987). In summary, when the lysed sample is mixed with phenol and chloroform, protein is denatured and dissolved in the phenol/chloroform. The nucleic acids, being a more polar molecule, dissolves in the water and the solution will separate, with the nucleic acid in the upper aqueous phase and the protein and other components in the interphase and lower organic phase. The aqueous phase, containing the nucleic acid of interest can then be carefully siphoned off using a pipette into a separate tube. This step can be tricky, the more of the aqueous phase that can be extracted the higher the recovery; however, any co-extraction of the organic phase will lead to contaminants that could inhibit the PCR. Products such as Phase Lock gel, are designed to make this step easier by creating greater separation between the organic and aqueous phase. The tube containing the aqueous phase can then be centrifuged to pellet the nucleic acid before precipitation by isopropanol or ethanol. Commercial kits using the above principles are those available from Invitrogen.

Caesium chloride ultracentrifugation

Caesium chloride ultracentrifugation (Meselson *et al.*, 1957; Radloff *et al.*, 1967; Maniatis *et al.*, 1982) is a method for isolating specific species of double-stranded DNA. The lysed sample is mixed with a DNA stain, such as ethidium bromide, and CsCl. The solution is then ultracentrifuged for at least 10 hours where the DNA separates due to relative density and physical form (e.g. linear or circular plasmid molecules) in a manner analogous to migration through agarose (acting as a molecular sieve) during agarose gel electrophoresis. Under ultraviolet light, the DNA bands can be visualized. To extract the species of interest a needle and syringe is used to puncture the side of the tube and extract the band. The DNA itself is then purified using isopropanol or ethanol precipitation.

Isopropanol or ethanol purification

Following isolation the nucleic acid maybe contaminated with salts (Zeugin and Hartley, 1985). The addition of sodium acetate and two to three volumes of 95% ethanol, creating a low polarity environment, causes exposure of the negatively charged phosphate back bone of nucleic acids to the positively charged Na^+ ions of sodium acetate. The nucleic acid is therefore neutralized making it much less hydrophilic and therefore less soluble in water. Centrifugation is then used to pellet the precipitated nucleic acid. A typical protocol then involves decanting the supernatant and resuspending the pellet in 70% ethanol by vortexing. Then once again centrifuge into a pellet, decant the supernatant and allow the pellet to air dry. If isopropanol is used instead of ethanol only a 1:1 volume is required, but a high centrifuge speed and longer drying time is needed, as isopropanol is less volatile. Over drying may cause a decreased yield. Finally resuspend the pellet in water or 1 × TE (Tris-EDTA) buffer.

Owing to the requirement to centrifuge and visualize bands, phase separation or pellets, the above techniques do not lend themselves well to simple, walk away, automated procedures.

Silica adsorption

The exact mechanism as to how nucleic acids adsorb to silica appears to be poorly understood and is still a matter of debate (Boom *et al.*, 1990). Particles of diatomaceous earth and other amorphous forms of silica dioxide are highly porous, thus presenting a very high surface area (800 m^2 per gram). As the lysate migrates through the micro pores in silica, in the presence of chaotropic salts, it interacts with the silica, binding to it and allowing the sample impurities to be washed away. Both the nucleic acid phosphorous backbone and silica have a negative charge and should, therefore, repel each other. There are several theories of how the nucleic acid adsorb onto the silica dioxide in the presence of chaotropic salts. One theory is that water forms a hydration shell around nucleic acid, preventing it from interacting with the silica dioxide. In the presence of chaotropic salts, this shell is disrupted allowing the formation of hydrogen bonds between the nucleic acid and the silica dioxide surface. Alternatively, following disruption of the hydration shell, cation bridges form between the nucleic acid and the silica dioxide surface. The ionic strength of the buffer can selectively cause adsorption that is specific to the nucleic acid species involved, so plasmid or genomic DNA can be selected by carefully optimized control of the ionic strength of the binding buffer containing chaotropic salts. For this reason, it is possible to have protocols and kits containing optimized binding and wash buffers, that are specifically designed to purify certain species of nucleic acids, even though the silica chemistry involved may be identical for all these protocols and kits. Other interactions are also possible; for example smaller nucleic acid species are able to migrate through the pores of the silica, whereas the larger genomic DNA will have more restricted access to such spaces. Ultimately, once binding has been established the excess chaotropic salts are washed away using ethanolic buffers with increasingly diluted ethanol throughout the protocol. Finally the nucleic acid is eluted using a low salt (i.e. Tris-EDTA) buffer or water. This elution can be understood in the context of the postulated hydration shell mechanism where the lower salt buffers allow reformation of the hydration shell, breaking the interaction between the silica and nucleic acid.

Silica adsorption technology is a cost-effective solution to provide purified nucleic acid for many downstream applications such as PCR and sequencing. However the purity achieved is not sufficient for more sensitive applications, such as transfecting into endotoxin-sensitive cells. The chemistry has also been criticized for producing low yields of typically 25–50%. It is generally accepted that this is not an adsorption issue, but an elution issue, due to the strength of the interaction between the nucleic acid and silica. Using elution buffer at elevated temperature (> 50°C) has been shown to improve nucleic acid yields. There has been recent advances in this chemistry which appear to show marked improvement in the yields, close to the yields often reserved to precipitation techniques.

From a practical point of view, the first iteration (Boom *et al.*, 1990) of using silica for purifying nucleic acids had only a slight advantage over isopropanol isolation, in that a phase separation step was not required. But the subsequent washing and centrifugation steps still required a high degree of manual intervention. Over time the method evolved to simplify the manual intervention by packing the silica into a column, priming the column, and then adding homogenized sample with binding buffer, to initiate adsorption. Gravity was used to allow passage of sample and buffers through the column. However, the time to achieve purification using gravity feed was not efficient. Eventually efficiency was addressed by further evolution to the current use of kits available. In kit form there appear to be two common technologies that use silica adsorption, beads and centrifuge columns. Centrifuge column technology was the first to market where the homogenized sample is mixed with binding buffer (optimized chaotropic salt solution) before adding the mix to columns containing a membrane (possibly also used in size exclusion) sandwiching the silica resin chemistry. The column is then centrifuged, and it is the centrifugal force that pushes the sample through the resin, and representing the first adsorption step described above. An alternative to centrifugation, which was designed to ease the automation of column technology, is the use of a vacuum manifold. In this application a vacuum is created to force the sample and buffers through the column. However, due to the physics of a disc (membrane/resin), the force exerted on the area of the membrane by a vacuum is concentrated on the centre, even more so than centrifugal force. Therefore the edges of the membrane may trap more of the sample or buffers. Another issue with column technology is that samples containing non/low-soluble PCR inhibitors can remain on the membrane, and some may be co-eluted with the nucleic acid in the final stages causing partial or complete inhibition of PCR. Automation, using columns, must involve liquid handling, due to specific volumes that need to be added to the columns. In a regulated environment this brings the additional burden of validating the pipetting performance, depending on the liquid type used. The DNeasy, and RNeasy family of kits, as sold by Qiagen, are an example of column kits using silica adsorption.

Silica magnetite beads are a more recent technology that utilize an iron core coated in silica. This technology was developed for the specific need of automating the silica binding chemistry. Here, the homogenized sample is again mixed with a binding buffer which is then mixed with the beads to initiate the adsorption step in the protocol. Bead technology can use liquid handling, by immobilizing the beads on the bottom of the tube/plate with a magnet under the tube/plate, and using the liquid handler to add and remove the wash and elution buffers. However a much simpler and neater technique is to move the beads themselves by using automation with simple magnetic rods to pickup and release the beads into the different wash buffers, leading ultimately to the elution buffer. Such automation is much simpler to maintain as the need to validate any liquid handling performance is negated. Another advantage that beads have over columns is the fact that the non/low-soluble inhibitors are not forced to interact with the chemistry, as the beads flow freely in solution, attracting the nucleic acids. For this reason bead chemistries tend to have equivalent yields to columns, but are also less problematic with regards to PCR inhibitors. Following the success of silica magnetite beads, the concept of encapsulating an iron core with the adsorption chemistry of interest, in order to easily automate the procedure, has been replicated for several applications, some of which are described in this chapter.

ChargeSwitch®

An increasingly common method for nucleic acid purification exploits the native negative charge of nucleic acid molecules and employs a charge–charge interaction to bind the target molecules. The chemistry is dependent on a low pH lysis and binding environment wherein the surface ligands sequester free protons that result in a positively charged surface. The negatively charged nucleic acids therefore bind and can be purified of cellular debris and proteins by aqueous wash steps. Liberation of the nucleic acids is then readily effected by the transfer of the bound nucleic acids into a high-pH environment, where the surface loses its charge and the bound molecules are readily washed free.

A number of other kits employing this type of technology are available on the market for the isolation of genomic and plasmid DNA as well as RNA. Such kits employ a relatively mild urea-based chemical lysis and are often free of organic solvent based buffers and concentrated salts that offer significant advantages. Processes involving high quantities of salts throughout may have a detrimental effect on the signal strength observed during subsequent real-time PCR analysis of the purified nucleic acids (Freeman *et al.*, 1999). For commercial processes, the use of ethanol within the process wash buffers introduces significant cost issues into supply chain management. The hygroscopic and corrosive nature of ethanol and its flammability contribute to the increased cost of the transportation. Further, the addition of guanidinium-based salts, while excellent chaotropes and highly effective in host lysis by sequestration of water, may lead to inhibition of the PCR if these salts are carried forward (Demeke and Jenkins, 2009).

Isolation using carboxyl derivatives

A number of proprietary and non-proprietary isolation methods employ particles or planar surfaces with chemistries featuring carboxyl functional groups. The method is centred on interaction between the carboxyl group and the nucleic acid in a high salt/PEG environment (Hawkins *et al.*, 1994). The method has been successfully employed for the separation of total nucleic acids from a background of cellular debris and proteins (Horak *et al.*, 2007), as well as for the graded separation of nucleic acids by molecular weight to differentiate, for example, plasmid DNA from genomic.

For the separation of DNA from other cellular contaminants, the method employs a combination of polyethylene glycol, of molecular weight 6000–10,000, and the chloride salt of lithium, sodium, magnesium, potassium, calcium, caesium or barium. Typically, however, the method is limited to PEG-8000, at a concentration of 5–15% w/v, and sodium chloride at a concentration of 0.5–5.0 M, which is the general range required for nucleic acids to bind to carboxyl functional groups. Specific concentrations of these reagents can be optimized from this general range; for example non-specific binding of DNA can be achieved by using PEG-8000 at 10% w/v and sodium chloride at 1.25 M.

The application of this method to bulk separations is addressed through a number of specialized kits currently on the research tools market. Most commonly these feature the use of solid-phase particles with derivatized surfaces exposing the nucleic acids in the bulk medium to the carboxyl-functional group chemistry designed to capture the molecules. The use of various particle formation methods to create particles with maximized surface area leads to highly efficient methods for nucleic acid isolation and often leads to high yields. Combination with appropriate methods for the treatment and washing of these beads can result in highly effective purification of the molecule of interest. It has further been postulated that the appropriate selection of polyethylene glycol and salt concentrations can act as a selective method for nucleic acids by molecular weight in a batch process method.

Chromatographic methods, with the potential for being adapted for operation in continuous mode, have also been reported employing this chemistry for the separation of specifically sized nucleic acid strands. These methods are of particular utility in the isolation of digestion products where fine resolution of product size is essential. Affinity resins crafted with optimized porosity and bed voidage allow differentiation of nucleic acid strands to a resolution of < 10 base pairs. Column based processes, with their low intrinsic shear effects, are often the desired process route for the purification of high molecular weight nucleic acids that are difficult to isolate in-tact through bead based processes. The high level of purification achievable is a significant advantage of these systems, but yield can often be compromised. Combination of a batch method to achieve the requisite yield of nucleic acids with a chromatographic method to achieve the required purity could therefore be the desired process route.

Cellulosic beads

Recent additions to the nucleic acids purification market include those employing cellulosic surfaces for the isolation of both DNA and RNA. A prominent example is the Promega Magazorb® kit that employs a chaotropic lysis (5 M guanidinium chloride, 8 M urea) followed by PEG-salt precipitation of the nucleic acids onto paramagnetic beads comprising carboxymethyl-cellulose and diethylaminoethyl cellulose (up to 10% w/w). The precise mechanism by which nucleic acids bind to the cellulosic surface is as yet unknown, but it is postulated that binding is driven by the presence of an aqueous-aqueous two phase system brought about by the concentrated sodium chloride and PEG-4000 present in the binding buffer (Lebreton *et al.*, 2002; Zhang *et al.*, 2006; Li *et al.*, 2010). In such an environment, the highly polar nucleic acids will tend towards the polar salt rich region and the apolar cellular contaminants or overall-neutral large proteins tend towards the PEG rich region. The high salt environment of the binding buffer (1.25 M NaCl) facilitates binding of the nucleic acids to the beads as well as sequestration of apolar contaminants (10% w/v PEG-8000). Subsequent washing under lower salt conditions (0.5 M NaCl) results in the removal of proteins that have lower overall polarity than the nucleic acids that are subsequently eluted in the effectively salt free environment of the nuclease free water.

Ion exchange purification

An emerging alternative to caesium chloride ultracentrifugation techniques is ion exchange chromatography based DNA purification (Yang *et al*,. 2008) which is the basis of a number of kits currently on the market for both plasmid and genomic DNA. The significant process advantage offered by such kits is the ability to perform ion exchange purification without the need for specialist chromatography equipment, which is often expensive and occupies a large spatial footprint. However, these kits are limited to gravity flow through columns with poor diffusional efficiency leading to concentration gradients across the bed. The loose packing required for gravity flow also leads to inefficient use of the binding sites within the individual beads. Ion exchange membranes have the advantages of high flow rate and thus the potential for high-throughput applications. Such membranes have been shown to have wide utility in the capture and purification of plasmid DNA (Endres *et al.*, 2003; Teeters *et al.*, 2003; Zhang *et al.*, 2003), oligonucleotide sequences (Deshmukh *et al.*, 2000) and many other non-nucleic acid biomolecules (Roper and Lightfoot, 1993; Knudsen *et al.*, 2001; Specht *et al.*, 2004).

As with many of the techniques described here, anion exchange techniques rely upon the negatively charged nucleic acid backbone interacting with the oppositely charged anion exchange surface. In its most simple format, the anion exchange technique employs a binding environment of near neutral pH and medium ionic strength to affect binding. The high density of surface groups for binding is exploited to achieve high levels of recovery in a relatively mild environment. For bulk capture and purification of nucleic acids, without distinguishing for example plasmid from genomic DNA, or RNA from DNA, such purification is highly effective both in terms of recovery and purity. However, for more selective capture the pH of the binding environment can be manipulated to affect the charge density of the matrix and the ionic strength of the buffers tuned to yield proteins, RNA or DNA specifically. In addition to PCR (Jones and Winistorfer, 1992; Jung *et al.*, 1992), due to its highly selective nature the technique has been employed for such applications as mammalian cell transfection (Chen *et al.*, 1991), DNA sequencing (Hall *et al.*, 1989; Voss *et al.*, 1990) and transfection (Kayne *et al.*, 1988; Luytjes *et al.*, 1989).

Elution

The final step in the isolation and purification of nucleic acids is the elution of the captured nucleic acids from the medium upon which they were adsorbed. The choice of elution buffer is dependent upon the mechanism of nucleic acid binding to the isolation medium but will often be an inert buffer with low ionic strength (low-salt) and slightly alkali pH (~ 8). Such a buffer serves the multiple purposes of eluting the target molecules, no intrusion on the PCR process and maintaining the integrity of the nucleic acids in a standard environment. Heating the elution buffer prior to use may also be beneficial to increase yield from capture methods as it aids stripping of the nucleic acid from the binding chemistry. Eluates from these processes can then be stored for short (at 4°C) or long periods (at –20°C) before analysis by PCR. Light, in particular ultra-violet, is highly damaging to nucleic acids, so exposure must be limited to further preserve their integrity.

Conclusion

This chapter has outlined popular methods from the literature for the isolation and purification of nucleic acids from a range of sample types. The evolution of some these techniques towards easy to use kits has made them an essential aspect of the nucleic acid analyst's tool box. In selecting the appropriate method particular care must be taken in identifying the target, such as genomic versus plasmid DNA or DNA versus RNA, and

Table 7.1 Summary of the relative strengths of the described sample extraction and isolation chemistries described

				Silica					Ion exchange	
Criteria	CsCl	Phe/Chl	Isopropanol	Beads	Columns	Charge switch beads	Carboxyl beads	Cellulosic beads	Beads	Column
Technical skill required	+++	++	++	+	+	+	+	+	++	++
Speed	+	++	++	+++	+++	+++	++	++	+++	++
Automatable	+	+	+	+++	+++	+++	++	++	+++	++
Yield	+++	+++	+++	++	++	++	++	++	++	++
Purity	+++	+++	+++	++	++	++	++	++	+++	+++

CsCl refers to caesium chloride ultracentrifugation, Phe/Chl refers to phenol–chloroform isolation.+= poor, ++=good, +++=excellent.

in a number of other considerations, such as the automatability of the technique at hand. Additionally, thought must be given to the future application of the method. If the method is destined for a QC environment, with higher throughput, and potentially lower expertise, then techniques which are easily automatable maybe more desirable, even if this requires a compromise in other areas. The techniques discussed have been summarized with some of these considerations in Table 7.1.

References

Boom R., Sol, C.J., Salimans, M.M., Jansen, C.L., Wertheim-van Dillen P.M., and van der Noordaa J. (1990). Rapid and simple method for purification of nucleic acids. J. Clin. Microbiol. *28*, 495–503.

Chen W., Andres, D., Goldstein, J., Russell, D., and Brown, M. (1991). cDNA cloning and expression of the peptide-binding beta subunit of Rat p21 farnesyltransferase, the counterpart of yeast DPR1/RAM1. Cell *66*, 327–334.

Chomczynski P., and Sacchi, N. (1987). Single-step method of RNA isolation by acid guanidinium thiocyanate-phenol-chloroform extraction. Anal. Biochem. *162*, 156–159.

Clewell D.B., and Helinski, D.R. (1970). Properties of a supercoiled deoxyribonucleic acid-protein relaxation complex and strand specificity of the relaxation event. Biochemistry *9*, 4428–4440.

Cullen, D.W., and Hirsch, P.R. (1998). Simple and rapid method for direct extraction of microbial DNA from soil for PCR. Soil Biol. Biochem. *30*, 983–993.

Demeke T., and Jenkins, G.R. (2009). Influence of DNA extraction methods, PCR inhibitors and quantification methods on real-time PCR assay of biotechnology-derived traits. Anal. Bioanal. Chem. *396*, 1977–1990.

Deshmukh R.R., Warner, T.N., Hutchison, F., Murphy, M., Leitch, W.E., De Leon P., Srivatsa, G.S., Cole, D.L., and Sanghvi, Y.S. (2000). Large-scale purification of antisense oligonucleotides by high-performance membrane adsorber chromatography. J. Chromatogr. A *890*, 179–192.

Endres H.N., Johnson, J.A., Ross, C.A., Welp, J.K., and Etzel, M.R. (2003). Evaluation of an ion exchange membrane for the purification of plasmid DNA. Biotechnol. Appl. Biochem. *128*, 132–149.

Engler H.D., Selepak, S.T., and Callaway, J. (1996). Effects of sonication of clinical specimens on detection of cytomegalovirus in shell vial cultures. Clin. Diag. Virology *5*, 67–71.

Erb, R.W., and Wagner-Döbler I. (1993). Detection of polychlorinated biphenyl genes in polluted sediments by direct DNA extraction and polymerase chain reaction. Appl. Environ. Microbiol. *59*, 4065–4073.

Freeman W.M., Walker, S.J., and Vrana, K.E. (1999). Quantitative RT-PCR: pitfalls and potential. BioTechniques *26*, 112–125.

Gross-Bellard M., Oudet, P., and Chambon, P. (1973). Isolation of high-molecular-weight DNA from mammalian cells. Eur. J. Biochem. *36*, 32–38.

Hall B., Betts, P., and Wootton, J. (1989). DNA sequence analysis of artificially evolved ebg enzyme and ebg repressor gene. Genetics *123*, 635–648.

Hawkins T.L., O'Connor-Morin T., Roy, A., and Santillan, C. (1994). DNA purification and isolation using a solid-phase. Nucleic Acids Res. *22*, 4543–4544.

Ho C.W., Tan, W.S., Kamarudin, S., Ling, T.C., and Tey, B.T. (2008). The release of hepatitis B core antigen from *Escherichia coli* by batch mode bead milling. Proc. Biochem. *43*, 206–212.

Horak D., Rittich, B., and Spanova, A. (2007). Carboxyl-functionalized magnetic microparticle carrier for isolation and identification of DNA in dairy products. J Magnetism Magentic Mat. *311*, 249–254.

Jones D., and Winistorfer, S. (1992). Recombinant circle PCR and recombinant PCR for site-specific mutagenesis without PCR product purification. Biotechniques *12*, 528–535.

Jung M., Dritschilo, A., and Kasid, U. (1992). Reliable and efficient direct sequencing of PCR double-stranded genomic DNA template. PCR Methods Appl. *1*, 171–174.

Kayne P., Kim, U., Han, M., Mullen, J., Yoshizaki, F., and Grunstein, M. (1988). Extremely conserved histone H4 N terminus is dispensible for growth but essential for repressing the silent mating loci in yeast. Cell *55*, 27–39.

Knudsen H.L., Fahmer, R.L., Xu, Y., Norling, L.A., and Blank, G.S. (2001). Membrane ion-exchange chromatography for process-scale antibody purification. J. Chromatogr. A *907*, 145–154.

Kuske, C.R., Banton, K.L., Adorada, D.L., Stark, P.C., Hill, K.K., and Jackson, P.J. (1998). Small-scale DNA sample preparation method for field PCR detection of microbial cells and spores in soil. Appl. Environ. Microbiol. *64*, 2463–2472.

Leuko S., Goh, F., Ibanez-Peral R., Burns, B.P., Walter, M.R., and Neilan, B.A. (2008). Lysis efficiency of standard DNA extraction methods for *Halococcus* species in an organic rich environment. Extremophiles: life under extreme conditions. *12*, 301–308.

Luytjes W., Krystal, M., Enami, M., Parvin, J., and Palese, P. (1989). Amplification, expression, and packaging of a foreign gene by influenza virus. Cell *59*, 1108–1113.

Maniatis, T., Fritsch, E.F., and Sambrock, J. (1982). Molecular Cloning. A Laboratory Manual (Cold Spring Harbor Laboratory Press, NY, USA), pp. 86–96.

Marra M.A., Kucaba, T.A., Hillier, L.W., and Waterston, R.H. (1999). High-throughput plasmid DNA purification for 3 cents per sample. Nucleic Acids Res. *27*, e37.

Meselson M., Stahl, F.W., and Vinograd J. (1957). Equilibrium sedimentation of macromolecules in density gradients. Proc. Natl. Acad. Sci. U.S.A. *43*, 581–583.

Miller D.N., Bryant, J.E., Madsen, E.L., and Ghiorse, W.C. (1999). Evaluation and optimization of DNA extraction and purification procedures for soil and sediment samples. Appl. Environ. Microbiol. *65*, 4715–4724.

Moré M.I., Herrick, J.B., and Silva, M.C. (1994). Quantitative cell lysis of indigenous microorganisms and rapid extraction of microbial DNA from sediment. Appl. Environ. Microbiol. *60*, 1572–1580.

Ogram, A., Sayler, G.S., and Barkay, T. (1987). The extraction and purification of microbial DNA from sediments. J. Microbiol. Methods *7*, 57–66.

Porteous L.A., Seidler, R.J., and Watrud, L.S. (1997). An improved method for purifying DNA from soil for polymerase chain reaction amplification and molecular ecology applications. Mol. Ecol. *6*, 787–791.

Radloff, R., Bauer, W., and Vinograd, J. (1967). A dye-buoyant-density method for the detection and isolation of closed circular duplex DNA in HeLa cells. Proc. Natl. Acad. Sci. U.S.A. *57*, 1514–1521.

Roper D.K., and Lightfoot, E.N. (1993). Separation of biomolecules using adsorptive membranes. J. Chromatogr. A *702*, 3–26.

Schupbach J., Tomasik, Z., Knuchel, M., Opravil, M., Gunthard, H.F., Nadal, D., and Boni, J. (2006). Optimized virus disruption improves detection of HIV-1 p24 in particles anduncovers a p24 reactivity in parients with undetectable HIV-1 RNA under long-term HAART. J. Med. Virol. *78*, 1003–1010.

Specht R., Han, B., Wickramasinghe, S.R., Carlson, J.O., Czermak, P., Wolf, A., and Reif, O.W. (2004). Densonucleosis virus purification by ion exchange membranes. Biotechnol. Bioeng. *88*, 465–473.

Teeters M.A., Conrardy, S.E., Thomas, B.L., Root, T.W., and Lightfoot, E.N. (2003). Adsorptive membrane chromatography for purification of plasmid DNA. J. Chromatogr. A *989*, 165–173.

Volossiouk T., Robb, E.J., and Nazar, R.N. (1995). Direct DNA extraction for PCR-mediated assays of soil organisms. Appl. Environ. Microbiol. *61*, 3972–3976.

Voss H., Zimmerman, J., Schwager, C., Stegemann, J., Erfle, H., Stuckey, K., and Ansorge, W. (1990). Automated fluorescent sequencing of cosmic DNA. Nucl. Acids Res. *18*, 1066.

Yang Y., Hebron, H.R., and Hang, J. (2008). High performance DNA purification using a novel ion exchange matrix. J. Biomol. Tech. *19*, 205–210.

Zeugin J.A., and Hartley, J.L. (1985). Ethanol Precipitation of DNA. Focus *7*, 1–2.

Zhang S., Krivosheyeva, A., and Nochumson, S. (2003). Large-scale capture and partial purification of plasmid DNA using anion-exchange membrane capsules. Biotechnol. Appl. Biochem. 37, 245–249.

Oligonucleotide Primers and Probes: Use of Chemical Modifications to Increase or Decrease the Specificity of qPCR

8

Scott D. Rose, Richard Owczarzy, Joseph R. Dobosy and Mark A. Behlke

Abstract

Although the vast majority of primers and probes employed in qPCR applications today are synthesized using unmodified DNA bases, selective use of chemically modified bases and non-base modifying groups can prevent primer dimer artefacts, improve specificity, and allow for selective amplification of sequences that differ by as little as a single base. A wide variety of chemical modifications have been characterized for use in qPCR. As a general class, the modifications that are in greatest use today increase the binding affinity of the oligonucleotides (i.e. increase the melting temperature, T_m). T_m-enhancing modifications allows both primers and probes to be shorter, improving the differential T_m ($\Delta T_m = T_{m\ match} - T_{m\ mismatch}$) between perfect match and mismatch hybridization. These modifications have widespread application in allele-specific PCR and in the detection of single nucleotide polymorphisms (SNPs). Conversely, a second class of base modifications are in common use that decrease specificity and improve duplex formation in the presence of base mismatches. Although these modifications lower T_m, they have less of an impact on primer stability than do actual mismatched bases. Universal bases permit use of primers and probes in polymorphic loci when it is desirable to detect all sequence variants and minimize mismatch discrimination.

Introduction

The primers and probes used in real-time quantitative PCR (qPCR) are synthetic oligonucleotides and can be manufactured using natural DNA bases or can comprise RNA bases, artificially modified bases, or a variety of pendant chemical groups not found in nature. While most needs are well met using unmodified oligodeoxynucleotides, the use of chemical modifications can improve qPCR performance in a number of areas. This chapter will cover three specific areas of interest, with a primary focus on increasing specificity and allelic discrimination methods:

1 increasing specificity, with a focus on SNP discrimination;
2 preventing primer-dimer or other PCR artefacts;
3 decreasing specificity using universal bases.

Increasing specificity

Specificity relates to the ability of the qPCR reaction to generate a positive signal from a desired nucleic acid sequence while simultaneously not generating a signal from any similar yet distinct sequences also present in the target nucleic acid sample. In some instances this involves discrimination of related sequences within a gene family where multiple base differences are present. In its most extreme form, this involves discrimination of sequences that differ by only a single base, i.e. a SNP. In broadest terms, qPCR approaches to SNP discrimination can be categorized as those methods which employ biased amplification, where one allele is selectively amplified over other alleles, and those methods which do not employ biased amplification, where all related sequences are amplified equally and a detection probe is employed to distinguish base identity.

SNP discrimination using unbiased amplification

PCR-based methods that discriminate between related sequences using unbiased amplification employ primers placed in regions of the sequence that are identical between alleles, positioning the SNP between the two primers. Sequence identity is usually interrogated by differential hybridization with a third probe oligonucleotide. Success is entirely dependent upon having a sufficiently high ΔT_m between the 'match' and 'mismatch' targets so that the 'match probe' hybridizes to the target site during amplification and the 'mismatch probe' does not. A variety of different assay formats exist that employ this strategy, most of which show improved performance using chemically modified oligonucleotides.

Linear dual-labelled probes

One of the more popular formats for quantitative sequence detection is the 5′-nuclease assay, which was first described by Gelfand and colleagues (Cetus Corporation) in 1991 (Holland *et al.*, 1991). This assay employs a fluorescence-quenched dual-labelled linear probe that binds to the target nucleic acid during the annealing phase of each cycle and is degraded during the extension phase of PCR. The probe has a reporter fluorophore on one end and a quencher on the other end, which absorbs fluorescence emission from the reporter dye (the intact probe is 'dark'). The reaction is cycled using a two-step programme with incubation at a single temperature during which time both the primer annealing and polymerase extension steps occur. The T_m of a sequence is the temperature at which 50% of the possible duplexes are formed and 50% remains in the single-stranded state. Generally, the probe is designed to have a T_m 5°C or more above the annealing temperature so that it is almost completely hybridized to the target template strand during DNA synthesis. During elongation, the DNA polymerase encounters the probe which is quickly degraded by the enzyme's 5′–3′ exonuclease activity. Degradation physically separates the reporter fluorophore from the quencher, resulting in the appearance of a detectable signal (the degraded probe is 'bright'). This assay format is often referred to as the TaqMan® assay.

The 5′-nuclease assay was initially developed to measure gene expression levels; it was quickly adapted for use in allelic discrimination. This method allows for a convenient, homogenous closed-tube assay where the same primers are employed to amplify both alleles. Two probes are present in the reaction that are labelled using detectably distinct fluorophores, one specific for allele 'A' and the other specific for allele 'B'. The presence of one or both alleles in a sample can be detected in real time by monitoring rise of signal in two separate fluorescence channels (Livak, 1999). The probe oligonucleotides usually need to be 20 bases or more in length to be completely hybridized to the target nucleic acid at the 60°C annealing/extension temperature (precise length will of course vary with sequence). Unfortunately, single base mismatches in probe sequences this long are often insufficient to produce a large enough ΔT_m between match and mismatch probes to obtain clean separation of signal between alleles. This problem is easily solved by using probes that are made with T_m-enhancing modifications. Modifications that increase T_m permit use of shorter probes at the same reaction temperature. A single base mismatch has a greater impact on duplex stability in short sequences than in long sequences, so the ΔT_m between match and mismatch improves as probe length decreases. Many factors contribute to the T_m and ΔT_m and it is important to be able to accurately predict these values when designing assays.

The ability of a probe oligonucleotide to distinguish between two species solely on the basis of hybridization is a thermodynamic problem that is influenced by the sequence context, the precise match versus mismatch base pair being studied, the length of the probe, and physical parameters of the hybridization reaction, including temperature, probe concentration, and buffer composition. As a general rule of thumb, a good place to start is to design the 'match probe' to have a T_m around 3–5°C above the reaction temperature and the 'mismatch probe' to have a T_m around 5–7°C below the reaction temperature, giving a ΔT_m of 10°C between the match and mismatch hybridization events. For reactions run at a standard temperature of 60°C, this means selecting sequences with T_m values around 64–65°C and 54–55°C for the match and mismatch probes respectively. Obviously having the ΔT_m as large as possible is desirable; however, it is generally not necessary to exceed 10°C and sometimes in actual practice probes with a lower ΔT_m may still perform well. While empiric testing of different probe designs is sometimes necessary, the use of computer assisted design tools that employ the latest thermodynamic parameters can often be used to predict probe sequences that discriminate well without further optimization, or at least greatly accelerates the process of finding good probes.

Typically, reactions are performed at around 60°C with 250 nM probe in a buffer that contains ~50 mM monovalent cation (sodium or potassium), ~3 mM divalent cation (magnesium), and 0.8 mM deoxyribonucleotide triphosphates (dNTPs). It is important to include the total dNTP concentration in thermodynamic calculations since dNTPs bind Mg^{2+} ions and reduce the concentration of free Mg^{2+} present in the reaction. Buffer composition can vary and unfortunately many vendors do not disclose the exact salt content of their PCR reagents, which complicates probe design. Thermodynamic parameters have been determined that allow for the accurate prediction of the T_m of short DNA oligonucleotides (SantaLucia *et al.*, 1996, 1998). In addition, the stabilizing effects of both monovalent and divalent cations are well characterized (Owczarzy *et al.*, 2004, 2008a), and the relative contribution of some fluorescent dyes and quencher have been measured (Moreira *et al.*, 2005). Sufficient detail is known about the thermodynamic effects of DNA base mismatches to model hybridization events of the kind needed to design SNP probes (SantaLucia and Hicks, 2004). A variety of algorithms and sequence analysis tools are available as part of commercial software packages or are freely accessible on the internet and that can help with design of both primers and probes for qPCR applications, including SNP detection (Owczarzy *et al.*, 2008b).

Early approaches to increase T_m and improve SNP discrimination through use of chemical modification focused on alterations of the sugar and base groups. Although 2′-*O*-methyl RNA (2′OMe RNA) is often thought of as being a T_m-increasing modification, it actually has increased affinity when binding to an RNA target while the T_m effects with a DNA target are variable. Use of this modification was not found to improve probe performance. In fact, extensive modification with 2′OMe RNA can impair nuclease cleavage of the probe and can be detrimental (Kuimelis *et al.*, 1997). Propynyl pyrimidines are a class of modified base where a propyne group is attached at the C5 position of the pyrimidine ring. This modification was first used to increase binding affinity of antisense oligonucleotides (Wagner *et al.*, 1993), but *in vivo* toxicity limited utility of this application. The propynyl pyrimidines, however, are quite effective in increasing T_m and improve performance of probes used in the 5′-nuclease assay in qPCR (Kuimelis *et al.*, 1997); each base substitution leads to a 0.5–1.5°C increase in T_m. Nearest neighbour parameters have not been reported for these bases; thus T_m predictions cannot be accurately made and empirical testing usually must be done when designing probes of this type. Unfortunately, these modifications are only available for pyrimidines, limiting benefit to pyrimidine-rich probe sequences. To be maximally useful, T_m-enhancing base modifications should be available for all four nucleotides. Structures of the various modifications discussed in this chapter are shown in Fig. 8.1.

Locked-nucleic acids (LNAs) are a T_m-enhancing modification simultaneously developed by Jesper Wengel and his group at the University of Southern Denmark and by Takeshi Imanishi and his group at the University of Osaka. LNAs have a methylene bridge that connects the 2′-O with the 4′-C of the ribose. This bicyclic structure locks the ribose in the 3′-*endo* conformation and increases T_m by 2–6°C per substitution (Kvaerno and Wengel, 1999; Obika *et al.*, 1997; Singh *et al.*, 1998; Singh and Wengel, 1998). LNA nucleotides are commercially available for all four nucleotides (guanine, adenosine, 5-methylcytosine, and thymidine) and have found widespread use in many applications, including antisense, RNA interference, and PCR. Substitution of LNA nucleotides for DNA in probes allows use of much shorter sequences at the standard 60°C reaction temperature and these shorter probes show a significant improvement in the ΔT_m between match and mismatch hybridization compared with parent unmodified (longer) sequences (Letertre *et al.*, 2003; Costa *et al.*, 2004; Johnson *et al.*, 2004; Ugozzoli *et al.*, 2004). The early reports from 2003–2004 which first demonstrated the utility of the LNA modification in dual-labelled hydrolysis probes substituted LNA nucleotides for DNA bases using a variety of different modification patterns with a primary focus on simply improving T_m and reducing probe length. This design strategy was effective; however, no systematic attempt was made to optimize the specific pattern of LNA modifications around the SNP site to maximize the ΔT_m between match and mismatch. In fact, the SNP site was often avoided for modification because it was uncertain if the increased binding affinity of the LNA nucleotide would improve or hurt SNP discrimination if placed directly at the site of mismatch. This question was recently addressed by Owczarzy and colleagues in 2006 (Integrated DNA Technologies), who carefully studied the effect that placement of LNA nucleotides at and around the SNP site had on T_m. Melting temperatures were experimentally measured for a large series of modified and unmodified oligonucleotides hybridized to perfect match or single base mismatch complements. The test set included all possible base variations at the SNP site and the adjacent bases, comparing the effects that LNA substitution had at these sites (You *et al.*, 2006). It was found that simply placing three consecutive LNA nucleotides spanning the SNP site (with the SNP positioned centrally) gave the maximum ΔT_m for all base combinations except for one, where the mismatch is a G:T pair and the guanine base is in the probe. In the single G:T exception, LNA nucleotides actually stabilized the

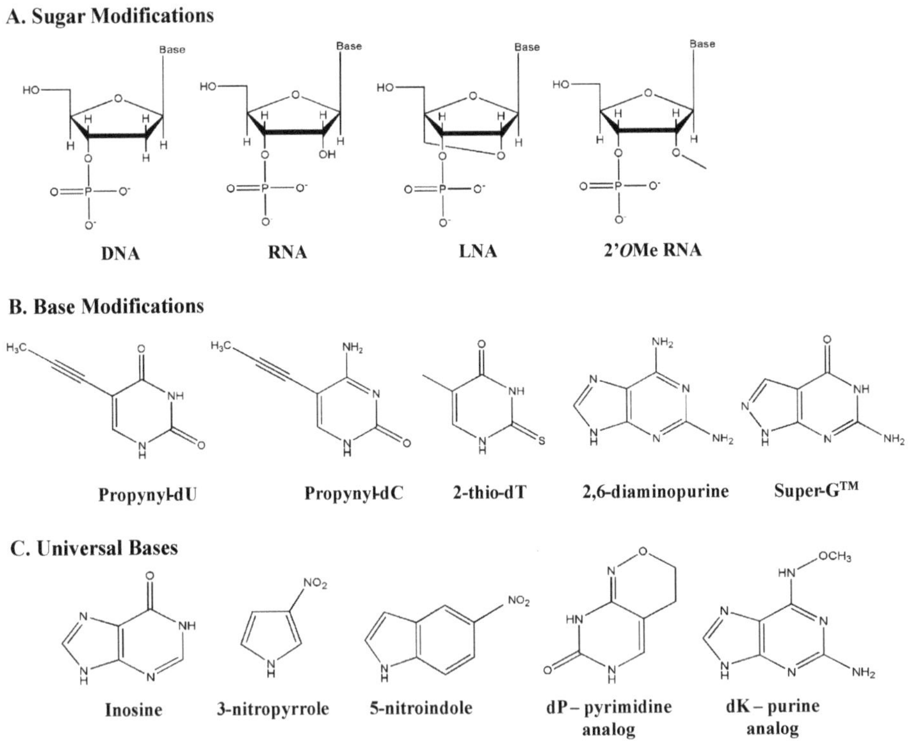

Figure 8.1 Oligonucleotide modifications. (A) Modifications to the ribose. DNA, RNA, LNA (locked nucleic acids), and 2′-*O*-methyl (2′*O*Me) are shown. (B) Base modifications. Propynyl-dU, propynyl-dC, 2-thiothymidine, 2,6-diaminopurine (2,6-DAP, or 2-amino-dA), and Super-G™ are shown. (C) Universal bases. Inosine, 3-nitropyrrole, 5-nitroindole (5-NI), dP, and dK are shown.

mismatch relative to the unmodified sequences, making SNP discrimination worse. LNA modifications can still be used to improve probes having a G:T mismatch by positioning the LNA nucleotides in flanking regions of the probe and reducing probe length. Alternatively, the probe can be designed to the other strand, in which case the mismatch is a T:G pair (with the thymidine base in the probe), and the standard 'LNA triplet' design strategy can be used.

When using the 'LNA triplet' modification pattern described above, it is important to balance the actual T_m of the modified probe so that 'match' hybridization stays around 64–65°C (i.e. 4–5°C above reaction temperature) by trimming bases from each end of the parent unmodified sequence to compensate for the increased binding affinity of the LNA nucleotides. Additional LNA nucleotides can be substituted in the flanking regions of the probe to allow further shortening of the probe sequence, which will continue to improve mismatch discrimination. Ideally, this process would be done using design software that calculates the T_m of different modification patterns and automatically optimizes design. Assistance with estimating the T_m of LNA-modified oligos and design of SNP-discriminating probes can be found on the Exiqon website at: http://lnatools.com/. The interactive tools on this site employ thermodynamic calculations that use experimentally determined parameters and unpublished proprietary rule sets for probe design. Unfortunately, enough inaccuracy exists in these calculations that empiric testing of different probe designs is often required. Experimentally determined nearest neighbour (NN) parameters were published for isolated LNA nucleotides in LNA/DNA hybrids by Celadon Laboratories in 2004 (McTigue *et al.*, 2004). Additional parameters are needed to enable predictions to be made with sufficient accuracy for the T_m and ΔT_m of LNA containing sequences to design probes for use in SNP assays, including parameters for sequential LNA nucleotides and LNA/DNA mismatches. These missing parameters were recently obtained from studies performed at Integrated DNA Technologies (IDT) and were published in 2011 (Owczarzy *et al.*, 2011); the resulting new algorithms enable the accurate calculation of the T_m of complex LNA-modified oligonucleotides of this type and can be accessed at: http://www.idtdna.com/analyzer/Applications/OligoAnalyzer/ or http://biophysics.idtdna.com.

A SNP in the human *SMAD7* gene (NM_005904) was employed here as a model site to demonstrate these principles. This SNP comprises a C/T variation (rs4939827) and the C-SNP is associated with increased risk of colorectal carcinoma (Tenesa *et al.*, 2008). Genomic sequence at this SNP site is shown in Fig. 8.2A. Probes were designed using existing DNA NN parameters and the new LNA match/mismatch parameters described above, keeping the T_m for the perfect match around 65°C. Calculations were made for a reaction mix having a probe concentration of 250 nM, 50 mM KCl, 3.0 mM $MgCl_2$, and 0.8 mM dNTPs (equivalent to a final free Mg^{++} concentration of 2.2 mM). Unmodified probes were designed as a control and modified probes were designed using the LNA 'triplet pattern' at the SNP site. Additional LNA residues were placed in flanking sequence as needed to allow the probes to be shortened from 20–22 bases (for the unmodified probes) to 14 bases; final LNA content ranged from three to four bases. Sequences of the probes designed are shown in Fig. 8.2B. Although not necessary for the sequences studied here, note that probes ending in a 5′-G base should have an additional non-complementary base inserted between the G and the reporter dye, to reduce the impact of G-quenching (which is particularly problematic for fluorescein) (Kaji *et al.*, 2009). Sequences of the SMAD7 probes are shown in Table 8.1. T_m values for the mismatch alleles were calculated using the new thermodynamic parameters and are also shown in Table 8.1. The predicted T_m values for the mismatch targets ranged from 50.9–55.5°C for the LNA-modified probes and resulted in a ΔT_m (match minus mismatch) of 10.0–15.7°C. In contrast, the unmodified probes showed predicted T_m's for the mismatch target ranging from 59.6–64.6°C, with a ΔT_m for match minus mismatch of only 1.5–5.6°C. Accuracy of the new T_m algorithms employed was tested by performing melting curve analysis on the SMAD7 probes. Experimental T_m values were obtained for both match and mismatch targets.

Melting experiments were conducted on a Beckman DU650 spectrophotometer (Beckman-Coulter) with a Micro T_m Analysis accessory, a Beckman High Performance Peltier Controller (to regulate the temperature), and 1 cm path-length cuvettes. UV-absorbance values at 268 nm wavelength were recorded using a PC interfaced to the spectrophotometer and temperature was increased at 0.1 degree increments. Pre-annealed duplexes were melted at 2 μM concentration in 10 mM Tris pH 8.3 (at 25°C), 50 mM KCl, and 2.2 mM $MgCl_2$ (equivalent to a PCR reaction buffer containing 3.0 mM $MgCl_2$ with 0.8 mM dNTPs). The melting profiles were analysed using methods that have been previously described (Owczarzy, 2005). Results are shown in Table 8.1. Experimental melting data were obtained using 2 μM duplex, which will raise T_m by 2–4°C compared

A) *SMAD7* SNP locus, rs4939827

```
GGGGCTGTCTCTAATCCACCATGCTCACAGCCTCATCCAAAAGAGGAAA(C/T)AG
GACCCCAGAGCTCCCTCAGACTCCTCAGGAAACACAGACAATGCTGGGGTTTAGAG

rs4939827 For
GGCTGTCTCTAATCCACCAT

rs4939827 Rev
CCCAGCATTGTCTGTGTTTC
```

B) Probe Alignments

```
5'         FAM-AGGAAACAGGACCC-FQ              3'
5'    FAM-AAAGAGGAAACAGGACCCCA-FQ             3'

5' CATCCAAAAGAGGAAACAGGACCCCAGAGCTCCCT 3'
3' GTAGGTTTTCTCCTTTGTCCTGGGGTCTCGAGGGA 5'

3'            MGB-CTTTGTCCTGGGGT-FAM          5'


5'         HEX-AGGAAATAGGACCC-FQ              3'
5'     HEX-AAGAGGAAATAGGACCCCAGAG-FQ          3'

5' CATCCAAAAGAGGAAATAGGACCCCAGAGCTCCCT 3'
3' GTAGGTTTTCTCCTTTATCCTGGGGTCTCGAGGGA 5'

3'            MGB-CTTTATCCTGGGGT-VIC          5'
```

Figure 8.2 Assays for the C/T SNP rs4939827 in the human *SMAD7* gene. (A) Genomic sequence of the region surrounding the C/T SNP site is shown. For and Rev primer sequences are shown and their binding sites are underlined. (B) Both strands of the genomic sequence of the 'C' and 'T' alleles are shown aligned with the different probe sequences that were tested. The SNP site is indicated in larger font. DNA bases are uppercase. LNA nucleotides are underlined. Reporter dyes employed (FAM, HEX, VIC) are indicated. FQ indicates the Iowa Black FQ dark quencher. MGB indicates a minor groove binder modification group linked to an Eclipse dark quencher.

Table 8.1 Predicted versus measured T_m values for SMAD7 probes

	Predicted T_m (°C) (0.25μM)			Measured T_m (°C) (2.0μM)		
Probe sequence	Match	Mismatch	ΔT_m	Match	Mismatch	ΔT_m
5′ FAM-AGGAAACAGGACCC FQ 3′	66.6	50.9	15.7	68.3	54.1	14.2
5′ FAM-AAAGAGGAAACAGGACCCCA-FQ 3′	65.2	59.6	5.6	65.5	56.4	9.1
3′-MGB-CTTTGTCCTGGGGT-FAM 5′	66.0	NA	NA	65.7	56.5	9.2
5′-HEX-AGGAAATAGGACCC-FQ 3′	65.5	55.5	10.0	67.6	59.1	8.5
5′-HEX-AAGAGGAAATAGGACCCCAGAG-FQ 3′	66.1	64.6	1.5	65.1	63.0	2.1
3′-MGB-CTTTATCCTGGGGT-FAM 5′	66.0	NA	NA	67.4	57.7	9.7

Probes specific for the 'C' allele were labelled with FAM at the 5′-end. Probes specific for the 'T' allele were labelled with HEX or VIC at the 5′-end, as indicated. The SNP site is indicated in larger, bold font. LNA nucleotides are underlined. FQ indicates the Iowa Black FQ dark quencher. MGB indicates a minor groove binder modification group linked to an Eclipse dark quencher. The MGB probes were designed to the opposite strand and are accordingly shown in reverse orientation. Note that predicted T_m values generated using a probe concentration of 0.25μM were used for probe design; measured T_m values are done at 2μM duplex concentration to accommodate needs for spectrometric methods. NA=mismatch predictions not available.

with the 0.25 μM duplex concentration used in qPCR (0.25 μM was used for the original T_m predictions when designing the probes). Measured T_m values were found to all be within a few degrees of the expected values for three of the four probes, well within expected experimental error for these methods. The unmodified 'C'-SNP probe, however, showed a lower than predicted T_m against the mismatched target, resulting in a ΔT_m of over 9°C.

The SMAD7 probes and primers shown in Fig. 8.2 and Table 8.1 were functionally tested in real-time qPCR. Reactions were run on a BioRad CFX384 in 384 well plates using 10 μl reaction volumes. Reactions consisted of either 1X Taqman Genotyping Master Mix (Life Technologies) or a homemade Immolase Master Mix (0.4 U Immolase DNA polymerase (Meridian Bioscience, Cincinnati, Ohio), 1 μl of 10× polymerase buffer (160 mM $(NH_4)_2SO_4$, 670 mM Tris-HCl pH 8.3, 0.1% Tween-20), 3 mM $MgCl_2$, 800 μM dNTPs), 250 nM of each probe, 500 nM of the forward and reverse primers, and 2 ng human genomic DNA. Cycling conditions included an initial 10 minutes soak at 95°C, followed by 40 cycles of 15 seconds at 95°C and 60 seconds at 60°C. Human genomic DNAs that were homozygous C/C or T/T or heterozygous C/T at SNP locus rs4939827 were obtained from the Coriell Institute for Medical Research Cell Repository (http://ccr.coriell.org/, GM07019 (T/T), GM07029 (C/T), GM07034 (C/C)). 'C'-SNP results were collected in the FAM fluorescence channel and 'T'-SNP results were collected in the HEX fluorescence channel. Amplification plots are shown in Fig. 8.3. The LNA-modified probes (Fig. 8.3a) resulted in unambiguous base

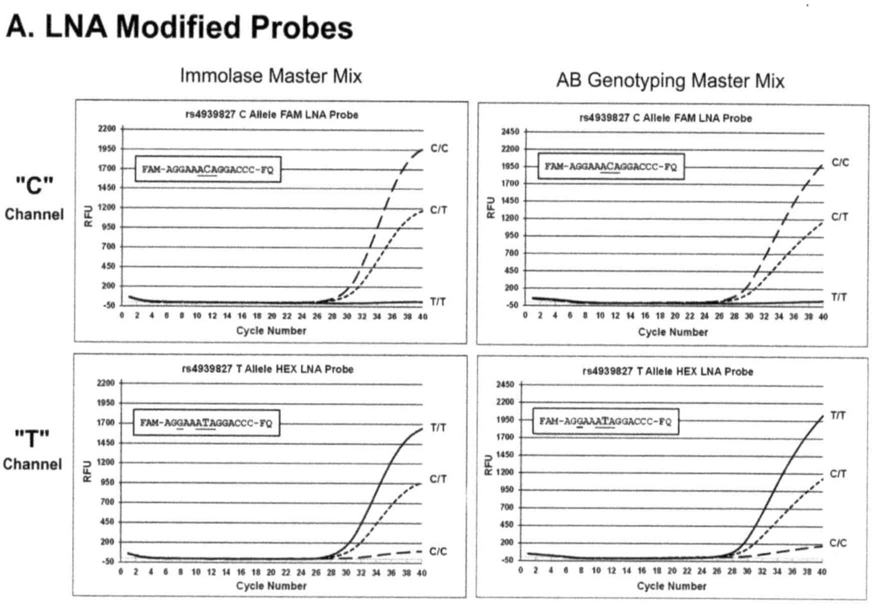

Figure 8.3 5′-Nuclease assay amplification plots for different probe designs at a SNP locus. The 5′-nuclease qPCR assay was performed using 2ng of human genomic DNA homozygous or heterozygous at the SMAD7 rs4939827 locus using probes of different design and two different buffer systems, the Immolase Master Mix or the Applied Biosystems (AB) Genotyping Master Mix. Reactions were run in multiplex format with probes specific for both alleles present in the reaction and were monitored in separate fluorescent channels. Probe

B. Std Unmodified Probes

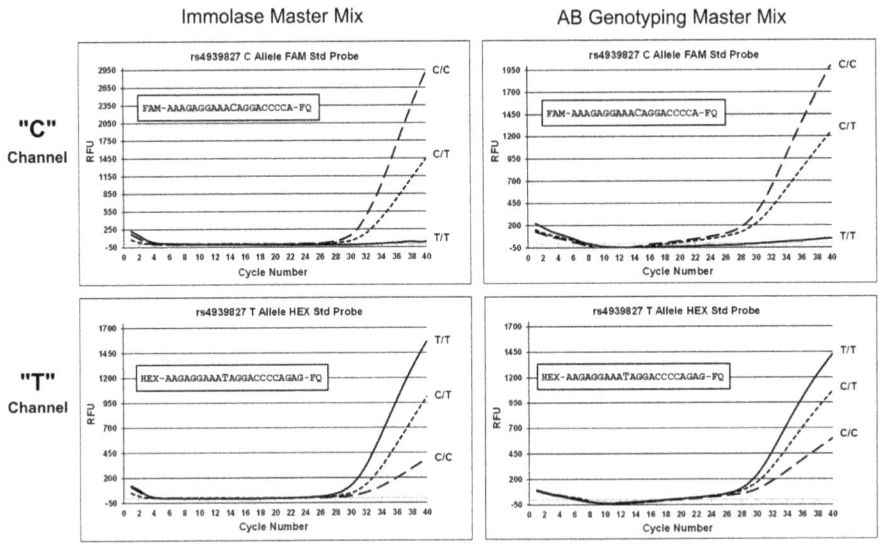

C. MGB Modified Probes

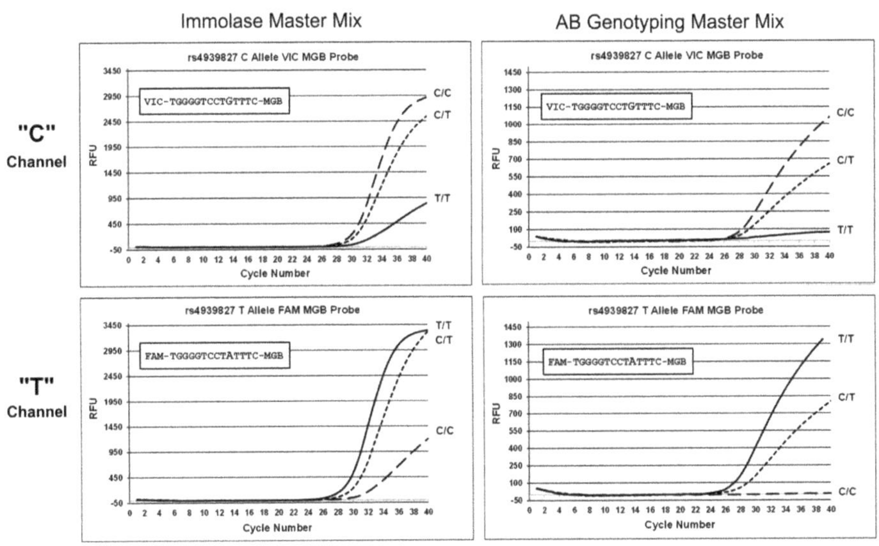

sequences are inset at the upper left corner of the amplification plots. Reporter dyes FAM, HEX, and VIC are indicated. FQ=Iowa Black FQ dark quencher. MGB=minor groove binder with Eclipse dark quencher. LNA nucleotides are underlined and the position of the SNP base is indicated in larger font. (A) Reactions using LNA-modified dual-labelled probes. (B) Reactions using unmodified DNA dual-labelled probes. (C) Reactions using MGB-modified dual-labelled probes.

discrimination, with the 'C'-probes only detecting amplification with C/C and C/T DNAs and the 'T'-probes only detecting amplification with T/T and C/T DNAs. Similar results were obtained in both of the master mix buffer systems. Importantly, these results were obtained without the need for any empiric optimization and probes were simply designed using the method described above.

The unmodified DNA dual-labelled probes were also tested and results are shown in Fig. 8.3b. The unmodified 'C'-SNP probe gave good discrimination but the unmodified 'T'-SNP probe gave poor discrimination. These results are consistent with the experimentally measured T_m and ΔT_m values obtained for match and mismatch pairs for the unmodified probes (Table 8.1). Interestingly, predictions made using nearest neighbour parameters suggested that both assays would show poor SNP discrimination, whereas in actual practice one worked well and the other failed. Nevertheless, both probes must perform well together, and genotypes could not be reliably called using the unmodified probe assay. Thus, use of ΔT_m-enhancing LNA nucleotides improved the ability of qPCR to distinguish between alleles at this locus, converting a failed assay into a successful assay.

Three additional modified bases were reported by Epoch Biosciences that can be used to improve performance of dual-labelled probes (and primers) in qPCR. The Super A™ and Super T™ bases represent modifications of adenosine and thymidine that increase stability when base paired with unmodified A or T. Substitution of these bases in a probe sequence will increase T_m, particularly for AT-rich sequences (Afonina *et al.*, 2002a; Dempcy *et al.*, 2003). Nearest neighbour parameters have not been reported for these bases, however, so T_m predictions cannot be accurately made to assist with design. The third base, Super G™ (8-aza-7-deazaguanine, or PPG) does not increase T_m but does disrupt the Hoogsteen-bonding potential for the N-7 position of the guanine ring (Kutyavin *et al.*, 2002). Hoogsteen-bonding is required for formation of G-tetrads (or G-quadruplexes), a structure that is naturally found in telomeres and can also cause unwanted aggregation of synthetic oligonucleotides (Simonsson, 2001; Reddy and Hardin, 2003). Substituting Super G™ in G-runs (sequences having four or more sequential guanine residues) can prevent G-tetrad formation and eliminate the various complications that arise from tetrad formation.

Another strategy to increase T_m, shorten probe length, and improve specificity is use of non-nucleoside T_m-enhancing modifying groups. Compounds have been known for many years that will bind DNA duplexes with high affinity, such as intercalators and minor groove binders (Bailly and Henichart, 1991). A novel compound of this class was developed by Epoch Biosciences in 1996 that binds the minor groove of a DNA duplex and significantly stabilizes hybridization of a short DNA oligonucleotide to its complementary sequence. This group is a complex molecule, 1,2-dihydro-(3*H*)-pyrrolo[3,2-*e*]indole-7-carboxylate ($CDPI_3$), which is more conveniently called 'MGB' (Kutyavin *et al.*, 1997). The thermodynamic properties of the MGB vary with sequence, length, and chemistry. When attached to the end of a short A:T 8mer (poly-dT_8:poly-dA_8), the MGB leads to greater than a 40°C increase in T_m. The increase in binding affinity is greater for AT-rich sequences than for GC-rich sequences. T_m changes for phosphorothioate DNA have similar magnitude as unmodified DNA; however, the impact is much less when the MGB is attached to a 2′OMe-modified oligonucleotide sequence, possibly because geometry of the minor groove is different in RNA (or RNA:DNA heteroduplexes) than DNA.

The first application proposed for this new chemistry was modification of the 5′-end of PCR primers to enable use of very short oligos (eight to ten bases long) to prime DNA synthesis (Afonina *et al.*, 1997). It was proposed that this would have particular utility in designing primers which were specific for short blocks of highly conserved sequence (for example, in amplifying a broad range of viral serotypes, it is necessary to place primers in very limited areas where sequence is sufficiently conserved between species). Importantly, the MGB group lies in the minor groove of the DNA helix and occupies space corresponding to around six bases, so if this group is positioned at the 5′-end of a primer it will not overlap or interfere with the 3′-terminus of an oligonucleotide that is even as short as eight to ten bases. It was quickly appreciated that the MGB group would also have significant utility if used for dual-labeled probes in the 5′-nuclease assay. In this format, the MGB group is placed at the 3′-end of the probe with a dark quencher positioned between the MGB and the 3′-DNA residue and a reporter fluorophore is placed at the 5′-end. Use of the MGB was demonstrated to permit shortening some probe sequences from 27 bases to as little as 12 bases in length while keeping a constant 65°C T_m, thanks to a 5–20°C contribution from the MGB (Kutyavin *et al.*, 2000, 2003). While probes of this design work well for general gene quantification purposes, the real value becomes apparent when performing allelic discrimination assays. It is thought that the group binds the 3′-end of the probe/target duplex, and, if a SNP site is positioned within the terminal roughly eight bases from the 3′-end (within or near the MGB binding domain), the resulting ΔT_m for match versus mismatch is often in the 10–20°C range, making for very robust SNP assays.

Use of the MGB modification at the 3′-end of a probe for 5′-hydrolysis format qPCR assays was exclusively licensed from Epoch Bioscience by Applied Biosystems and is the basis for the current line of TaqMan® probes. Probes of this design are in widespread use today for all qPCR applications; this format is most often used for low to mid-volume assay throughput needs (De la Vega *et al.*, 2005). The PrimerExpress™ assay design software from Applied Biosystems will provide an estimate of T_m for MGB-modified probes. However, no estimates are given for the ΔT_m when paired with mismatched targets. Applied Biosystems (now Life Technologies) also offers a probe design service via their website and has predesigned or validated probes available for many targets. Predesigned MGB-TaqMan assays are available for most genes and many SNPs from Applied Biosystems and are easily searched/ordered using tools on their website. However, these probes are only provided as fully formulated 'assays', where all probes and the primers come pre-mixed in a single tube, making it difficult to study the performance of individual probes. Further, the sequences of the probes are not provided when ordered via this route. Therefore, instead of ordering pre-designed assays, MGB-Taqman probes for the SMAD7 SNP locus were designed using the PrimerExpress™ software and these sequences were obtained from Applied Biosystems as a custom synthesis. Experimental T_m values were determined for the MGB probes using identical methods and buffers as was previously used for the LNA probes. Results are reported in Table 8.1. The ΔT_m comparing match versus mismatch hybridization was 9–10°C, a value grossly similar to that obtained using the LNA modified probes. The SMAD7 MGB-probes were used in allelic discrimination qPCR assays, using the same methods as outlined previously for the unmodified and LNA-modified probes. The results are shown in Fig. 8.3c. Performance of the MGB probes was excellent and complete discrimination of the C/C and T/T genotypes was achieved when using Applied Biosytems genotyping mastermix. However, performance of the MGB probes was very different when reactions were done using a different buffer (the Immolase Master Mix). Anneal/extension temperatures were varied with little benefit (not shown).

It is not clear why the 'C' allele probe showed good functional discrimination while the 'T' allele probe did not, even though both showed similar ΔT_m values (in a buffer having similar ionic composition to the Immolase Master Mix). These results illustrate the important point that buffer composition is a critical variable when performing demanding applications such as SNP discrimination in qPCR. Some commercial reagents contain proprietary components and do not reveal ionic composition, making it difficult to predict performance. Empiric comparison of different reaction mixes may be needed to optimize performance.

The MGB group can also be added to the 5′-end of the probe. In this configuration, 5′-nuclease activity of the polymerase is blocked and the probe instead functions in a hybridization format with fluorescence signal resulting from the reduced quenching that occurs upon hybridization (Afonina *et al.*, 2002a,b). In single stranded form, the end-to-end distance in the random coil conformation is short and the reporter dye is effectively quenched. When hybridized to target, the duplex exhibits more rigid-rod conformation than single strands and the end-to-end length increases, reducing quenching. Fluorescence emission therefore increases from the reporter dye in the absence of probe degradation (Livak *et al.*, 1995).

Other chemical groups can be used to increase T_m and improve probe performance. Behr and colleagues from the University of Strasbourg developed methods to attach cationic spermine groups to oligonucleotides using a novel spermine phosphoramidite. Multiple spermine groups can be joined in serial fashion, creating a polycation. Attachment of this positively charged polymer to one of the ends of an oligonucleotide helps overcome the electrostatic repulsion between two annealing anionic DNA strands, increasing binding affinity (i.e. increasing T_m) (Noir *et al.*, 2008). Methods employing this technology in molecular biology assays have been commercialized by PolyPlus-Transfection, Inc. This modification has been called 'ZNA', for 'zip nucleic acids', when used in synthetic oligonucleotides. Like the effects described previously for the MGB group, addition of a 5′-ZNA modification permits short primers to be used in PCR (Moreau *et al.*, 2009). Similarly, attachment to the 3′-end of a dual-labelled probe is compatible with use in the 5′-nuclease assay and increases probe T_m, enabling use of shorter probes with better performance in allelic discrimination assays than unmodified probes (Paris *et al.*, 2010). Unlike the MGB group, which is usually positioned at the very 3′-end, ZNA probes position the quencher at the 3′-end and typically four spermine groups are placed between the quencher and the 3′-most base of the nucleic acid sequence. This poly-spermine tail flips over and binds the DNA nucleotides near the 3′-end of the probe, which fortuitously also results in the quencher being positioned closer to the reporter fluorophore (at the 5′-end). Forcing a shorter dye-quencher distance increases quenching, lowering background fluorescence and further improves probe performance.

Another variant of a linear hybridization probe for SNP discrimination was described by Yamane (Wakunaga Pharmaceutical Co., Japan) and were called 'MagiProbes' (Yamane, 2002). In this case, a reporter fluorophore and a pyrene quencher were positioned centrally in the probe near the SNP site in adjacent or nearly adjacent positions. When hybridized to a perfect match target, the pyrene group intercalates into the DNA duplex resulting in loss of quencher function and fluorescence emission seen from the reporter. Conversely, when hybridized to a mismatch target, the pyrene group remains external to the duplex and continues to quench the reporter.

Hairpin dual-labelled probes

Hairpin fluorescent dual-labelled probes, called 'Molecular Beacons', were first described by Tyagi and Kramer (Public Health Research Institute, New York, NY, USA) in 1996 (Tyagi and Kramer, 1996). These probes have a reporter fluorophore on one end and a quencher on the other end; a target hybridizing domain occupies the central loop portion of the probe sequence and two short sequences are placed on either end which are self-complementary (and non-complementary to target). In the absence of target, the Molecular Beacon forms a hairpin, which forces the reporter and quencher into close proximity. As a result, quenching efficiency is very high and probes of this design have a very good signal-to-noise ratio. Further, hairpin formation competes with target binding and improves the thermodynamics of match/mismatch discrimination, increasing the relative ΔT_m. With careful adjustment of the hairpin stem length, good SNP discrimination can sometimes be achieved without additional chemical modification (Tyagi *et al.*, 1998; Bonnet *et al.*, 1999; Tsourkas *et al.*, 2003b). As with linear probes, LNA nucleotides can be included within the target binding domain which will allow for use of shorter probe sequences, further improving the ΔT_m. Non-base modifiers (like MGB or ZNA), however, cannot easily be used in Molecular Beacons as these groups would primarily increase stability of the stem and would be detrimental to target binding. Synthesizing probes with 2′-O-Me-RNA in place of DNA increases nuclease stability and permits use of Molecular Beacons in live cells to directly visualize mRNA by fluorescence microscopy (Bratu *et al.*, 2003); note that shorter stem domains must be used with this chemistry as the T_m of the stem is much higher for 2′O-Me–2′O-Me duplexes (Tsourkas *et al.*, 2003a). The use of a 2′-O-Me sugar backbone, however, does not confer any benefit for use in qPCR applications. A version of Molecular Beacons called 'Scorpions' was described by Brown and colleagues (University of Southampton, UK) where the probe is tethered to a PCR primer and configured so that target binding can occur at specific points in the reaction cycle (Thelwell *et al.*, 2000). This design showed good sensitivity and specificity compared with traditional probes and in particular worked well with fast cycling times.

Enzymatic interrogation of SNP identity

The SNP discrimination methods discussed above all rely upon differential hybridization of a probe oligonucleotide with different target nucleic acids to distinguish base identity at the site of interest. It is also possible to perform SNP discrimination using probes with high affinity that bind well to both match and mismatch targets that include modification(s) which permit the probes to interact with an enzyme that is capable of discriminating between paired and unpaired bases at the SNP site. The best example of this class of SNP assay uses RNase H enzymes to selectively cleave probes. RNases H are a family of enzymes that specifically cleave at RNA bases in an RNA:DNA heteroduplex. The enzymes do not cleave single stranded RNA or DNA. Two major classes of RNase H enzymes exist, RNase H1 and RNase H2 (Ohtani *et al.*, 1999). RNase H1 enzymes require a stretch of at least four consecutive RNA bases to show maximal enzyme activity and cleave between RNA residues (Hogrefe *et al.*, 1990). In contrast, RNase H2 enzymes will cleave a substrate having only a single RNA residue in an otherwise DNA sequence and cleavage occurs at the 5′-side of the RNA base at what is actually a DNA linkage (Eder and Walder, 1991; Eder *et al.*, 1993). Cleavage patterns for RNase H1 and RNase H2 enzymes are illustrated in Fig. 8.4. RNase H enzymes have been isolated from a variety of prokaryotic species that are thermostable

(Kanaya, 2001), some of which can be used at elevated temperatures in molecular biology assays or even be included in PCR reactions.

Peter Duck (Meiogenics, MD, USA) and Joseph Walder (University of Iowa and Integrated DNA Technologies) described 20 years ago a method to use chimeric RNA:DNA oligonucleotides as probes to detect complementary sequences with isothermal signal amplification. Probes were made with DNA bases at both ends and contained a stretch of at least four RNA residues in the central region. When hybridized to a DNA target, RNase H1 will cleave within the RNA domain, cutting the probe in half. The shortened probe dissociates from the target, permitting a new probe molecule to hybridize and repeat the cycle (isothermal cycling); cleavage products accumulate and can be detected by a variety of methods (Duck *et al.*, 1990; Walder and Walder, 1995). Harvey and Han (Eximus Biotech, Maryland) adapted this method to a fluorescence format by adding a dye and a quencher to the chimeric probe molecule on opposite sides of the cleavable RNA domain, so that the probe is dark in the intact state and bright after RNase H cleavage (Harvey *et al.*, 2004). Cleavage by RNase H1 is less efficient if a mismatch is present in the RNA cleavage domain between the probe and the DNA target. This method can be linked to PCR by using a thermostable RNase H1 enzyme and the resulting assay can be used to discriminate SNPs (Harvey *et al.*, 2008). In this case, a commercially available RNase H1 enzyme from a *Thermophilus* sp. was employed (Epicenter, Madison, WI).

While this method works, more than one cleavable linkage exists within the 4-RNA central domain and it is not entirely clear where it is best to position the SNP when designing the assay. If an RNase H2 enzyme is employed, the same assay strategy can be used with a probe that has only a single RNA base (i.e. a single cleavable linkage). Hou and colleagues (Shanghai Jiaotong University, China) characterized the ability of the RNase H2 enzyme from *Chlamydia pneumoniae* to discriminate mismatches when a probe containing a single RNA residue was hybridized to targets having mismatches at different positions and

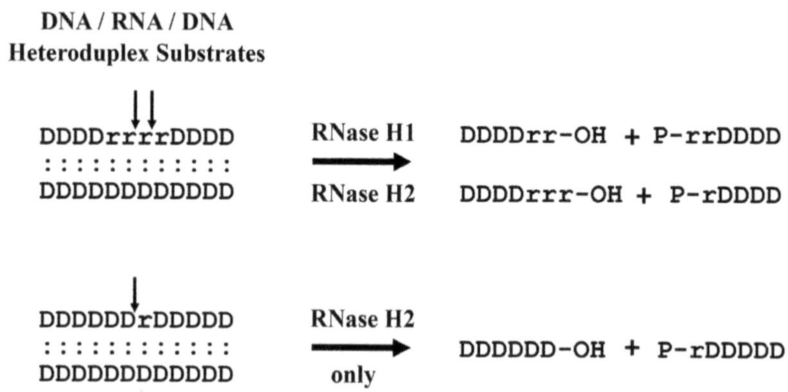

Figure 8.4 RNase H cleavage patterns. Two classes of RNA:DNA heteroduplex substrates are shown. The top row shows a substrate having four consecutive RNA bases and the bottom row shows a substrate having a single RNA base. Arrows indicate expected sites for RNase H cleavage for each substrate. Expected products following cleavage by RNase H1 and/or RNase H2 are shown on the right. D=DNA, r=RNA, OH=3′-hydroxyl, P=5′-phosphate.

found that discrimination was best when the SNP was placed at the RNA base; the bases on either side of the RNA base (at the '+1' and '–1' positions) also showed high selectivity (Hou *et al.*, 2007a). They adapted this assay for use with Molecular Beacons, positioning the RNA base centrally within the target binding loop domain of the probe. Since the *C.p.* RNase H2 enzyme is not thermostable, the Molecular Beacon hybridization/cleavage steps were run post-amplification as an end-point assay (Hou *et al.*, 2007b,c). The same group later improved this method by using a thermostable RNase H2 enzyme from *Thermus thermophilus*, which permitted a homogeneous assay format where the probe and cleaving enzyme were present and functioned in real time during PCR (Liu *et al.*, 2010).

SNP discrimination using biased amplification

PCR-based methods that discriminate between related sequences using biased amplification lead to differential accumulation of product with one allele being preferentially amplified over the other allele. If the different alleles are present at roughly similar levels in the test nucleic acid sample, SNP discrimination methods that use biased or unbiased amplification can perform equally well. However, if one allele is under-represented in the population, unbiased amplification is often only capable of detecting the more prevalent species. Detection of a rare allele in the presence of a high background of wild-type sequence requires use of a method that employs biased amplification, especially if that species is present at levels of 10^{-2} to 10^{-6} or lower. Unlike unbiased amplification, biased amplification normally requires that one of the primers overlaps the SNP site. The system is designed so that the match template supports amplification while the mismatch template does not. Sequence identity is determined by the presence or absence of amplification. Probe-based methods discussed previously usually rely upon differential hybridization between match versus mismatch targets (i.e. thermodynamic discrimination). For biased amplification, primers typically bind to both alleles and discrimination is based upon enzymatic events, e.g. whether or not the oligonucleotide will prime DNA synthesis or not. Most of the methods to perform biased amplification with enzymatic interrogation employ chemically modified oligos, and the chemical modifications enable the process.

Allele-specific PCR

The simplest system to perform allele-specific PCR (ASPCR) employs primers where one primer overlaps the SNP, which is positioned as the 3′-end base of the primer. Differences in the priming efficiency on match versus mismatch templates determine the level of selectivity. In practice, actual discrimination can vary from very low to fairly high and depends largely on the sequence context and precise bases present as the mismatch pair at the SNP site. Also, different polymerases show different degrees of selectivity using the same primers and template (Mendelman *et al.*, 1990). A survey was performed at IDT where we examined the relative discrimination seen in a synthetic amplicon using allele-specific PCR. Four unmodified primers were made having the four possible bases at the 3′-position; four templates were made having the four possible bases at the SNP site. All 16 pairwise combinations were tested in a SYBR® Green format, and reactions were tested individually (not in a multiplex format). The ΔCq (the cycle number at which signal was first detected for the mismatch minus the cycle number at which signal was first detected for the match) ranged from 1.2 to 10.7 cycles, with an average ΔCq of 5.4 for the 12 mismatch pairings (Dobosy *et al.*, 2011). In other words, sometimes this method works well and sometimes it does not. Careful optimization of primer design, reaction temperature, buffer composition, and enzyme choice can improve results. Traditionally, these reactions are performed individually in separate tubes, as there is no simple way to distinguish between the two different allele products when using unmodified primers. Using modified dye labelled primers (such as the hairpin fluorescence-quenched Amplifluor™ system), it is possible to perform assays for both alleles in a single tube. Interestingly, results are often significantly improved in multiplex format and it appears that kinetic competition between alleles increases specificity and reliability (Myakishev *et al.*, 2001).

Use of chemically modified bases can further improve the specificity of ASPCR. In particular, placing an LNA nucleotide at the 3′-end of the allele-specific primer will decrease priming on mismatched templates without compromising efficiency of PCR on matched templates. Usually only a single LNA modification is used on each primer, which is placed at the 3′-end with the LNA nucleotide pairing with the SNP (Latorra *et al.*, 2003; Nakitandwe *et al.*, 2007). This strategy works best when using a DNA polymerase that does not have 3′–5′ exonuclease activity. If a high fidelity DNA polymerase is used that has proofreading function (3′–5 exonuclease activity), then the enzyme can remove the terminal 3′-base from the primer (error correction) and amplification will proceed whether a mismatch is present or not. In this scenario, it is preferred to position the LNA nucleotide at the 'N-2' position, the base adjacent to the 3′-end (Di Giusto and King, 2004). In this design, the LNA modification confers structure to the primer that makes this last linkage relatively nuclease resistant so that the 3′-base is retained in the primer and can contribute to specificity as intended.

The 'amplification refractory mutations system' (ARMS) and 'mismatch amplification mutation assay' (MAMA) are methods that introduce an artificial second mismatch into allele-specific PCR primers as a way to increase selectivity (Newton *et al.*, 1989; Cha *et al.*, 1992). As with standard ASPCR, the SNP under interrogation is positioned at the 3′end of the primer. A second mismatch is introduced in the primer directly adjacent to the 3′-base, resulting in the 'match' primer having a single mismatch with target (the artificial mismatch) and the 'mismatch' primer having two mismatches with the target (the artificial mismatch plus the SNP). At this position, the single internal mismatch is permissive and the 'match' primer does not show any significant decrease in priming efficiency. However, the double mismatch is very inefficient when priming on the mismatch target and ΔCqs of ten or more cycles can often be achieved, even for base mismatches that are unfavourable for standard ASPCR. The primers used in the original versions of these assays are unmodified. In many cases, the specificity of ARMS or MAMA primers can be further enhanced by placing an LNA modification at the SNP site located at 3′ end.

Biased amplification using blocked primers

Another approach to perform biased amplification employs primers that are blocked at the 3′-end which cannot prime DNA synthesis until the block is removed. The blocked 3′-OH group can be removed by cleavage using either an endonuclease (internal cleavage) or a 3′-exonuclease; cleavage activates the primer, allowing PCR to proceed. To function as a SNP assay, the 3′-deblocking step must be selective and only occur when a perfect match is present between the primer and the target at the cleavage site. Specificity of the cleavage process is critical to success.

One successful version of this assay strategy was developed by Liu and Sommer (City of Hope National Medical Center, Duarte, CA, USA) and is called pyrophosphorolysis-activated polymerization, or PAP (Liu and Sommer, 2000, 2002). Like traditional ASPCR primers, PAP primers have the SNP site positioned at the 3′-end of the primer sequence; however, PAP primers are also blocked at the 3′-end. The 3′-end block is removed by the polymerase enzyme via pyrophosphorolysis, which is essentially a reverse polymerization reaction that occurs in the presence of sufficiently high levels of inorganic pyrophosphate (Deutscher and Kornberg, 1969). High levels of inorganic pyrophosphate inhibit DNA polymerases, so careful buffer composition optimization must be done. PAP reactions also employ lower levels of deoxynucleotide triphosphates than are typically employed in PCR (e.g. 50 μM of each dNTP with 300 mM sodium pyrophosphate). Pyrophosphorolysis occurs only when the terminal base of the primer is base-paired with a complementary base on the target; a mismatch base will not function as a substrate for this enzymatic reaction. A variant of the PAP method uses both blocked forward and blocked reverse primers, called Bi-PAP (bidirectional pyrophosphorolysis-activated polymerization). The SNP is positioned at the 3′-end of both primers. As a result, the primers overlap by one base, but this short overlap is not sufficient to allow primer-dimer amplification (the primers are blocked – after deblocking, the primers no longer overlap). For amplification to occur, successful PAP must occur on each primer independently, greatly increasing specificity of the final assay. Bi-PAP has been shown to be capable of rare allele detection at levels as low as 10^{-7}–10^{-8} (Shi *et al.*, 2007). A wide variety of 3′-blocking strategies have been tested and the best results are obtained when the 3′-base is a dideoxynucleotide (ddN). Widespread use of PAP therefore requires the availability of primers with a 3′-terminal dideoxy base for each of the four bases; ddA and ddC synthesis reagents are available as commercial solid supports, so synthesis of primers ending in an A or C base is straightforward. Unfortunately, ddT and ddG supports are not available and primers ending in a T or G base have to be made using expensive reverse synthesis or by enzymatic methods, limiting the general utility of this approach.

Biased amplification using RNase H and blocked-cleavable primers

Another method in this class of assays employs blocked primers that have a single RNA base positioned near the 3′-end of the primer. Following hybridization to a complementary sequence, the terminal blocking group is removed by action of a thermostable RNase H2 enzyme (in contrast with PAP, this time the 'deblocking enzyme' acts via internal endonuclease cleavage). RNA-containing blocked-cleavable primers can be used in PCR and provide improved specificity when compared with traditional priming methods, resulting in lower background, reduced mispriming, and elimination of primer-dimer formation (Walder *et al.*, 2009; Dobosy *et al.*, 2011). PCR done using blocked-cleavable primers has been called 'rhPCR'. Many of the features of RNase H enzymes were reviewed in the previous section. It is important to note that the cleavage reaction requires that the primer must be duplexed with a DNA target. RNase H2 will not degrade single-stranded nucleic acids; it requires an RNA–DNA heteroduplex substrate. When a blocked primer of this design hybridizes to the target, the substrate has a single RNA base embedded in an otherwise DNA sequence. RNase H2 cleaves at the 5′-side of the RNA base at the DNA linkage, leaving a 3′-hydroxyl at this end and a cleavage product which is an activated, functional primer. The second cleavage product contains a 5′-phosphate, the single RNA base, a few DNA bases, and the 3′-blocker; this is an inactive fragment that is short, dissociates from the target, and is 'disposable'. Importantly, the cleavage reaction is sensitive to correct base pairing such that primer deblocking is inhibited by the presence of a base mismatch in the vicinity of the cleavage site. The basic scheme for performing rhPCR with blocked-cleavable primers and RNase H2 is outlined in Fig. 8.5.

Ideally the RNase H2 enzyme employed in rhPCR will be very thermostable. A number of suitable enzymes have been isolated from archaeal sources, including *Pyrococcus furiosus* (Sato *et al.*, 2003), *Methanococcus jannaschii* (Lai *et al.*, 2003), *Thermococcus kodakaraensis* (Haruki *et al.*, 1998; Mukaiyama *et al.*, 2004), and *Pyrococcus abyssi* (Walder *et al.*, 2009; Le Laz *et al.*, 2010; Dobosy *et al.*, 2011). These enzymes require divalent cations for catalysis and can employ either Mg^{++} ions (usually in the 2–5 mM range) or Mn^{++} ions (usually in the 0.3–1.5 mM range). Turnover rate is fast enough that primer annealing and cleavage can be done in real time during PCR using the same primer anneal/extend times that are usually employed (30–60 seconds dwell time) with no loss of priming efficiency. Cycling should be performed as a two-step PCR (e.g. 60°C–95°C steps), using the same temperature for both primer annealing and extension reactions. A three-step PCR (e.g. 60°C–72°C–95°C steps) will be inefficient as primer annealing and deblocking cannot occur during the 72°C extension phase.

For a primer/target heteroduplex to be an efficient substrate for RNase H2 cleavage, a minimum amount of nucleic acid must be present in duplex form both 5′ and 3′ to the RNA base. These requirements have been studied for *Pyrococcus abyssi* RNase H2 and 8–10 DNA bases are needed 5′-to the cleavable cleavage site and four or more DNA bases are needed 3′-to the cleavage site for maximal activity. A variety of methods can be used to block the 3′-end of the primer, including dideoxy nucleotides or non-base modifiers placed attached to the 3′-end of the terminal DNA residue. A C3 spacer group (1,3-propanediol) is inexpensive and works well as a blocking group (Walder *et al.*, 2009; Dobosy *et al.*, 2011). Using these design parameters, a typical rhPCR primer will have a 'DNA primer domain' towards the 5′-end (20–25 bases), a single RNA base, four DNA bases on the 3′-side of the RNA, and a terminal 3′-C3 spacer (or other blocking group). This basic 'first generation' rhPCR design primer will prevent primer-dimer formation and improve specificity. In addition to primer hybridization, mismatches at or near the RNA cleavage site will reduce the rate of cleavage and prevent efficient PCR. It is not necessary that the cleavage event be 'all or none' at the

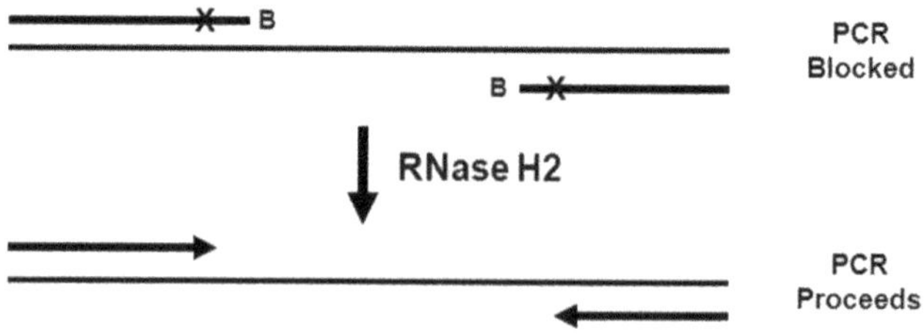

Figure 8.5 rhPCR using blocked-cleavable primers. Blocked primers are aligned against template on the top. X=scissile linkage, which in rhPCR is a single RNA base. B=blocking group, most commonly a C3-spacer (1,3-propanediol). Following cleavage, the RNA base and 3′-nucleic acids are removed and fully active primers remain having an unblocked terminal 3′-hydroxyl group (arrowheads).

RNA base to have a dramatic effect on PCR. If a mismatch reduces primer deblocking from 100% to 50% efficiency, then half of potential PCR priming events never happen and PCR amplification reduces from the standard exponential 2^n (where n = cycle number) to 1.5^n. Over 30 cycles, this represents a 5600-fold reduction in amplification.

The relative ability of a base mismatch to reduce cleavage has been studied for the RNase H2 enzymes from *Thermococcus kodakaraensis* (Liu *et al.*, 2010) and *Pyrococcus abyssi* (Walder *et al.*, 2009; Dobosy *et al.*, 2011). The best sites to place a mismatch (i.e. a SNP) are at positions complementary to the RNA base (position '0') or complementary to the DNA base which is immediately 5′-to the RNA base (position '–1'). RNase H2 cleavage occurs at the linkage between these two bases. From an assay standpoint, however, it is preferable to position the SNP at the RNA base. Since cleavage occurs on the 5′-side of the RNA base, if an unwanted cleavage event does occur in the presence of a mismatch, then the resulting primer will have this base excised and the first base of primer extension by the DNA polymerase will be encoded by the SNP. This ensures that the daughter product will faithfully reproduce the actual sequence present in the target nucleic acid. If the mismatch was positioned at the DNA base (position '–1'), cleavage still occurs immediately 5′-to the RNA base and then the SNP base encoded by the primer remains in the primer and is locked into the extension product, converting the PCR product to the primer sequence.

When the 'first generation' rhPCR primers are used in SNP assays, the observed Δ*Cq*s (mismatch – match) typically range from 5–15 cycles depending on the sequence context and precise base mismatches involved. While it is easy to distinguish the correct genotype present from these results, this level of discrimination is lower than would be expected given the known selectivity of the RNase H2 cleavage event. The basis for this 'lower than expected selectivity' was investigated. Reaction products were cloned from rhPCR done using mismatched templates and sequence was determined. It was unexpectedly found that the amplicons had converted to the primer sequence, which should not have been possible if primer cleavage was only occurring at the 5′-side of the RNA base (the SNP was positioned at this site) (Walder *et al.*, 2009). It appears that a very low frequency (less than 1 in 1000) of cleavage events occur at a non-canonical cleavage position, such that the enzyme cuts on the 3′-side of the RNA base. Cutting at this position results in the primer-encoded SNP base being incorporated in the amplicon, which converts the PCR product to the primer sequence. In subsequent cycles, this product will amplify at 100% efficiency as it is now perfect match to the primer. Thus the frequency of non-canonical cleavage ultimately determines the specificity of rhPCR.

Various patterns of chemical modification were tested in the rhPCR primers to decrease the frequency of non-canonical cleavage on the 3′-side of the RNA base. Substituting non-base groups in the short DNA domain on the 3′-side of the RNA base creates substrates that are less easily cleaved by RNase H2 (such primers require 10- to 100-fold higher enzyme concentrations to achieve reaction efficiency similar to what is seen using a 'first generation' primer); however, the frequency of non-canonical cleavage is reduced and Δ*Cq* values of 10–20+ cycles are easily achieved using this 'second generation' primer design. Currently the favoured 'second generation' primer design incorporates a DNA base, two C3-spacers, and a final DNA base on the 3′-side of the RNA residue. No 3′-blocking group is required with this primer design as the C3-spacers reduce or totally eliminate the ability of the terminal DNA base to prime DNA synthesis. Even if priming did occur, the spacers prevent the primer from functioning as a template for DNA synthesis so second strand synthesis would terminate at the primer, further blocking PCR. This 'second generation' primer design should be employed for allelic discrimination assays; however, the 'first generation' primer design is still preferred for other applications since lower concentrations of enzyme are needed to perform each reaction.

Examples of SNP discrimination using biased amplification

A series of SNP discrimination assays were performed using several of the methods included in the 'biased amplification' section using the same SNP in the *SMAD7* gene previously used in probe studies in this chapter (NM_005904, C/T variation, rs4939827). This site is a relatively challenging SNP assay for these kinds of methods, since one of the mismatch pairings is T:G. The T:G/G:T mismatch pairs are thermodynamically fairly stable and as a result are difficult to distinguish. Primers used included: standard ASPCR, 3′-LNA ASPCR, ARMS, 3′-LNA ARMS, and both the first generation and second generation rhPCR blocked-cleavable primers. Primers are shown in Fig. 8.6 aligned on the target SNP site.

The SNP discrimination assays were run in singleplex format on a Roche LightCycler® 480 in 384 well plates using 10 µl reaction volumes. Reactions consisted of 5 µl of 2′ BIO-RAD iQ™ SYBR® Green Supermix (BIO-RAD, Hercules, CA), 200 nM each of the forward and reverse primers, and 2 ng human genomic DNA. Cycling conditions included an initial 5 minutes incubation at 95°C, followed by 80 cycles of 10 seconds at 95°C and 30 seconds at 60°C (note – reactions were run 80 cycles only to demonstrate the selectivity of the blocked primers; a standard 40–45 cycle run would be used in a typical assay). Human genomic DNAs that were homozygous C/C or T/T at SNP locus rs4939827 were obtained from the Coriell Institute for Medical Research Cell Repository (http://ccr.coriell.org/, GM18562 and GM18537). The quantification cycle number (*Cq*) was determined using the absolute quantification/2nd derivative method. All reactions were run in triplicate and data reported in Fig. 8.6 represents an average of the three data points. The rhPCR reactions included recombinant *Pyrococcus abyssi* RNase H2 in the reaction mix (2.6 mU per 10 µl reaction for the first generation primers and 50 mU per 10 µl reaction for the second generation rC-containing primer and 200 mU for the rU containing primer) (Walder *et al.*, 2009).

At the locus, unmodified ASPCR primers were ineffective and showed < 1 cycle different in *Cq* values seen between match and mismatch targets. Adding a single LNA nucleotide to the 3′-end of this primer design boosted the ΔCq to six to nine cycles. Unmodified ARMS primers performed even better, showing a ΔCq of 9–11 cycles. Adding a 3′-LNA nucleotide to the ARMS primers increased discrimination to 10–13 cycles; however, a three cycle delay in the match *Cq* was observed for the 'T' allele. The first generation blocked-cleavable primers (rhPCR) gave a ΔCq of around 11 cycles and the second generation version of this assay gave the best results in this survey with ΔCq values of 28–30 cycles. It is important to note that these assays were performed in singleplex format. The ASPCR approach often shows improved specificity when run in a multiplex format (which requires modification to add the appropriate dyes/quenchers to the primers to permit multiplex fluorescence detection) where kinetic competition between primers for each allele occurs. Nevertheless it is clear that such biased amplification methods can work well, even at difficult SNP sites like the rs4939827 locus used in the present example.

Assay		Primer Seq	Cq match	Cq mismatch	ΔCq
Control	5′	CAGCCTCATCCAAAAGAGGAAA	26.4	--	--
ASPCR	5′	CAGCCTCATCCAAAAGAGGAAAC	26.7	27.4	0.7
3′-LNA	5′	CAGCCTCATCCAAAAGAGGAAAC	25.9	32.3	6.4
ARMS	5′	CAGCCTCATCCAAAAGAGGAATC	26.5	35.6	9.1
ARMS+LNA	5′	CAGCCTCATCCAAAAGAGGAATC	27.7	41.3	13.6
rhPCR, Gen1	5′	CAGCCTCATCCAAAAGAGGAAAcAGGAx	27.2	38.4	11.2
rhPCR, Gen2	5′	CAGCCTCATCCAAAAGAGGAAAcAxxA	27.2	55.7	28.5
	5′	GCTCACAGCCTCATCCAAAAGAGGAAACAGGACCCC 3′			
	3′	CGAGTGTCGGAGTAGGTTTTCTCCTTTGTCCTGGGG 5′			

Assay		Primer Seq	Cq match	Cq mismatch	ΔCq
Control	5′	CAGCCTCATCCAAAAGAGGAAA	26.4	--	--
ASPCR	5′	CAGCCTCATCCAAAAGAGGAAAT	26.6	27.0	0.4
3′-LNA	5′	CAGCCTCATCCAAAAGAGGAAAT	25.9	35.1	9.2
ARMS	5′	CAGCCTCATCCAAAAGAGGAATT	26.7	37.5	10.8
ARMS+LNA	5′	CAGCCTCATCCAAAAGAGGAATT	30.6	40.6	10.0
rhPCR, Gen1	5′	CAGCCTCATCCAAAAGAGGAAAuAGGAx	27.2	38.1	10.9
rhPCR, Gen2	5′	CAGCCTCATCCAAAAGAGGAAAuAxxA	28.6	59.3	30.7
	5′	GCTCACAGCCTCATCCAAAAGAGGAAATAGGACCCC 3′			
	3′	CGAGTGTCGGAGTAGGTTTTCTCCTTTATCCTGGGG 5′			

Figure 8.6 Biased amplification assays at a C/T SNP in the human *SMAD7* gene. Genomic sequence of the region surrounding the C/T SNP site is shown. SNP discrimination is based upon amplification from the For primer sequences, which are shown aligned with the genomic target. Assays tested include unmodified allele-specific PCR primers (ASPCR), 3′-LNA (variant of ASPCR), ARMS (note second mutation added at the next to last base from the 3′-end, an A–T-change), ARMS with 3′-LNA, first generation rhPCR blocked-cleavable primer with RNase H2 cleavage, and second generation rhPCR blocked-cleavable primer with RNase H2 cleavage. A control primer was also used which did not include the SNP base under interrogation. DNA bases are uppercase. LNA nucleotides are underlined. RNA bases are lowercase. 'X' is a C3 spacer (1,3-propanediol group). Reactions were performed in singleplex format using SYBR® Green detection using 2ng input homozygous C/C or T/T DNA. Cq values obtained for the match and mismatch templates are shown as well as the D*Cq* (mismatch *Cq* minus match *Cq*).

Biased amplification using template-blocking agents

Unlike the methods to perform biased amplification previously discussed, this version uses primers that do not overlap the SNP site and are non-discriminatory. A high affinity oligonucleotide or peptide nucleic acid (PNA) is made that is complementary to the target and overlaps the SNP site, which is positioned between the primers. Like a hybridization probe, the blocking agent must bind to the match target but not bind to the mismatch target under the reaction conditions employed. Unlike fluorescent probes, here the binding must be stable enough to block processivity of the DNA polymerase, interrupting amplification. Both PNA oligomers and LNA oligonucleotides have been shown to work in this application (Shinozaki *et al.*, 2007; Tatsumi *et al.*, 2008). It is essential that either the blocking agent is nuclease resistant or that the DNA polymerase employed lacks a 5′–3′ exonuclease activity. It can be challenging to find oligomers that bind with sufficient affinity to block the polymerase yet do not bind in the presence of a single mismatch. As a result, this method has not found widespread use.

Preventing primer–dimer formation

In spite of efforts to design primers that do not have significant overlap with themselves or each other, sometimes it is necessary for biological reasons to position primers at locations in a sequence which are not ideal and permits primer-dimer formation. 'Hot start' PCR methods have decreased some of the problems seen with primer-dimer formation; however, this does not solve all problems and sometimes it is necessary to use chemically modified primers that have reduced this risk. The blocked-cleavable rhPCR primers discussed above offer one method to prevent primer-dimer formation and will not be discussed further in this section.

Another method to minimize primer-dimer formation was described by Howard Gamper and colleagues from Epoch Pharmaceuticals (Kutyavin *et al.*, 1996; Gamper *et al.*, 2006) and relies upon the base modifications 2,6-diaminopurine (2,6-DAP, or 2-amino-dA) and 2-thiothymine (2-thio-dT). The modified base 2,6-DAP is an adenosine analogue and will pair with thymine bases in the normal fashion. Similarly 2-thio-dT is a thymine analogue which forms a stable base pair with native adenosine. However, 2-thio-dT will not pair with 2,6-diaminopurine. Thus primers made using these modified bases will hybridize to their intended targets and can support PCR with equal efficiency as unmodified primers; however, the modified primers cannot easily hybridize to each other as the 2,6-DAP and 2-thio-dT bases (A and T replacements) will not base pair. These modifications are fairly expensive and this method has not found widespread use, even though it is biochemically quite elegant and works well in practice.

A different approach was described by Gerald Zon, Richard Hogrefe, and colleagues from TriLink BioTechnologies (Lebedev *et al.*, 2008). Instead of using a base modification, the internucleotide linkage was modified at the 3′-most and/or next to last linkage with a 4-oxo-1-pentyl (OXP) phosphotriester (PTE) group. This modification impairs the ability of the oligonucleotides to prime DNA synthesis; however, the modifying group is heat labile and rapidly dissociates from the primers during the initial 95°C heat denaturation step, after which the primers revert to unmodified form and function normally. This prevents primer-dimer formation during the initial warming phase of PCR and is a supplement to standard hot-start methods. In particular, this kind of primer modification should be very useful in one-tube RT-PCR chemistries where the primers are present with an active polymerase and traditional 'hot start' is not employed. This method is inexpensive and primers with this chemistry can be obtained directly from TriLink BioTechnologies (San Diego, CA).

Decreasing specificity using universal bases

The previous sections of this chapter focused on methods to increase specificity. Sometimes it is also necessary to decrease specificity. For example, the SNPs that occur in outbred populations can interfere with assays intended to amplify all variants. This problem can be particularly difficult with infectious disease targets, where different strains or serotypes within the same bacterial species or the same virus may have widely varying sequences. This greatly complicates attempts to make a single assay capable of detecting all important relatives. One approach to deal with this problem is to introduce mixed base sites, where synthetic oligonucleotides are made having a mixture of bases inserted at a specific position so that within the population of oligos some subpopulation will always have a perfect match to the target. This strategy works well if the amount of variability is limited. However, as target variability increases, then the number of different sequence variants of the primers or probes that exist in the mixed population can become very high with the result that the actual molar concentration of perfect match primers is too low to support PCR without having to employ unacceptably high concentrations of oligonucleotide.

A solution to reduce complexity of the oligonucleotides that will bind multiple targets is to incorporate modified bases which can pair with more than one of the natural bases, or at least are minimally destabilizing when paired with different natural bases. An excellent review of this topic was published by Loakes in 2001 (Loakes, 2001). The first 'inert' base used for this purpose was inosine, a naturally occurring hypoxanthine variant. Inosine is commonly used today, but is an inferior 'universal base' because it has very different effects on T_m depending on which base it is mispaired with (Martin *et al.*, 1985). Inosine most stably pairs with cytosine and behaves as a guanine analogue that simply does not have as strong a mismatch thermodynamic cost when paired with other bases as does guanine. The approximate rank order of stability for inosine base paring is C > A > G > T (Kawase *et al.*, 1986). Unlike other universal bases that may show more uniform thermodynamic effects when paired with any native base, inosine has the advantage that it can be inserted in primers and supports template function. When present in a nucleic acid which is used as a template in DNA synthesis, inosine directs incorporation of cytosine, consistent with it being considered a 'G analogue'.

The first artificial base that seemed to have true 'universal base' properties was 3-nitropyrrole; however, this modification was fairly destabilizing when used in primers and was quickly replace with 5-nitroindole (5-NI) (Nichols *et al.*, 1994; Loakes *et al.*, 1995). 5-NI destabilizes short duplexes by 2–5°C per insertion depending on location but shows reasonably little difference between templates having different bases mispaired with it. 5-NI is a good choice to use in hybridization probes as a universal base; however it does not function well in primers. Most polymerases do not recognize 5-NI as a base and chain termination usually occurs when 5-NI is encountered in a template and thus 5-NI containing oligonucleotides do not support PCR.

Additional compounds have been identified that behave as universal purine or pyrimidine bases. The synthetic base analogue 6H,8H-3,4-dihydropyrimido[4,5-c][1,2]oxazin-7-one (called 'p') is a pyrimidine replacement that will pair with the purine bases A or G. The synthetic base analogue N6-methoxy-2,6-diaminopurine (called 'k') is a purine analogue that will pair with the pyrimidine bases C or T (Bergstrom *et al.*, 1997). These bases are particularly useful as they support template function when used in a primer. The 'p' base directs incorporation of dA (60%) and dG (40%) and the 'k' base directs incorporation of dT (87%) and dC (13%) (Hill *et al.*, 1998). A variety of other base analogues have been synthesized and described in the literature; however, the set described here are commercially available and can be obtained from a variety of sources.

Conclusions

A wide variety of chemical modification can be introduced into synthetic oligonucleotides that alter their properties as PCR primers or probes. Modifications can be made to any part of the nucleic acid molecule, including the nucleobases, the internucleotide linkages, and the ribose ring. The modifications can increase binding affinity (T_m) and are commonly used to shorten primers and probes, improving specificity when this is desired. Conversely, modified bases are available which decrease specificity and are permissive for mismatch pairing. Modifications can also be used to prevent formation of primer-dimers by interfering with the ability of the oligonucleotides to prime DNA synthesis or interact with each other. All of the modifications discussed in this chapter are easily obtained from commercial sources. New modifiers and applications for their use are likely to continue to be developed in the future.

Acknowledgements

The authors thank all members of the research staff at Integrated DNA Technologies for their contributions over the past decade to our understanding of the fine details of qPCR and the impact that chemical modifications have on assay performance. In particular, we thank Derek Thomas, Ashley Dvorak, Andrey Tautarov, and Michael Christodoulou for assistance with the present manuscript. All original work discussed herein was supported by funding from Integrated DNA Technologies, Inc.

References

Afonina, I., Zivarts, M., Kutyavin, I., Lukhtanov, E., Gamper, H., and Meyer, R.B. (1997). Efficient priming of PCR with short oligonucleotides conjugated to a minor groove binder. Nucleic Acids Res. *25*, 2657–2660.

Afonina, I., Belousov, Y.S., Metcalf, M., Mills, M., Mills, A., Sanders, S., Kutyavin, I., Walburger, D.K., Gorn, V.V., Shishkina, I., *et al.* (2002a). Single nucleotide polymorphism detection with MGB Eclipse assays. J. Clin. Ligand Assay *25*, 268–275.

Afonina, I.A., Reed, M.W., Lusby, E., Shishkina, I.G., and Belousov, Y.S. (2002b). Minor groove binder-conjugated DNA probes for quantitative DNA detection by hybridization-triggered fluorescence. Biotechniques *32*, 940–944, 946–949.

Bailly, C., and Henichart, J.P. (1991). DNA recognition by intercalator-minor-groove binder hybrid molecules. Bioconjug. Chem. *2*, 379–393.

Bergstrom, D.E., Zhang, P., and Johnson, W.T. (1997). Comparison of the base pairing properties of a series of nitroazole nucleobase analogs in the oligodeoxyribonucleotide sequence 5′-d(CGCXAATTYGCG)-3′. Nucleic Acids Res. *25*, 1935–1942.

Bonnet, G., Tyagi, S., Libchaber, A., and Kramer, F.R. (1999). Thermodynamic basis of the enhanced specificity of structured DNA probes. Proc. Natl. Acad. Sci. U.S.A. *96*, 6171–6176.

Bratu, D.P., Cha, B.J., Mhlanga, M.M., Kramer, F.R., and Tyagi, S. (2003). Visualizing the distribution and transport of mRNAs in living cells. Proc. Natl. Acad. Sci. U.S.A. *100*, 13308–13313.

Cha, R.S., Zarbl, H., Keohavong, P., and Thilly, W.G. (1992). Mismatch amplification mutation assay (MAMA): application to the c-H-ras gene. PCR Methods Appl. *2*, 14–20.

Costa, J.M., Ernault, P., Olivi, M., Gaillon, T., and Arar, K. (2004). Chimeric LNA/DNA probes as a detection system for real-time PCR. Clin. Biochem. *37*, 930–932.

De la Vega, F.M., Lazaruk, K.D., Rhodes, M.D., and Wenz, M.H. (2005). Assessment of two flexible and compatible SNP genotyping platforms: TaqMan SNP Genotyping Assays and the SNPlex Genotyping System. Mutat. Res. *573*, 111–135.

Dempcy, R.O., Podyminogin, M.A., and Reed, M.W. (2003). Synthesis, characterization, and application of substituted pyrazolopyrimidine nucleosides. Curr. Protoc. Nucleic Acid Chem. *Chapter 1*, Unit 1.8.

Deutscher, M.P., and Kornberg, A. (1969). Enzymatic synthesis of deoxyribonucleic acid. 28. The pyrophosphate exchange and pyrophosphorolysis reactions of deoxyribonucleic acid polymerase. J. Biol. Chem. *244*, 3019–3028.

Di Giusto, D.A., and King, G.C. (2004). Strong positional preference in the interaction of LNA oligonucleotides with DNA polymerase and proofreading exonuclease activities: implications for genotyping assays. Nucleic Acids Res. 32, e32.

Dobosy, J.R., Rose, S.D., Beltz, K.R., Rupp, S.M., Powers, K.M., Behlke, M.A., and Walder, J.A. (2011). RNase H-dependent PCR (rhPCR): improved specificity and single nucleotide polymorphism detection using blocked cleavable primers. BMC Biotechnol. *11*, e80.

Duck, P., Alvarado-Urbina, G., Burdick, B., and Collier, B. (1990). Probe amplifier system based on chimeric cycling oligonucleotides. Biotechniques 9, 142–148.

Eder, P.S., and Walder, J.A. (1991). Ribonuclease H from K562 human erythroleukemia cells. Purification, characterization, and substrate specificity. J. Biol. Chem. *266*, 6472–6479.

Eder, P.S., Walder, R.Y., and Walder, J.A. (1993). Substrate specificity of human RNase H1 and its role in excision repair of ribose residues misincorporated in DNA. Biochimie 75, 123–126.

Gamper, H.B., Jr., Arar, K., Gewirtz, A., and Hou, Y.M. (2006). Unrestricted hybridization of oligonucleotides to structure-free DNA. Biochemistry 45, 6978–6986.

Haruki, M., Hayashi, K., Kochi, T., Muroya, A., Koga, Y., Morikawa, M., Imanaka, T., and Kanaya, S. (1998). Gene cloning and characterization of recombinant RNase HII from a hyperthermophilic archaeon. J. Bacteriol. *180*, 6207–6214.

Harvey, J.J., Lee, S.P., Chan, E.K., Kim, J.H., Hwang, E.S., Cha, C.Y., Knutson, J.R., and Han, M.K. (2004). Characterization and applications of CataCleave probe in real-time detection assays. Anal. Biochem. *333*, 246–255.

Harvey, J.J., Brant, S.R., Knutson, J.R., and Han, M.K. (2008). SNP analysis using catacleave probes. J. Clin. Lab. Anal. *22*, 192–203.

Hill, F., Loakes, D., and Brown, D.M. (1998). Polymerase recognition of synthetic oligodeoxyribonucleotides incorporating degenerate pyrimidine and purine bases. Proc. Natl. Acad. Sci. U.S.A. *95*, 4258–4263.

Hogrefe, H.H., Hogrefe, R.I., Walder, R.Y., and Walder, J.A. (1990). Kinetic analysis of *Escherichia coli* RNase H using DNA-RNA-DNA/DNA substrates. J. Biol. Chem. *265*, 5561–5566.

Holland, P.M., Abramson, R.D., Watson, R., and Gelfand, D.H. (1991). Detection of specific polymerase chain reaction product by utilizing the 5′–3′ exonuclease activity of *Thermus aquaticus* DNA polymerase. Proc. Natl. Acad. Sci. U.S.A. *88*, 7276–7280.

Hou, J., Liu, X., Pei, D., and Liu, J. (2007a). RNase HII from Chlamydia pneumoniae discriminates mismatches incorporation into DNA-rN1-DNA/DNA duplexes. Biochem. Biophys. Res. Commun. *356*, 988–992.

Hou, J., Liu, X., Wang, J., Liu, J., and Duan, T. (2007b). Molecular beacons for isothermal fluorescence enhancement by the cleavage of RNase HII from Chlamydia pneumoniae. Anal. Biochem. *371*, 162–166.

Hou, J., Liu, X., Zheng, Y., and Liu, J. (2007c). A method for HLA genotyping using the specific cleavage of DNA-rN(1)-DNA/DNA with RNase HII from *Chlamydia pneumoniae*. Oligonucleotides *17*, 433–444.

Johnson, M.P., Haupt, L.M., and Griffiths, L.R. (2004). Locked nucleic acid (LNA) single nucleotide polymorphism (SNP) genotype analysis and validation using real-time PCR. Nucleic Acids Res. *32*, e55.

Kaji, T., Ito, S., Iwai, S., and Miyasaka, H. (2009). Nanosecond to submillisecond dynamics in dye-labeled single-stranded DNA, as revealed by ensemble measurements and photon statistics at single-molecule level. J. Phys. Chem. B *113*, 13917–13925.

Kanaya, S. (2001). Prokaryotic type 2 RNases H. Methods Enzymol. *341*, 377–394.

Kawase, Y., Iwai, S., Inoue, H., Miura, K., and Ohtsuka, E. (1986). Studies on nucleic acid interactions. Stabilities, I. of mini-duplexes (dG2A4XA4G2-dC2T4YT4C2) and self-complementary d(GGGAAXYTTCCC) containing deoxyinosine and other mismatched bases. Nucleic Acids Res. *14*, 7727-7736.

Kuimelis, R.G., Livak, K.J., Mullah, B., and Andrus, A. (1997). Structural analogues of TaqMan probes for real-time quantitative PCR. Nucleic Acids Symp. Ser., 255–256.

Kutyavin, I.V., Rhinehart, R.L., Lukhtanov, E.A., Gorn, V.V., Meyer, R.B., Jr., and Gamper, H.B., Jr. (1996). Oligonucleotides containing 2-aminoadenine and 2-thiothymine act as selectively binding complementary agents. Biochemistry 35, 11170–11176.

Kutyavin, I.V., Lukhtanov, E.A., Gamper, H.B., and Meyer, R.B. (1997). Oligonucleotides with conjugated dihydropyrroloindole tripeptides: base composition and backbone effects on hybridization. Nucleic Acids Res. 25, 3718–3723.

Kutyavin, I.V., Afonina, I.A., Mills, A., Gorn, V.V., Lukhtanov, E.A., Belousov, E.S., Singer, M.J., Walburger, D.K., Lokhov, S.G., Gall, A.A., *et al.* (2000). 3′-minor groove binder-DNA probes increase sequence specificity at PCR extension temperatures. Nucleic Acids Res. 28, 655–661.

Kutyavin, I.V., Lokhov, S.G., Afonina, I.A., Dempcy, R., Gall, A.A., Gorn, V.V., Lukhtanov, E., Metcalf, M., Mills, A., Reed, M.W., *et al.* (2002). Reduced aggregation and improved specificity of G-rich oligodeoxyribonucleotides containing pyrazolo[3,4-d]pyrimidine guanine bases. Nucleic Acids Res. *30*, 4952–4959.

Kutyavin, I., Lokhov, S., Lukhtanov, E., and Reed, M.W. (2003). Chemistry of minor groove binder–oligonucleotide conjugates. Curr. Protoc. Nucleic Acid Chem. *Chapter 8*, Unit 8.4.

Kvaerno, L., and Wengel, J. (1999). Investigation of restricted backbone conformations as an explanation for the exceptional thermal stabilities of duplexes involving LNA (Locked Nucleic Acid):[dagger] synthesis and evaluation of abasic LNA. Chemical Communications, 657–658.

Lai, B., Li, Y., Cao, A., and Lai, L. (2003). Metal ion binding and enzymatic mechanism of Methanococcus jannaschii RNase HII. Biochemistry *42*, 785–791.

Latorra, D., Campbell, K., Wolter, A., and Hurley, J.M. (2003). Enhanced allele-specific PCR discrimination in SNP genotyping using 3′ locked nucleic acid (LNA) primers. Hum. Mutat. *22*, 79–85.

Le Laz, S., Le Goaziou, A., and Henneke, G. (2010). Structure-specific nuclease activities of Pyrococcus abyssi RNase HII. J. Bacteriol. *192*, 3689–3698.

Lebedev, A.V., Paul, N., Yee, J., Timoshchuk, V.A., Shum, J., Miyagi, K., Kellum, J., Hogrefe, R.I., and Zon, G. (2008). Hot start PCR with heat-activatable primers: a novel approach for improved PCR performance. Nucleic Acids Res. *36*, e131.

Letertre, C., Perelle, S., Dilasser, F., Arar, K., and Fach, P. (2003). Evaluation of the performance of LNA and MGB probes in 5′-nuclease PCR assays. Mol. Cell. Probes *17*, 307–311.

Liu, Q., and Sommer, S.S. (2000). Pyrophosphorolysis-activated polymerization (PAP): application to allele-specific amplification. Biotechniques *29*, 1072–1076, 1078, 1080 passim.

Liu, Q., and Sommer, S.S. (2002). Pyrophosphorolysis-activatable oligonucleotides may facilitate detection of rare alleles, mutation scanning and analysis of chromatin structures. Nucleic Acids Res. *30*, 598–604.

Liu, X.P., Hou, J.L., and Liu, J.H. (2010). A novel single nucleotide polymorphism detection of a double-stranded DNA target by a ribonucleotide-carrying molecular beacon and thermostable RNase HII. Anal. Biochem. *398*, 83–92.

Livak, K.J. (1999). Allelic discrimination using fluorogenic probes and the 5′ nuclease assay. Genet. Anal. *14*, 143–149.

Livak, K.J., Flood, S.J., Marmaro, J., Giusti, W., and Deetz, K. (1995). Oligonucleotides with fluorescent dyes at opposite ends provide a quenched probe system useful for detecting PCR product and nucleic acid hybridization. PCR Methods Appl. *4*, 357–362.

Loakes, D. (2001). Survey and summary: The applications of universal DNA base analogues. Nucleic Acids Res. *29*, 2437–2447.

Loakes, D., Brown, D.M., Linde, S., and Hill, F. (1995). 3-Nitropyrrole and 5-nitroindole as universal bases in primers for DNA sequencing and PCR. Nucleic Acids Res. *23*, 2361–2366.

McTigue, P.M., Peterson, R.J., and Kahn, J.D. (2004). Sequence-dependent thermodynamic parameters for locked nucleic acid (LNA)-DNA duplex formation. Biochemistry *43*, 5388–5405.

Martin, F.H., Castro, M.M., Aboul-ela, F., and Tinoco, I., Jr. (1985). Base pairing involving deoxyinosine: implications for probe design. Nucleic Acids Res. *13*, 8927–8938.

Mendelman, L.V., Petruska, J., and Goodman, M.F. (1990). Base mispair extension kinetics. Comparison of DNA polymerase alpha and reverse transcriptase. J. Biol. Chem. *265*, 2338–2346.

Moreau, V., Voirin, E., Paris, C., Kotera, M., Nothisen, M., Remy, J.S., Behr, J.P., Erbacher, P., and Lenne-Samuel, N. (2009). Zip Nucleic Acids: new high affinity oligonucleotides as potent primers for PCR and reverse transcription. Nucleic Acids Res. *37*, e130.

Moreira, B.G., You, Y., Behlke, M.A., and Owczarzy, R. (2005). Effects of fluorescent dyes, quenchers, and dangling ends on DNA duplex stability. Biochem. Biophys. Res. Commun. *327*, 473–484.

Mukaiyama, A., Takano, K., Haruki, M., Morikawa, M., and Kanaya, S. (2004). Kinetically robust monomeric protein from a hyperthermophile. Biochemistry *43*, 13859–13866.

Myakishev, M.V., Khripin, Y., Hu, S., and Hamer, D.H. (2001). High-throughput SNP genotyping by allele-specific PCR with universal energy-transfer-labeled primers. Genome Res. *11*, 163–169.

Nakitandwe, J., Trognitz, F., and Trognitz, B. (2007). Reliable allele detection using SNP-based PCR primers containing Locked Nucleic Acid: application in genetic mapping. Plant Methods *3*, 2.

Newton, C.R., Graham, A., Heptinstall, L.E., Powell, S.J., Summers, C., Kalsheker, N., Smith, J.C., and Markham, A.F. (1989). Analysis of any point mutation in DNA. The amplification refractory mutation system (ARMS). Nucleic Acids Res. *17*, 2503–2516.

Nichols, R., Andrews, P.C., Zhang, P., and Bergstrom, D.E. (1994). A universal nucleoside for use at ambiguous sites in DNA primers. Nature *369*, 492–493.

Noir, R., Kotera, M., Pons, B., Remy, J.S., and Behr, J.P. (2008). Oligonucleotide–oligospermine conjugates (zip nucleic acids): a convenient means of finely tuning hybridization temperatures. J. Am. Chem. Soc. *130*, 13500–13505.

Obika, S., Nanbu, D., Hari, Y., Morio, K.-i., In, Y., Ishida, T., and Imanishi, T. (1997). Synthesis of 2′-*O*,4′-*C*-methyleneuridien and -cytidine. Novel bicyclic nucleotsides having a fixed C3,-*endo* sugar puckering. Tetrahedron Lett. *38*, 8735–8738.

Ohtani, N., Haruki, M., Morikawa, M., and Kanaya, S. (1999). Molecular diversities of RNases H. J. Biosci. Bioeng. *88*, 12–19.

Owczarzy, R. (2005). Melting temperatures of nucleic acids: discrepancies in analysis. Biophys. Chem. *117*, 207–215.

Owczarzy, R., You, Y., Moreira, B.G., Manthey, J.A., Huang, L., Behlke, M.A., and Walder, J.A. (2004). Effects of sodium ions on DNA duplex oligomers: improved predictions of melting temperatures. Biochemistry *43*, 3537–3554.

Owczarzy, R., Moreira, B.G., You, Y., Behlke, M.A., and Walder, J.A. (2008a). Predicting stability of DNA duplexes in solutions containing magnesium and monovalent cations. Biochemistry *47*, 5336–5353.

Owczarzy, R., Tataurov, A.V., Wu, Y., Manthey, J.A., McQuisten, K.A., Almabrazi, H.G., Pedersen, K.F., Lin, Y., Garretson, J., McEntaggart, N.O., *et al.* (2008b). IDT SciTools: a suite for analysis and design of nucleic acid oligomers. Nucleic Acids Res. *36*, W163–169.

Owczarzy, R., You, Y., Groth, C.L., and Tataurov, A.V. (2011). Stability and mismatch discrimination of locked nucleic acid-DNA duplexes. Biochemistry *50*, 9352–9367.

Paris, C., Moreau, V., Deglane, G., Voirin, E., Erbacher, P., and Lenne-Samuel, N. (2010). Zip nucleic acids are potent hydrolysis probes for quantitative PCR. Nucleic Acids Res. *38*, e95.

Reddy, M.S., and Hardin, S.H. (2003). Features in short guanine-rich sequences that stimulate DNA polymerization *in vitro*. Biochemistry *42*, 350–362.

SantaLucia, J., Jr. (1998). A unified view of polymer, dumbbell, and oligonucleotide DNA nearest-neighbor thermodynamics. Proc. Natl. Acad. Sci. U.S.A. *95*, 1460–1465.

SantaLucia, J., Jr., and Hicks, D. (2004). The thermodynamics of DNA structural motifs. Annu. Rev. Biophys. Biomol. Struct. *33*, 415–440.

SantaLucia, J., Jr., Allawi, H.T., and Seneviratne, P.A. (1996). Improved nearest-neighbor parameters for predicting DNA duplex stability. Biochemistry *35*, 3555–3562.

Sato, A., Kanai, A., Itaya, M., and Tomita, M. (2003). Cooperative regulation for Okazaki fragment processing by RNase HII and FEN-1 purified from a hyperthermophilic archaeon, Pyrococcus furiosus. Biochem. Biophys. Res. Commun. *309*, 247–252.

Shi, J., Liu, Q., and Sommer, S.S. (2007). Detection of ultrarare somatic mutation in the human TP53 gene by bidirectional pyrophosphorolysis-activated polymerization allele-specific amplification. Hum. Mutat. *28*, 131–136.

Shinozaki, M., O'Day, S.J., Kitago, M., Amersi, F., Kuo, C., Kim, J., Wang, H.J., and Hoon, D.S. (2007). Utility of circulating B-RAF DNA mutation in serum for monitoring melanoma patients receiving biochemotherapy. Clin. Cancer Res. *13*, 2068–2074.

Simonsson, T. (2001). G-quadruplex DNA structures – variations on a theme. Biol. Chem. *382*, 621–628.

Singh, S.K., and Wengel, J. (1998). Universality of LNA-mediated high-affinity nucleic acid recognition. Chem. Commun., 1247–1248.

Singh, S.K., Koshkin, A. A., Nielsen, P., and Wengel, J. (1998). LNA (locked nucleic acids): synthesis and high-affinity nucleic acid recognition. Chem. Commun., 455–456.

Tatsumi, K., Mitani, Y., Watanabe, J., Takakura, H., Hoshi, K., Kawai, Y., Kikuchi, T., Kogo, Y., Oguchi-Katayama, A., Tomaru, Y., *et al.* (2008). Rapid screening assay for KRAS mutations by the modified smart amplification process. J. Mol. Diagn. *10*, 520–526.

Tenesa, A., Farrington, S.M., Prendergast, J.G., Porteous, M.E., Walker, M., Haq, N., Barnetson, R.A., Theodoratou, E., Cetnarskyj, R., Cartwright, N., *et al.* (2008). Genome-wide association scan identifies a colorectal cancer susceptibility locus on 11q23 and replicates risk loci at 8q24 and 18q21. Nat. Genet. *40*, 631–637.

Thelwell, N., Millington, S., Solinas, A., Booth, J., and Brown, T. (2000). Mode of action and application of Scorpion primers to mutation detection. Nucleic Acids Res. *28*, 3752–3761.

Tsourkas, A., Behlke, M.A., and Bao, G. (2003a). Hybridization of 2′-O-methyl and 2′-deoxy molecular beacons to RNA and DNA targets. Nucleic Acids Res. *31*, 5168–5174.

Tsourkas, A., Behlke, M.A., Rose, S.D., and Bao, G. (2003b). Hybridization kinetics and thermodynamics of molecular beacons. Nucleic Acids Res. *31*, 1319–1330.

Tyagi, S., and Kramer, F.R. (1996). Molecular beacons: probes that fluoresce upon hybridization. Nat. Biotechnol. *14*, 303–308.

Tyagi, S., Bratu, D.P., and Kramer, F.R. (1998). Multicolor molecular beacons for allele discrimination. Nat. Biotechnol. *16*, 49–53.

Ugozzoli, L.A., Latorra, D., Puckett, R., Arar, K., and Hamby, K. (2004). Real-time genotyping with oligonucleotide probes containing locked nucleic acids. Anal. Biochem. *324*, 143–152.

Wagner, R.W., Matteucci, M.D., Lewis, J.G., Gutierrez, A.J., Moulds, C., and Froehler, B.C. (1993). Antisense gene inhibition by oligonucleotides containing C-5 propyne pyrimidines. Science *260*, 1510–1513.

Walder, J.A., and Walder, R.Y. (1995). Nucleic acid hybridization and amplification method for detection of specific sequences in which a complementary labeled nucleic acid probe is cleaved. US Patent 5,403,711.

Walder, J.A., Behlke, M.A., Rose, S.D., and Dobosy, J.R. (2009). RNase H-based assays utilizing modified RNA monomers. US Patent Application 2009/0325169.

Yamane, A. (2002). MagiProbe: a novel fluorescence quenching-based oligonucleotide probe carrying a fluorophore and an intercalator. Nucleic Acids Res. *30*, e97.

You, Y., Moreira, B.G., Behlke, M.A., and Owczarzy, R. (2006). Design of LNA probes that improve mismatch discrimination. Nucleic Acids Res. *34*, e60.v

Real-time PCR Arrays

9

Nick A. Saunders

Abstract
Real-time PCR arrays are tools that allow convenient testing of samples in many assays concurrently, parallel testing of many samples or testing of multiple samples and targets simultaneously. It is desirable to standardize and automate primer and probe selection due to the large number of assays that must be designed. Furthermore, it is useful to use probe selection techniques that increase the robustness of the individual assays since this will increase the level of compatibility between the assays and decrease the complexity of interpretation of the outputs.

A simple approach to creating real-time PCR arrays is to use microtitre plates which currently have capacities of 96, 384 or 1536 features. Such arrays can be populated with user designed assays or with tests selected from a menu of over one million that are commercially available. A primary application of such arrays has been to verify gene expression data obtained using hybridization.

Cramming additional features into a device of manageable scale has led to the introduction of nanolitre volume arrays that diverge from the microtitre plate pattern. Several thousand different reactions can now be included in a single real-time PCR array. The reduction in scale also has advantages in terms of the volumes of materials required.

As real-time arrays are miniaturised the number of pipetting steps required increases and it is often necessary to pre-configure them commercially leading to relative inflexibility. This limitation has prompted the development of arrays that include microfluidic channels and valves. These 'chips' can be loaded via relatively few liquid handling steps to create custom applications.

Introduction

In a conventional PCR application a collection of samples might be tested under identical reaction conditions, i.e. with the same primers. PCR arrays represent a simple inversion of this set-up in which samples are split between parallel reactions each with a different primer set. Using PCR arrays it is possible to obtain multiple data points from a sample in a single run. Although some non-real-time PCR array applications exist real-time detection is generally required.

The common applications of real-time PCR arrays can be broadly divided into typing and gene expression analyses. For example, typing by SNP profiling can now be performed conveniently using arrays comprised of libraries of individual assays brought together in a microtitre tray and one major supplier now has a collection of over 4.5 million pre-designed human SNP assays of which 170,000 have been validated (Abi website https://products.appliedbiosystems.com). These assays can be combined as required by the investigator. In the same way, assays designed to measure the levels of specific mRNAs may be brought together to give a transcript profile and the same supplier has over 1 million assays covering genes from 19 species.

The first real-time PCR arrays built up from these assays were designed to use standard instruments and 96 well microtitre trays. However, the power of arrays lies in the opportunity to combine parallel assays. Advances in instrumentation and the associated consumables for real-time PCR have led to the development of platforms that allow several thousand nanolitre sized reactions to be run concurrently. Developments in microfluidics are also contributing to the convenience of PCR array based assays and it is possible to run multiple samples on a single device with rapid automated reaction set-up.

Handling the design of large numbers of sets of primers and probes for use in arrays is greatly assisted by using software that applies layered rules to the available sequence data. Data handling tools are also helpful for analysis of the greater quantum of information from the array platforms. The utility of such software needs to increase in step with the growing requirements of the field.

This chapter is designed to present available and emerging technologies associated with real-time PCR arrays. Some of the applications, strengths and weaknesses of different PCR array formats will be discussed.

Assay design

The performance of real-time PCR assays depends upon the suitability of the primers and probes chosen. For sensitivity the primers must match sequences in all target genomes or transcriptomes sufficiently for priming to occur and for specificity the primers and probe(s) should mismatch all other sequences that might yield a signal. Additional requirements for sensitivity are that amplification should proceed with reasonable efficiency, that the formation of primer artefacts occurs infrequently and that the probe(s) have a decent fluorescent yield on interacting with the amplicon.

Real-time assays that include hydrolysis (TaqMan) probes are very extensively used in PCR array type applications mainly because of the relative ease of data interpretation (De la Vega *et al.*, 2005). These assays have well understood but somewhat strict design requirements that lend themselves to being configured using computer programmes that follow a simple set of thermodynamic and heuristic rules recommended by Applied Biosystems. PrimerExpress® (Applied Biosystems, Foster City, CA, USA) is the best known software intended for this task. The primary requirement is the availability of a sequence that is conserved throughout the taxa that the assay is being designed to detect. The software then provides possible primer/probe sets that can be configured to allow single nucleotide polymorphism (SNP) detection. A BLAST (Altschul *et al.*, 1990) search is performed to ensure that the primers are not hitting unintended targets. Finally the assay is tested in the laboratory although this may be minimal depending on the application. However, for microbial targets testing may involve analysis of nucleic acids from many strains.

Approaches designed to improve assay robustness applicable to arrays

Real-time PCR arrays involve the implementation of multiple assays that must be run in parallel. To facilitate this requirement it is important to use design techniques intended to provide probes and primers that give robust real-time PCR assays.

In general robust primer design for arrays has depended upon adherence to the generally accepted parameters governing the compatibility of primer pairs. It is then possible to use a single set of amplification conditions for all assays on the array although this involves some degree of compromise, i.e. a default annealing temperature and magnesium ion concentration.

Various methods exist for the detection of SNPs and other mutations by real-time PCR. A widely used strategy is allele-specific PCR in which one or both primers anneal at variant sites (Wu *et al.*,1989). Ideally, a primer matching a specific variant should amplify only that variant. However, in practice, amplification of related sequences occurs although the breakthrough rate varies over many orders of magnitude (Huang *et al.*, 1992). Commonly, the 3′ end of the mutant allele-specific primer is positioned directly over a mutant base in order to selectively amplify the mutant template. This strategy greatly reduces the breakthrough rate for amplification of non-matching sequences but the variability of non-specific amplification may necessitate empirical testing to find highly selective assays (Newton *et al.*,1989). Mismatch amplification mutation assay (MAMA) where an additional mismatch is introduced at the penultimate base is effective in decreasing the breakthrough rate (Li *et al.*, 2004) as is the use of locked nucleic acid residues at the SNP (Latorra *et al.*, 2003).

Other methods include allele-specific competitive blocker PCR and blocker–PCR (Seyama *et al.*, 1992; Orou *et al.*, 1995; Prix *et al.*, 2002; Amicarelli *et al.*, 2007; Morlan *et al.*, 2009) in which oligonucleotides that cannot be extended by the polymerase are used to inhibit amplification of a specific sequence variant by annealing to the primer site, the inter-primer sequence or both. Locked nucleic acid hydrolysis probes can also be used to enhance the impact of a SNP (Ugozzoli *et al.*, 2004).

Co-amplification at lower denaturation temperature-PCR (COLD-PCR) uses a modified protocol to enrich the chosen alleles (minority population) from a mixture so that they can be detected reliably (Li *et al.*, 2008). Additional post-denaturation steps are introduced into every cycle. The reaction is held at a temperature above that required for primer annealing to allow the amplicons to re-anneal. The temperature is then increased to an intermediate point set so that any heteroduplexes present are denatured but homoduplexes remain in double stranded form. At the end of this additional process, designed to enrich the proportion of minority population sequence in the single-stranded population, the cycle is resumed at the annealing step. The cumulative effect of many cycles with the additional steps is to increase the proportion of the minority sequence close to a 1:1 ratio.

Another approach to robust, sensitive detection of SNPs is to exploit the high template specificity of DNA ligase. Circularizable ligation probes (Padlock probes) developed and applied by Nilsson and colleagues (1994) have been adapted for use in real-time PCR (van Doorn *et al.*, 2007) and named PRI-lock probes (Fig. 9.1). The circularised probes are amplified via primer pairs that are specific to each probe sequence that acts like a barcode. Each PRI-lock probe consists of terminal regions that are complementary to the target separated by a linker. Within the linker are the opposed unique primer sequences, a generic sequence and a desthiobiotin modified base that allows capture and release. Following ligation of the terminal probe sequences in the presence of the target the priming sites are in the correct orientation to allow formation of an amplimer. Many PRI-lock probes can be combined in a single assay provided the primers are designed to amplify the amplimer under the same conditions. For a multiplex assay in which many SNPs are to be detected

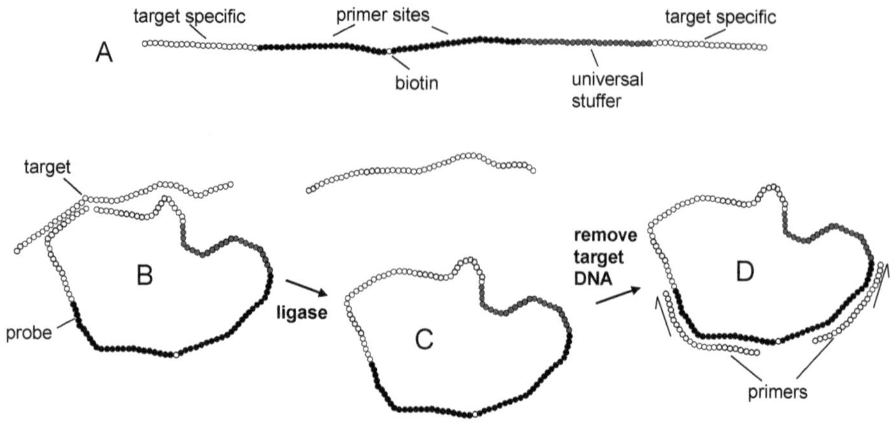

Figure 9.1 PRI-lock probes for real-time PCR. Each PRI-lock probe (A) consists of terminal regions that are complementary to the target separated by a linker. Within the linker are a set of primer sequences, a universal stuffer sequence and a desthiobiotin modified base to aid purification. Following ligation of the terminal probe sequences which requires that the target sequence hybridizes to both target specific regions of the probe (B) the target DNA is removed (C) by binding the probe to a streptavidin modified solid support via the biotin residue. The priming sites in the circular probe are in the correct orientation (D) to allow formation of an amplicon in a PCR with suitable primers. Each target sequence is coded to a different primer pair.

the probes are ligated on the test DNA recovered using the desthiobiotin moiety, treated with exonuclease to destroy non-ligated probe and assayed by real-time PCR on an array (OpenArray™).

Real-time PCR arrays in conventional plates

PCR arrays based on standard reaction plates (i.e. microtitre tray format; see Fig. 9.2) represent a straightforward application of readily available technology. Each well, or set of wells, is loaded with a different PCR but all wells receive an aliquot of the same sample. The primers and probes of each PCR reaction are generally designed to function with identical cycling conditions and reagents. This is a convenient but non-essential feature of plate real-time arrays which could, for example, be run on a gradient block.

Hydrolysis probe (TaqMan®) based assays are well suited for use in real-time PCR arrays partly because of the convenience of the format and partly because of the existence of large numbers of off-the-shelf designs and associated reagents. However, it is realistic to use lower cost assays based on SYBR Green detection and a wide range of off-the-shelf assays (SuperArray Biosciences now Qiagen) are available. SNP genotyping and expression profiling assays are available commercially and the individual tests are combined to form supplier designed off-the-shelf arrays or user specified custom arrays.

Potential applications of SNP arrays include their use for human genotyping for forensic purposes, the detection of human mutations associated with drug sensitivity, infectious agent genotyping and detection of markers for genetic modified foods (Mano *et al.*, 2009). Although, these applications have potential, TaqMan® arrays in conventional plates are not often the method of choice for work at this scale due to the availability of alternatives such

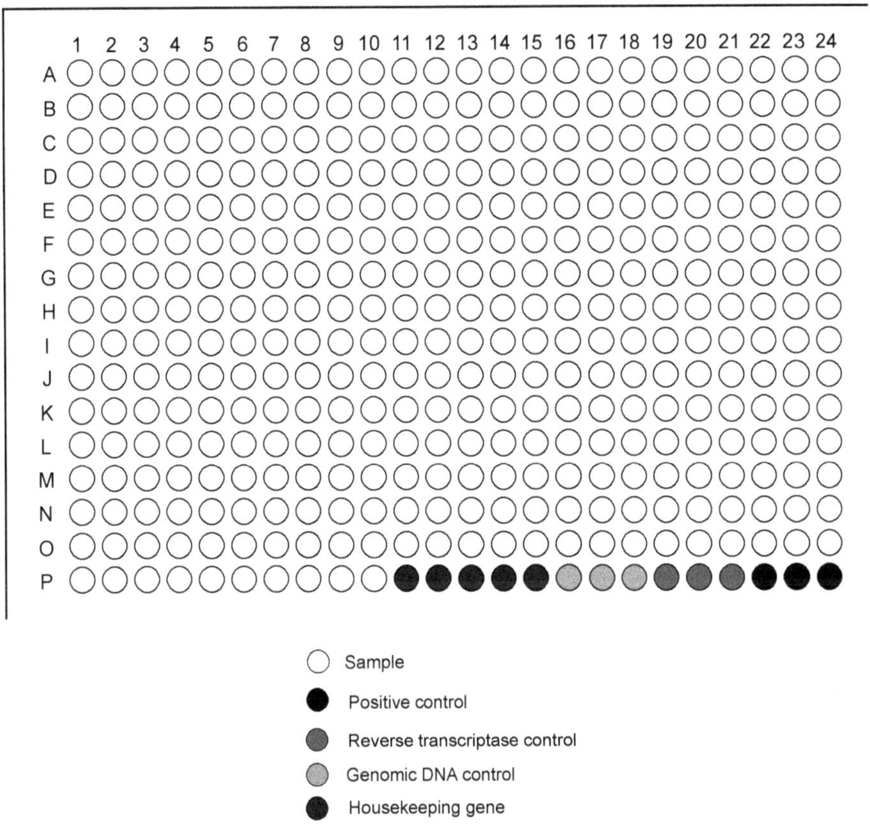

Figure 9.2 Real-time PCR on a plate array. The diagram shows a typical 384 well plate format. Here up to 370 PCR assays are included for measurement of the mRNA levels of target genes. A number of positive and negative controls can easily be accommodated. In this example assays for five housekeeping genes (i.e. genes expected to be expressed in the sample), three positive controls (i.e. a DNA target is spiked into the well) and three reverse transcriptase controls (i.e. a RNA target is spiked into the well) are included. The three negative genomic DNA controls detect genomic contamination (i.e. assay for a non-expressed sequence).

as multiplexing which can accommodate a larger number of samples (Tobler *et al.*, 2005; Tomas *et al.*, 2008).

Expression profiling arrays based on conventional plate formats have been more widely reported than SNP arrays. The key application has been validation of results obtained in microarray experiments. Validation is especially vital in the context that it may be very difficult to correlate any microarray results with real clinical or biological effects (Miklos and Maleszka, 2004). Reference RNA samples analysed in the MicroArray Quality Control study (The MicroArray Quality Control Study: http://www.fda.gov/nctr/science/centres/toxicoinformatics/maqc/index.htm) have been tested on a range of microarray formats (Shi *et al.*, 2006) and in real-time PCR arrays using TaqMan® probes (Shi *et al.*, 2006) or SYBR Green (Arikawa *et al.*, 2008). Although these studies have shown the general reliability of the microarray platforms in identifying genes that are differentially expressed it is also acknowledged that the PCR arrays provide the most sensitive and accurate data with a greater dynamic range than can currently be achieved using microarray formats (Wang *et al.*, 2006; Gaj *et al.*, 2008). Czechowski *et al.* (2004) used the additional sensitivity of the PCR array approach to overcome the detection and dynamic range limits inherent to hybridization based array methods of transcriptome analysis. The PCR array developed measured the levels of transcripts for 1465 Arabidopsis transcription factors (TFs) and was based on closely spaced gene-specific primer pairs and SYBR Green to monitor amplification. Transcript levels of 83% of all target genes could be measured with quantification over six orders of magnitude and a detection limit equivalent to one transcript molecule in 1000 cells. In comparison, Affymetrix chips detected less than 55% of TF transcripts in the same samples and the range of transcript levels was compressed by a factor more than 100. Thus the PCR array was able to reveal root-specific and shoot-specific TF genes that had not been identified as organ-specific previously.

Real-time PCR arrays are well suited for the measurement of the expression of a defined subset of genes. For example, SuperArray Biosciences supply a range of pre-formulated arrays. An example is the 'Human Toll-Like Receptor (TLR) Signalling Pathway RT^2 Profiler PCR Array', which can be used to measure the expression of 84 genes including pattern recognition receptors involved in TLR-mediated signal transduction and innate immunity to a wide range of bacteria, viruses, fungi and parasites. West and Damania (2008) used this array to demonstrate up-regulation of the TLR3 pathway by Kaposi's sarcoma-associated herpes virus during primary infection.

The determination of DNA methylation status at multiple sites is an interesting application of real-time DNA arrays. The technology is suited to the analysis of gene promoter sequences that become hypermethylated in defined biological settings. Hypermethylation may be correlated with biological phenotypes or disease outcomes. Restriction enzymes that either require or are inhibited by methylation are used to enrich a sample for target DNA with a defined methylation status (Herman *et al.*, 1996). The remaining DNA is then quantified in real-time PCR using specific gene PCRs. For example, the commercial Methyl-Profiler DNA Methylation qPCR array (SuperArray Biosciences Corp., Frederick, MD, USA) has been used to confirm results of a genome wide promoter CpG methylation array that uses the bisulfite conversion method (Kibriya *et al.*, 2011) It has also been used to analyse genomic DNA isolated from nasal epithelial cells of smokers and non-smokers for methylation in the promoter sites of human genes (Jaspers *et al.*, 2010). The relative concentrations of differentially methylated DNA are determined by comparing the amount of each digest with that of a mock digest.

Nanolitre plate arrays

Plate arrays that use nanolitre-scale reaction volumes are an interesting prospect for commercial development and some examples are now on the market. The advantages of moving to the nanolitre scale are the reductions in the use of samples and reagents together with an advance in the number of targets or samples that can be measured concurrently leading to potentially very high throughputs. An advantage of the array format is that if necessary each assay in a panel can be independently replaced without the need to re-optimize the entire assay set. However, current nanotitre plate array assay configurations are generally set by the commercial suppliers due to the need for specialized liquid handling and the use of large numbers of custom reagents (primers and probes) required. Array re-configuration is therefore likely to be relatively costly. Pre-formatted, high-throughput assays are likely to be very useful in a variety of applications. However, they lack the flexibility to address rapidly changing needs and cannot be expected to satisfy all requirements for niche assays. Consequently, systems with the facility for users to develop new applications hold a significant advantage.

As discussed in previous sections a significant advantage of the real-time PCR array approach compared to conventional arrays (i.e. those that depend on hybridization of labelled probes without prior amplification of the target) is high sensitivity. Further, at present qPCR provides a more reliable means of target quantification than conventional arrays and gives the potential for high sample throughput. However, compared to microarrays and macroarrays real-time arrays suffer the disadvantage that it is not feasible to achieve very high levels of multiplexing. Part of the motivation for introduction of nanolitre scale real-time PCR arrays is to reduce this relative performance deficit.

The OpenArray® Real-Time PCR System (Applied Biosystems) is designed to provide either genotyping or gene expression analysis. The arrays are constructed by micromachining holes into a stainless steel sheet with the dimensions of a standard microscope slide. The holes have a volume of 33 nl and the interior of each is treated to give a covalently bound hydrophilic coating of polyethylene glycol. The slide surface is given a hydrophobic fluoroalkyl coating. The OpenArray® holes are grouped in 48 subarrays each comprised of 64 wells giving a total number of 3072. For convenience in the robotic application of the reaction components from microtitre plates the subarray layout has the same pitch as a 384 well plate (Fig. 9.3). Primers are pre-loaded and dried down in user specified wells. The wells function to metre sample and master mix added by the loader via capillary action (i.e. the volume delivered, 33 nl, is determined by the hole capacity). The hydrophilic well and hydrophobic surface coating effectively ensures that the sample remains isolated when placed in a glass-walled cassette filled with an immiscible perfluorinated liquid (Fluorinert; 3M, St. Paul, MN, USA) hermetically sealed with epoxy.

Each well of the array can be used for an independent PCR assay although by design each subarray is used for analysis of a single sample in 64 PCRs. The system supports detection based on intercalation (SYBR Green) or hydrolysis probes. In addition, detection either in real-time or following cycling is possible.

Roberts and colleagues (2009) have reported use of OpenArrays® for high-throughput genotyping of human DNA based on assays for single nucleotide polymorphisms (SNPs) at 32 loci. SNPs represent the most convenient, abundant and accessible polymorphisms for recording genetic variation in a population. The attractions of SNP genotyping have increased with the emergence of a growing number of SNP databases, e.g. www.ncbi.nlm.nih.gov/projects/SNP. There are a number of significant applications of SNP genotyping including the diagnosis of human disease, pharmacogenetics, blood typing, forensics, paternity testing, fisheries management, pathogen identification and plant breeding (Brenan *et al.*, 2009). A panel of 90 DNA samples from the Centre d'Etude du Polymorphisme Humain (Fondation Jean Dausset-CEPH) was used for evaluation of the SNP panel on this platform. End-point analysis was used for the bi-allelic tests (most human SNPs are biallelic) based on pairs of hydrolysis probes labelled with VIC and FAM. Good rates of calling with high accuracy were obtained with from 0.1–1 ng human DNA per well although at the lower concentrations whole genome amplification (Dean *et al.*, 2001) was required to maintain the call rate for a proportion of the assays.

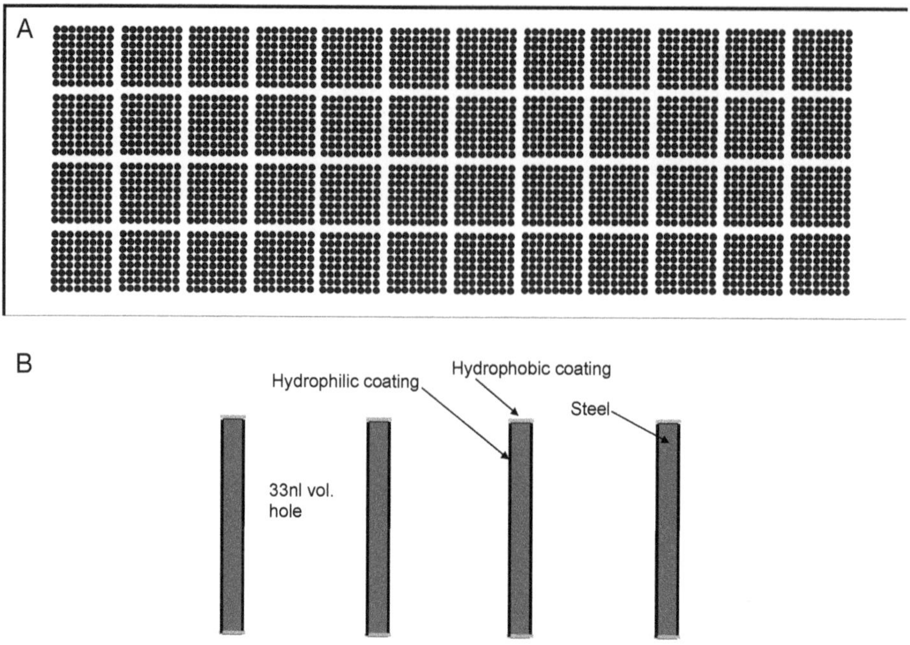

Figure 9.3 The OpenArray® slide. (A) Layout of 48 subarrays on an OpenArray® slide. The subarrays are spaced with the same pitch as a 384 well microtitre plate and each includes 64 holes (320µm diameter holes in an 8×8 array). (B) Cross-section of part of a slide with three complete holes included. The holes, formed by perforating a 0.3mm steel sheet the size of a microscope slide, have a 33nl capacity and are coated internally with a hydrophilic layer. The outer surface of the steel sheet has a hydrophobic coating.

OpenArrays® have also been exploited for the quantitative detection of multiple plant pathogens (van Doorn *et al.*, 2007). This paper introduced the concept of quantifiable PRI-lock probes (a variant of Padlock probes) as discussed in the section above 'Approaches designed to improve assay robustness applicable to arrays' and shows the potential application of the arrays to diagnosis of human and animal infection, food safety, biothreat detection and environmental pest management. The use of real-time arrays overcomes the major drawback of PCR multiplexing which is that the performance (i.e. particularly the sensitivity and accuracy of quantification) of the component assays is compromised by the number of target sequences included. In addition to the specificity conferred by the ligation based assay chemistry the use of the ligation probes confers the advantage that the array can be factory configured with a single set of primer pairs that can be applied universally.

The application to plant pathogens described by Van Doorn and colleagues (2007) provides a model for the development of similar detection systems offering high specificity and throughput with robust target quantification. In their study all targets were detected and no false-positives were recorded for non-target organisms with very similar ligation target sites, including alleles differing at a single nucleotide, indicating high specificity of the ligation reaction (Fig. 9.1). It was demonstrated that the presence of multiple targets had no statistically significant influence on either specificity or sensitivity. Indicating that this multiplex detection system could provide independent detection of multiple targets without inhibition due to competition.

The SmartChip™ Real-Time PCR System (WaferGen Biosystems) is a platform for rapid quantitative gene-expression analysis (Gelfand and Kastury, 2009). The SmartChip™ arrays consist of 5000–30,000 wells each of which can contain a different PCR reaction, in a volume as low as 100 nl. The arrays are run on a platform capable of rapid-cycling and real-time detection. The platform is capable of performing thousands of nanoscale quantitative PCR assays in a single run. This provides the capability to perform screens normally associated with microarrays but combined with the excellent sensitivity inherent in real-time PCR. The SmartChips can be configured either to measure a large number of target sequences on a single sample or in a multiarray format (i.e. with a smaller number of targets analysed for up to 384 samples on a single array). A dedicated dispenser is available to ensure reliable delivery of samples to the arrays. It is claimed that use of the system will allow the generation of expression results many times faster and at greatly reduced cost compared to conventional methods.

Current SmartChips are factory configured with panels of assays that target genes relevant to a particular application, e.g. cancer studies and toxicology. In addition, WaferGen offer custom arrays with user designed assay panels. Although this platform is promising it is yet to find significant levels of application. A problem is that its laboratory niche is currently occupied by platforms such as the high-density arrays manufactured by Affymetrix Inc and bead arrays such as those of Illumina Inc (Saunders, 2011).

Microfluidic arrays

The use of microfluidic devices is one solution to the problem of how to overcome one of the major potential bottlenecks hampering the use of real-time PCR arrays in high-throughput applications, i.e. the need to transfer samples into multiple test wells with minimal pipetting steps. Recent years have seen encouraging progress in the development of such devices and commercial products are now available.

The development of microfluidic devices for use in PCR has been an area of activity since the early 1990s and the developments have been reviewed by Zhang and colleagues (2006, 2007). For example, Belgrader and colleagues (1998) reported the development of a microchip PCR instrument with an array of ten silicon reaction chambers with thin-film resistive heaters and solid-state optics and showed that it could be applied to rapid pathogen detection. However, this approach to array construction, i.e. building an array from individual devices, is limited in its cost-effective scalability, compared with the alternative approach of building an array of reaction chambers into a single unit.

The TaqMan® Low Density Array (Applied Biosystems) is a consumable device consisting of 384 wells connected by a series of micro fluidic channels (Fig. 9.4) designed for custom or off-the-shelf configuration with TaqMan® assays. User-selected assays are factory loaded into each well. Up to eight cDNA samples per array are mixed with Master Mix and loaded into the eight ports. A single sample reaction mix is delivered to 48 wells via a centrifugal filling process. As for the plate arrays, low density microfluidic arrays are useful either as the primary tool for gene expression analysis of a limited number of genes (for example Rom *et al.*, 2009) or as a confirmatory assay for selected genes identified using high-density arrays (for example Hartman *et al.*, 2008). Although, these arrays were designed primarily as a

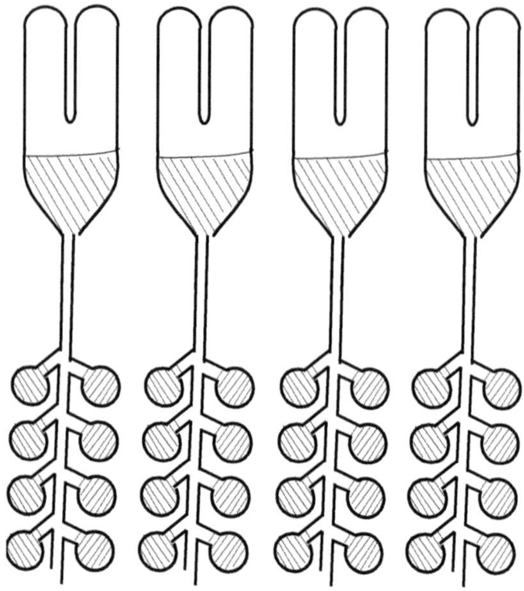

Figure 9.4 Low density microfluidic array. Each of the 384 wells is pre-loaded (Applied Biosystems) with dried primers and probes. Potentially each well can be loaded with a different assay though more usually each set of 48 wells is configured identically. For use each of the eight reservoirs is loaded with sample and this is distributed to the associated wells by centrifugation. After centrifugation each well is isolated. The wells of the card are configured as on a 384 well microtitre tray.

low-density alternative to microarrays for the analysis of gene expression they can also be employed for detection of multiple targets in parallel. For example, Kodani and colleagues (2011) have applied the array cards to the simultaneous detection of respiratory pathogens.

Fluidigm microfluidic chips are similar in capability to the TaqMan® Low Density Arrays. However, the array dimensions (i.e. the number of samples and different assays) of the devices are generally larger and all of the reaction components, including the probes and primers, may be added to the chip in the user laboratory so that the system is extremely flexible. A range of Fluidigm chips is now available which differ in the number of sample and assay (i.e. the probes and primers) inlets. The available formats are 192 samples × 24 assays, 48 samples × 48 assays and 96 samples × 96 assays. For most applications the chips are single use due to the difficulty of cleaning to the standard necessary to avoid carry-over. However, a reusable 48 samples × 48 assays chip is available that can be cleaned to the level necessary for use in SNP genotyping. The chips are mounted on a plastic carrier with the sample inlets and detector inlets together with interface and containment accumulators that can be pressurized to close the valves (i.e. NanoFlex™ valves) that are the central design feature of these arrays. The dimensions and positions of the inlet ports conform to those of a 384 well microtitre plate. A network of fluid lines, valves and reaction chambers are included on the chips. The valves deflect under pressure to seal and regulate the flow of liquids in the device. Each reaction chamber is connected to a sample and detector reservoir via a containment valve. The sample and detector sides of each reaction chamber are separated by a separately controllable interface valve (Fig. 9.5). For loading, pressure is applied to

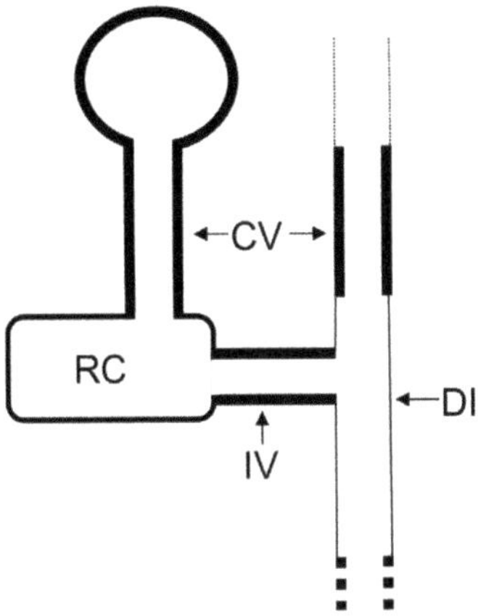

Figure 9.5 NanoFlex™ valves associated with a single reaction chamber. Each reaction chamber (RC) has two associated control valves (CV). One of the control valves is in the line leading to the detector and the other is in the sample line. The reactants are loaded down each line with the control valves open and the interface valve (IV) closed. This delivers sample to the reaction chamber and detector to the detector inlet (DI). For mixing the control valves are shut and the interface valve is opened allowing the detector and sample solutions to meet.

push the sample and detector fluids down separate fluid lines. Sample goes to the individual reaction chambers and detector solution collects in an inlet proximal to each well. At this point mixing of the two fluids is prevented by the closed interface valve. For mixing, the two containment valves associated with each reaction chamber are closed and then the interface valve is opened allowing the reagents in the detector inlet access to the reaction chamber so that mixing can occur. When mixing is complete the interface valve is closed and the chip is ready for PCR. The loading process can be completed in 1 hour.

Spurgeon and colleagues (2008) have evaluated the performance of the 48 × 48 Fluidigm array, which accommodates 2304 individual reactions. Gene expression data was compared with publicly available data (GeneNote database http://bioinfo2.weizmann.ac.il/cgi-bin/genenote/home_page.pl) obtained by microarray and conventional real time PCR. The reproducibility of the microfluidic array was good for replicates both within and between chips and the results were essentially identical in quality the conventional real-time RT PCR data and superior to the microarray data for RNA extracted from the same tissue type.

An interesting application of a 48×48 array has been described by Guo and colleagues (2010), who examined the levels of 48 gene transcripts in single-cell extracts from cells at various stages of progression from zygote to blastocyst. The cells were separated manually and lysed to release mRNA for reverse transcription and sequence-specific preamplification of the cDNA before loading onto the array. It was possible to show that in cells of embryos at the morula stage (consisting of about 30 cells) each expressed transcription factors at the same levels but that by the 64-cell stage three cell types were clearly distinguished by transcription factor expression levels. Fluidigm arrays have been further developed to enable single cells to be isolated, extracted and amplified within the same device (White *et al.*, 2011). These authors described an integrated microfluidic device for RT-qPCR analysis of 50 single cells from six independent samples (i.e. 300 cells in total). All steps including single-cell capture, lysis, reverse transcription, and qPCR can be executed on the device.

Digital PCR

PCR arrays are generally applicable to digital PCR in which DNA is partitioned to a level at which individual target molecules are isolated and can be individually amplified (Sykes *et al.*, 1992). The application of digital PCR to target quantification, the original driver of the concept, has now been rendered largely redundant by real-time PCR. However, digital arrays may be an alternative to real-time arrays in various applications. For example, haplotyping of individual samples can be achieved using digital arrays that probe individual sample DNA molecules. Fluidigm Digital Arrays™ have been used for this application since they allow individual target molecules to be isolated, amplified and detected within the device (Menzel *et al.*, 2010).

References

Altschul, S.F., Gish, W., Miller, W., Myers, E.W., and Lipman, D.J. (1990). Basic local alignment search tool. J. Mol. Biol. *215*, 403–410.

Amicarelli, G., Shehi, E., Makrigiorgos, G.M., and Adlerstein, D. (2007). FLAG assay as a novel method for real-time signal generation during PCR: application to detection and genotyping of KRAS codon 12 mutations. Nucleic Acids Res. *35*, e131.

Arikawa, E., Sun, Y., Wang, J., Zhou, Q., Ning, B., Dial, S., Guo, L., and Yang, J. (2008). Cross-platform comparison of SYBR(R) Green real-time PCR with TaqMan PCR, microarrays and other gene expression measurement technologies evaluated in the MicroArray Quality Control (MAQC) study. BMC Genomics *9*, 328.

Belgrader, P., Benett, W., Hadley, D., Long, G., Mariella, R., Jr., Milanovich, F., Nasarabadi, S., Nelson, W., Richards, J., and Stratton, P. (1998). Rapid pathogen detection using a microchip PCR array instrument. Clin. Chem. *44*, 2191–2194.

Brenan, C.J., Roberts, D., and Hurley, J. (2009). Nanoliter high-throughput PCR for DNA and RNA profiling. Methods Mol. Biol. *496*, 161–174.

Czechowski, T., Bari, R.P., Stitt, M., Scheible, W.R., and Udvardi, M.K. (2004). Real-time RT-PCR profiling of over 1400 Arabidopsis transcription factors: unprecedented sensitivity reveals novel root- and shoot-specific genes. Plant J. *38*, 366–379.

De la Vega, F.M., Lazaruk, K.D., Rhodes, M.D., and Wenz, M.H. (2005). Assessment of two flexible and compatible SNP genotyping platforms: TaqMan SNP Genotyping Assays and the SNPlex Genotyping System. Mutat. Res. *573*, 111–135.

Dean, F.B., Nelson, J.R., Giesler, T.L., and Lasken, R.S. (2001). Rapid amplification of plasmid and phage DNA using Phi 29 DNA polymerase and multiply-primed rolling circle amplification. Genome Res. *11*, 1095–1099.

van Doorn, R., Szemes, M., Bonants, P., Kowalchuk, G.A., Salles, J.F., Ortenberg, E., and Schoen, C.D. (2007). Quantitative multiplex detection of plant pathogens using a novel ligation probe-based system coupled with universal, high-throughput real-time PCR on OpenArrays™. BMC Genom. *8*, 276.

Gaj, S., Eijssen, L., Mensink, R.P., and Evelo, C.T. (2008). Validating nutrient-related gene expression changes from microarrays using RT(2) PCR-arrays. Genes Nutr. *3*, 153–157.

Gelfand, D., and Kumar, K. (2009). High-throughput nanovolume qPCR: next-generation platform combines microarray throughput with sensitivity. Genet. Eng. Biotechnol. News 42–43.

Guo, G., Huss, M., Tong, G.Q., Wang, C., Li Sun, L., Clarke, N.D., and Robson, P. (2010). Resolution of cell fate decisions revealed by single-cell gene expression analysis from zygote to blastocyst. Dev. Cell *18*, 675–685.

Hartman, A.L., Ling, L., Nichol, S.T., and Hibberd, M.L. (2008). Whole-genome expression profiling reveals that inhibition of host innate immune response pathways by Ebola virus can be reversed by a single amino acid change in the VP35 protein. J. Virol. *82*, 5348–5358.

Herman, J.G., Graff, J.R., Myöhänen, S., Nelkin, B.D., and Baylin, S.B. (1996). Methylation-specific PCR: a novel PCR assay for methylation status of CpG islands. Proc. Natl. Acad. Sci. U.S.A. *93*, 9821–9826.

Huang, M.M., Arnheim, N., and Goodman, M.F. (1992). Extension of base mispairs by Taq DNA polymerase: implications for single nucleotide discrimination in PCR. Nucleic Acids Res. *20*, 4567–4573.

Jaspers, I., Horvath, K.M., Zhang, W., Brighton, L.E., Carson, J.L., and Noah, T.L. (2010). Reduced expression of IRF7 in nasal epithelial cells from smokers after infection with influenza. Am. J. Respir. Cell Mol. Biol. *43*, 368–375.

Mano, J., Shigemitsu, N., Futo, S., Akiyama, H., Teshima, R., Hino, A., Furui, S., and Kitta, K. (2009). Real-time PCR array as a universal platform for the detection of genetically modified crops and its application in identifying unapproved genetically modified crops in Japan. *J. Agric. Food Chem. 57*, 26–37.

Kodani, M., Yang, G., Conklin, L.M., Travis, T.C., Whitney, C.G., Anderson, L.J., Schrag, S.J., Taylor, T.H., Jr., Beall, B.W., Breiman, R.F., *et al.* (2011). Application of TaqMan low-density arrays for simultaneous detection of multiple respiratory pathogens. J. Clin. Microbiol. *49*, 2175–2182.

Kibriya, M.G., Raza, M., Jasmine, F., Roy, S., Paul-Brutus, R., Rahaman, R., Dodsworth, C., Rakibuz-Zaman, M., Kamal, M., and Ahsan, H. (2011). A genome-wide DNA methylation study in colorectal carcinoma. BMC Med. Genomics *4*, 50.

Latorra, D., Campbell, K., Wolter, A., and Hurley, J.M. (2003). Enhanced allele-specific PCR discrimination in SNP genotyping using 39 locked nucleic acid (LNA) primers. Hum. Mutat. *22*, 79–85.

Li, B., Kadura, I., Fu, D., and Watson, D.E. (2004). Genotyping with TaqMAMA. Genomics *83*, 311–320.

Li, J., Wang, L., Mamon, H., Kulke, M.H., Berbeco, R., and Makrigiorgos, G.M. (2008). Replacing PCR with COLD-PCR enriches variant DNA sequences and redefines the sensitivity of genetic testing. Nature Med. *14*, 579–584.

Menzel, S., Qin, J., Vasavda, N., Thein, S.L., and Ramakrishnan, R. (2010). Experimental generation of SNP haplotype signatures in patients with sickle cell anaemia. PLoS One *5*, e13004.

Miklos, G.L., and Maleszka, R. (2004). Microarray reality checks in the context of a complex disease. Nat. Biotechnol. *22*, 615–621.

Morlan, J., Baker, J., and Sinicropi, D. (2009). Mutation detection by real-time PCR: a simple, robust and highly selective method. PLoS One *4*, e4584.

Newton, C.R., Graham, A., Heptinstall, L.E., Powell, S.J., Summers, C., Kalsheker, N., Smith, J.C., and Markhamet, A.F. (1989). Analysis of any point mutation in DNA. The amplification refractory mutation system (ARMS). Nucleic Acids Res. *17*, 2503–2516.

Nilsson, M., Malmgren, H., Samiotaki, M., Kwiatkowski, M., Chowdhary, B.P., and Landegren, U. (1994). Padlock probes: circularizing oligonucleotides for localized DNA detection. Science *265*, 2085–2088.

Orou, A., Fechner, B., Utermann, G., and Menzel, H.J. (1995). Allele-specific competitive blocker PCR: a one-step method with applicability to pool screening. Hum. Mutat. *6*, 163–169.

Prix, L., Uciechowski, P., Böckmann, B., Giesing, M., and Schuetz, A.J. (2002). Diagnostic biochip array for fast and sensitive detection of K-ras mutations in stool. Clin. Chem. *48*, 428–435.

Roberts, D.G., Morrison, T.B., Liu-Cordero, S.N., Cho, J., Garcia, J., Kanigan, T.S., Munnelly, K., and Brenan, C.J. (2009). A nanoliter fluidic platform for large-scale single nucleotide polymorphism genotyping. Biotechniques *46*, Suppl. ix–xiii.

Rom, J., von Minckwitz, G., Marmé, F., Ataseven, B., Kozian, D., Sievert, M., Schlehe, B., Schuetz, F., Scharf, A., Kaufmann, M., *et al.* (2009). Phase I study of apoptosis gene modulation with oblimersen within preoperative chemotherapy in patients with primary breast cancer. Ann. Oncol. *20*, 1829–1835.

Saunders, N.A. (2011). Applications of nanomaterials to arrays for infectious disease diagnosis. Nanomed. *6*, 271–280.

Seyama, T., Ito, T., Hayashi, T., Mizuno, T., Nakamura, N., and Akiyama, M. (1992). A novel blocker-PCR method for detection of rare mutant alleles in the presence of an excess amount of normal DNA. Nucleic Acids Res. *20*, 2493–2496.

Shi, L., Reid, L.H., Jones, W.D., Shippy, R., Warrington, J.A., Baker, S.C., Collins, P.J., de Longueville, F., Kawasaki, E.S., Lee, K.Y., *et al.* (2006). The MicroArray Quality Control (MAQC) project shows inter- and intraplatform reproducibility of gene expression measurements. Nat. Biotechnol. *24*, 1151–1161.

Spurgeon, S.L., Jones, R.C., and Ramakrishnan, R. (2008). High throughput gene expression measurement with real time PCR in a microfluidic dynamic array. PLoS One *3*, e1662.

Sykes, P.J., Neoh, S.H., Brisco, M.J., Hughes, E., Condon, J., and Morley, A.A. (1992). Quantitation of targets for PCR by use of limiting dilution. Biotechniques *13*, 444–449.

Tobler, A.R., Short, S., Andersen, M.R., Paner, T.M., Briggs, J.C., Lambert, S.M., Wu, P.P., Wang, Y., Spoonde, A.Y., Koehler, R.T., *et al.* (2005). The SNPlex genotyping system: a flexible and scalable platform for SNP genotyping. J. Biomol. Tech. *16*, 398–406.

Tomas, C., Stangegaard, M., Børsting, C., Hansen, A.J., and Morling, N. (2008). SNPforID Consortium. Typing of 48 autosomal SNPs and amelogenin with GenPlex SNP genotyping system in forensic genetics. Forensic Sci. Int. Genet. 3, 1–6.

Ugozzoli, L.A., Latorra, D., Puckett, R., Pucket, R., Arar, K., and Hamby, K. (2004). Real-time genotyping with oligonucleotide probes containing locked nucleic acids. Anal. Biochem. *324*, 143–152.

Wang, Y., Barbacioru, C., Hyland, F., Xiao, W., Hunkapiller, K.L., Blake, J., Chan, F., Gonzalez, C., Zhang, L., and Samaha, R.R. (2006). Large scale real-time PCR validation on gene expression measurements from two commercial long-oligonucleotide microarrays. BMC Genom. *7*, 59.

White, A.K., Van Insberghe, M., Petriv, O.I., Hamidi, M., Sikorski, D., Marra, M.A., Piret, J., Aparicio, S., and Hansen, C.L. (2011). High-throughput microfluidic single-cell RT-qPCR. Proc. Natl. Acad. Sci. U.S.A. *108*, 13999–14004.

West, J., and Damania, B. (2008). Upregulation of the TLR3 pathway by Kaposi's sarcoma-associated herpesvirus during primary infection. J. Virol. *82*, 5440–5449.

Wu, D.Y., Ugozzoli, L., Pal, B.K., and Wallace, R.B. (1989). Allele-specific enzymatic amplification of beta-globin genomic DNA for diagnosis of sickle cell anemia. Proc. Natl. Acad. Sci. U.S.A. *86*, 2757–2760.

Zhang, C., and Xing, D. (2007). Miniaturized PCR chips for nucleic acid amplification and analysis: latest advances and future trends Nucl. Acids Res. *35*, 4223–4237.

Zhang, C., Xu, J., Ma, W., and Zheng, W. (2006). PCR microfluidic devices for DNA amplification. Biotechnol. Adv. *24*, 243–284.

The Validation of Real-time PCR Assays for Infectious Diseases 10

Melvyn Smith

Abstract

The real-time polymerase chain reaction is now established as one of the core technologies for diagnosing infectious diseases. The early stages of the technique's development were followed by a dramatic increase in the number of diagnostic assays being published, together with the introduction of commercially produced tests. Each of the numerous publications showed a number of differences in the approach to validating the newly produced assays and in the quality and quantity of the data supporting their validation. As a result, many workers have, at times, found it difficult to reproduce the published results from other laboratories. These difficulties can arise from, for example, a lack of information in the publication, differences in equipment between laboratories, the use of different extraction methods and sequence variations in the pathogen being detected. Over the years a number of authors have voiced their concerns over the subject of what constitutes a properly validated assay, highlighting the issues of basic scientific good practice and the responsibilities of journals in publishing full validation data. This chapter summarizes the recent work covering validation and verification methodology in order to provide a practical guide to help inform and standardize the process.

Introduction

The PCR (Saiki *et al.*, 1985; Mullis *et al.*, 1986) and in particular, real-time PCR, (Higuchi *et al.*, 1992, 1993) is now one of the most widely used methods for the routine diagnosis of infectious diseases (Weile and Knabbe, 2009). This is particularly so for viral infections, where, in many cases, the pathogen cannot be grown or is difficult to grow in culture. As the technology has developed, the range of molecular tests has expanded to include qualitative assays to detect a single analyte or target, quantitative assays used to monitor the effectiveness of antimicrobial drug therapy and multiplex assays capable of detecting two or more pathogens in a single sample.

For many years, the development of assays took place in the laboratory carrying out the tests, the so-called 'in-house' or laboratory-developed assay. Although more and more commercially developed tests have become available, novel assays continue to be developed in academic hospital and diagnostic laboratories, which are more able to respond quickly to new and re-emerging infections. This was the case with the coronavirus causing severe acute respiratory syndrome (SARS), first recognized as an atypical pneumonia in the Guangdong province of China in November 2002 (Mahony *et al.*, 2004). The full sequence of the virus was obtained within 3 weeks of the start of an international cooperative project involving teams from the USA and Canada (Marra *et al.*, 2003). This was shortly followed by the first in-house reverse transcription real-time PCR assay (Drosten *et al.*, 2003); it was not until the following year that two commercial kits became available (Drosten *et al.*, 2004). Similarly, the appearance of swine influenza A H1N1 virus in the spring of 2009 was swiftly followed by the release of three in-house assays from the Health Protection Agency in the UK, the Centers for Disease Control and Prevention in America and the National Virus Reference Laboratory in Ireland (Ellis *et al.*, 2009).

Although often considerably more expensive than in-house assays, commercial kits do, however, enable the rapid introduction of new tests into laboratories because design and optimization have already been completed. Commercial assays also benefit from consistent and standardized manufacturing procedures, resulting in more reproducible results. Perhaps the single most advantageous aspect of using commercial diagnostic kits is that they are CE marked or FDA approved, providing some form of reassurance to the laboratory. However, CE marking is only a declaration of compliance with European legislative requirements, it does not necessarily guarantee the rigorous validation of the assay, or give any details of how and when a test should be used (RCP, 2006). Furthermore, commercial assays, by definition, have to be commercially attractive. Therefore, specialist, small-scale tests for rarely occurring infectious pathogens will not be cost-effective from a commercial standpoint. Thus, there will continue to be a need for laboratory-developed assays.

The need for validation

In-house assays

Over the years many laboratories have established a workable methodology for validating in-house assays. However, a study of the literature covering the dramatic growth of both research and diagnostic real-time PCR assays continues to show a rather scattered approach to the problem. Some of the difficulties encountered in the literature include; a lack of detail in the optimization of extraction techniques, no details of the methods used in primer and probe design, no evidence of amplicon sequencing to confirm assay specificity, imprecise estimates of sensitivity and specificity and assays that do not include internal or extraction controls (Apfalter *et al.*, 2005; Bustin *et al.*, 2009; Murphy and Bustin, 2009). Clearly a lack of detailed experimental information in the original publication will make assessing the clinical utility of the assay difficult. This is a particular problem in the developing world, where clinical care can be compromised by the lack of regulatory controls and where clinicians are sometimes faced with having to select tests on published data that can originate from inadequately designed studies (Peeling *et al.*, 2008). Some of these problems lie with the scientific literatures' approach to publishing, where space limitations often restrict the amount of detail permitted in a paper. These difficulties have led to a number of papers calling for an improvement in the standards of reporting diagnostic assays, including the STARD initiative (standards for reporting of diagnostic accuracy; Bossuyt *et al.*, 2003), DEEP (Diagnostic Evaluation Expert Panel; Peeling *et al.*, 2008) and the proposed MIQE guidelines (the minimum information for publication of quantitative real-time PCR experiments; Bustin *et al.*, 2009).

In addition to the fundamental requirements of good scientific practice, there has been an ongoing debate in Europe regarding the validation of in-house assays. The public consultation on the revision of Directive 98/79/EC of the European Parliament and of the Council on 27 October 1998 on *In Vitro* Diagnostic Medical Devices has once again raised the prospect of CE marking in-house diagnostic assays. Provision was made in the Directive for exemption for devices used on the premises of the original manufacturer (or the immediate vicinity), referring to the in-house diagnostic assay. The revision sets out to remove or modify this exemption to ensure suitable safety standards and prevent unfair competition between in-house assays and CE marked *in vitro* diagnostic tests. While there is insufficient space here to cover the complexities associated with the existing and potential legislation, there is clearly a need for change. This need is recognized not only by government bodies, but also researchers working in the discipline (Apfalter *et al.*, 2005; Bustin *et al.*, 2009; Raymaekers *et al.*, 2009; Burd, 2010; Cohen, 2010; TDR, 2010). However, the imposition of rigorous controls on the development of assays must not stifle innovation and the ability of front-line laboratories to respond quickly to new and emerging threats. Under these circumstances it makes sense for laboratories developing and publishing assays to ensure that they produce rigorously documented experimental proof to support any potential CE-marking/IVDD requirements that may be necessary in the future.

Commercial assays

Although commercial assays are generally CE marked or FDA approved (typically for very specific sample types), this does not necessarily mean the assay has been validated to a defined standard for all applications. Furthermore, although the assay may perform acceptably in the commercial developer's laboratory, a number of factors can affect the assay's performance elsewhere. There may be issues with staff competences and workflow systems. For example, not all laboratories have separate working areas recommended to reduce contamination. There may also be differences in equipment maintenance schedules, which can include anything from freezers and pipettes to thermal cyclers and can fundamentally affect the performance of an assay. Although manufacturers of commercially produced FDA or IVD-CE marked kits are responsible for the performance of the assays, it is still necessary for the implementing laboratory to verify the stated criteria (Raymaekers *et al.*, 2009; Burd, 2010). Although there are no formal requirements in the UK for laboratories to evaluate newly introduced commercial assays (Price and Christenson, 2008), in the USA, FDA-approved, FDA-modified [i.e. tests modified from the manufacturer's instructions but with performance characteristics determined by the user laboratory in a manner consistent with Clinical Laboratory Improvement Act (CLIA) requirements] and laboratory-developed assays, require verification before introducing the test into the clinical laboratory (Sloan, 2007). CLIA stipulates that prior to implementation of an FDA-cleared test, laboratories must verify the following manufacturer's performance specifications; accuracy, precision, reportable range and reference intervals. Where a FDA-modified test or laboratory-developed test is to be introduced, CLIA stipulates that, together with the above characteristics, analytical sensitivity and specificity must be established (Sloan, 2007).

In an attempt to clarify the situation, it is the intention in this Chapter to provide a set of basic working guidelines that form the basis of the validation process, culled from some of excellent publications available and experiences in our laboratory. These guidelines can be applied throughout the continuous process of maintaining the validated status of an assay. The key concepts in the process of validating an assay are illustrated in Fig. 10.1.

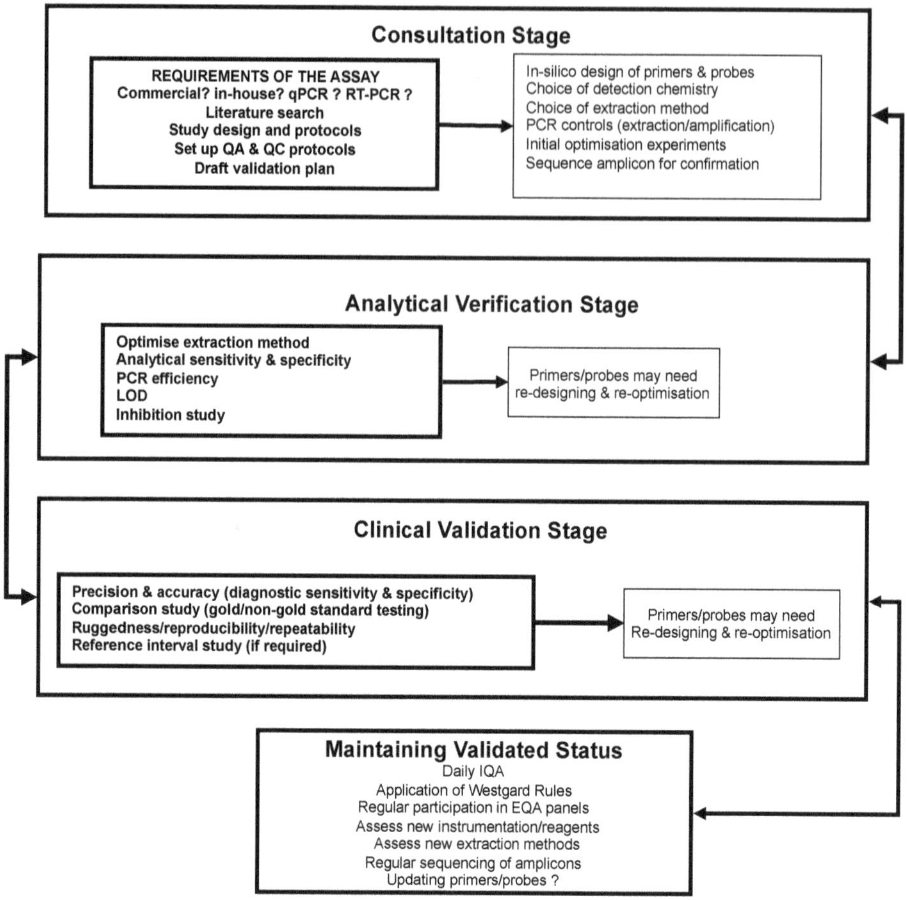

Figure 10.1 Assay validation – key components.

Some definitions

Accuracy

A term, now replaced by 'trueness' used to describe the level of agreement between the test being developed and another comparative reference method or the gold standard (Sloan, 2007; Burd, 2010). Trueness is defined by the ISO as 'the closeness of agreement between the average value obtained from a large series of tests results and an accepted reference value' (Hauk *et al.*, 2007). Further definitions include, the difference between actual and experimentally measured values expressed as fold-changes or estimates of copy numbers (Bustin *et al.*, 2009). Experiments to determine trueness are fundamental to the clinical validation of an assay.

Contamination

Contamination in a clinical laboratory is a potentially serious problem. Reporting false-positive results can lead to issues with the management of patients, protracted delays and significant financial costs in the laboratory until the problem is resolved. It is crucially important to ensure that as many safeguards are in place to reduce the risk of contamination to an absolute minimum. While completely eliminating the problem is impossible, a number of measures can be taken to minimize both its occurrence and subsequent impact. These measures can be divided into three main areas: (i) the layout of the laboratory, (ii) the design of the assay itself and (iii) training staff in the use of the assay.

The two major sources of difficulties are with cross-contamination from positive to negative samples and cross-contamination of amplified target from previous reactions. These problems can be minimized by carefully planning the layout of the laboratory and implementing good laboratory practice (GLP). There have been numerous publications dealing with the subject (Prince and Andrus, 1992; Hartley and Rashtchian, 1993; Miffin, 1995; Espy *et al.*, 2006; Chen *et al.*, 2007). There is insufficient space here for a detailed discussion on the subject; however, it is worth presenting a brief outline of the issues. The basic requirements are to separate the processes involved into the three main functions, which are:

- a clean laboratory for reagent preparation;
- a specimen laboratory for extraction and making up the final PCR reaction mixes; and
- an amplification laboratory for carrying out the PCR.

Post amplification analysis must take place outside the amplification laboratory; this is much easier to accomplish if the amplification equipment is networked and the software loaded onto remote computers, located in separate authorization areas. The workflow-pattern proceeds from the clean area to the specimen/extraction area to the amplification area and on to the analysis of results. Separate lab coats and gloves must be provided for each area and it is useful to use specific colour-coded gloves and lab coats for each area. Strict GLP must be enforced in each area, i.e. hand-washing before leaving each area, changing gloves frequently, particularly in the specimen laboratory, regular cleaning of all surfaces with hypochlorite solutions (Prince and Andrus, 1992) and the use of PCR hoods with ultraviolet lighting.

The design of the assay can have a significant impact on reducing potential contamination. The use of a suitable number and carefully placed negative controls (including environment controls) will give an indication of contamination before it becomes a serious issue. Using one-step reverse transcription PCR, rather than two-step, will reduce the number of times tubes containing potential contaminants are opened. Similarly, if possible, avoid using nested PCR; this will significantly reduce the potential for amplicon carryover. The technique, an early development of standard PCR, was used in a number of studies to increase sensitivity and specificity. Amplicons are re-amplified in a second-round reaction, using an inner, nested set of primers. The amplicons for the second round of PCR have to be transferred from the opened tubes from the first round reaction, providing a perfect opportunity for amplicon to contaminate the laboratory. Regular, documented use of UV-equipped PCR cabinets will help to control contamination.

Designing assays employing uracil DNA glycosylase (UNG) can minimize potential amplicon carryover into subsequent reactions (Longo *et al.*, 1990). UNG *in vivo* removes uracil from DNA by cleaving the N-glycosidic bond between the base and sugar phosphate backbone and the resulting apyrimidinic sites prevent replication by DNA polymerase. Substituting deoxyuridine for deoxythymidine in a PCR, results in amplicons containing dU residues. By treating subsequent reactions with UNG any potential amplicon contaminants will be degraded and cannot be amplified. The use of pre-mixed, small aliquot, single use reagents simplifies set-up procedures and reduces the number of samplings from stocks. Regular environmental swabbing of all work areas is recommended and will potentially provide an indication of the problem before contamination becomes a serious issue.

Training and assessing the competency of staff carrying out the tests is a major step in reducing the potential for contamination. Staff involved in molecular diagnostic work must be trained in all aspects of the procedure, from setting up the reactions to reporting the results. Staff training and competencies must be documented and regularly reviewed by running tests with previous positive and negative specimens. The introduction of regular bleaching of work areas before and after use and regular glove changes is recommended and the procedure should be documented as part of the laboratories GLP guidelines.

Crossing point/crossing threshold/threshold cycle

These are some of the terms used in the literature to describe the same feature, i.e. the point at which the amplification reaction enters the exponential phase, or the fractional cycle used for quantification. For quantification assays (qPCR) the term 'quantification cycle' (Cq) should be used (Bustin *et al.*, 2009: Lefever *et al.*, 2009). Where necessary, for qualitative assays C_t can be used.

Inhibition

There are two types of inhibition to be considered, that due to interfering substances in the sample post-extraction and poor sensitivity of the primers and/or probe. Inhibition can be monitored by including an internal amplification control (IAC) that is co-amplified with the target. If, in a sample, the IAC fails to amplify, or shows an increase of three or more cycles in amplification compared with the average cycle number in the negative controls, that sample is considered inhibitory to the PCR. The rationale behind this is that an increase of approximately three cycles represents a one log reduction between amplification of the two controls, a figure sufficient to indicate inhibition. This represents a rather simplistic view of the problem, since the effects of inhibition can be considerably more subtle than a straightforward elimination or suppression of the PCR in a sample. The reduction of amplification efficiency can have significant effects on quantitative results; a fall in amplification efficiency of just 5% can result in a three-fold difference in the amount of amplicon after 25 cycles (Sisti *et al.*, 2010). Where the amplification efficiency of a sample is less than expected and is not due to problems of inhibitory substances, it may be due to sequence variation in the primer/probe binding region, in which case the amplicon must be sequenced. This type of ongoing analysis of an assay's efficiency must form part of the long-term validation process. This is particularly so in the case of retroviruses which are prone to mutation due to the lack of proofreading activity in the viral RNA polymerase.

Linear dynamic range/reportable range

These are interchangeable terms describing the range of target concentrations over which the PCR is linear, i.e. the lowest to highest copy number of template detected, established from the calibration curve determinations (Bustin *et al.*, 2009). The calibration curve should be based on a log-dilution series of four to six orders (four as a minimum); the correlation coefficient or coefficient of determination (r^2) must be reported together with confidence intervals (CIs) for the full dynamic range. A reportable range does not need to be determined for qualitative assays. For quantitative assays only results within the linear dynamic range can be reported.

Normalization

A crucial process in quantification assays, where normalization controls for extraction efficiencies, the yield from reverse-transcription and amplification efficiency. This enables comparisons of mRNA across different samples. The most common method of providing a normalization base-line measure is through the use of reference genes.

PCR controls

There are a number of controls that must be used each time a diagnostic PCR assay is run. These include a positive control, usually a positive sample that can be diluted for long-term use. It should be titrated to run at a level that can be consistently amplified, but at the lower limit of the assay. A number of negative or no-template controls (NTCs) should be run to monitor for contamination. In our laboratory, we run one NTC after every eight samples. In addition, the NTC should also be set up as an environment control, where the tube is left open in the PCR laboratory and is used to monitor the presence of contaminating nucleic acid in the environment. An internal amplification control (IAC) is a non-related amplification target, added to each reaction to monitor the PCR reagents and the thermal cycling programme. This control can be constructed using standard molecular biology methods, incorporating primer-binding sites, ligated to the ends of a non-homologous sequence that is amplified with equal efficiency to the target by the same primers, but detected with its own sequence-specific probe. An extraction control is added to the sample and co-extracted with it. The type of system used depends on whether a DNA or RNA virus is being assayed. For reverse transcription assays, an RNA control must be used to monitor the RT step following extraction. A number of different approaches have been employed, using T4 bacteriophage as a control with the DNA viruses HSV-1 and HSV-2 (Gerriets *et al.*, 2008) and MS2 bacteriophage with RNA viruses such as norovirus (Rolfe *et al.*, 2007). The advantage of using this approach rather than spiking naked nucleic acid into the sample prior to extraction is that the lysis steps in the procedure are fully controlled. This is a more accurate replication of the conditions that the target pathogen is subjected to during extraction.

PCR efficiency

Amplification efficiency may be measured by the use of calibration curves; this is the most effective method of easily establishing a rapid and reproducible indication of mean PCR efficiency (Bustin *et al.*, 2009). A set of log-diluted calibration standards can be used for this (as used in analytical sensitivity-testing) and efficiency determined from the slope of the calibration curve, where a theoretical maximum value of 1.00 (or 100%) indicates the amount of product doubles with each cycle of amplification.

Precision

A measure of the variation in an assay, which can be accounted for by a number of factors, such as temperature fluctuations affecting annealing fidelity and/or denaturation, pipetting errors affecting the concentration of reaction components and stochastic variation. Precision typically varies with the concentration of the template, reducing as template copy numbers fall. Technical variance in precision measurements should be expressed as SD error bars or CIs on calibration curves with replicate samples of varying concentrations (Bustin *et al.*, 2009). Biological replicates are used to demonstrate statistical variations between groups or treatments and it may also be necessary to report interassay precision between the different laboratories involved in the validation exercise.

Quantification standards

A set of externally amplified templates, typically constructed using standard cloning techniques from which copy numbers of the target in a sample can be compared and therefore, estimated. Careful validation of the amplification efficiencies for the target and calibration standards needs to be carried out to ensure that relevant comparisons can be made. This is because many standards are constructed by cloning a similar, but non-related (to the target pathogen) sequence into a plasmid and such closed-circle constructs do not usually amplify with the same efficiency as a linear template. If problems are found, linearizing the plasmid clone with a suitable restriction enzyme may help to improve amplification efficiency to that of the target pathogen. It may also be necessary to combine the standards in a matrix that replicates the genetic complexity of the samples, where a non-related genome can be used for this purpose, typically salmon or herring sperm DNA.

Reference genes/housekeeping genes

The term housekeeping gene is no longer recommended, reference gene should be used instead (Bustin *et al.*, 2009). For normalization, the ratio of the concentration of the reference genes to those of the target genes of interest is calculated and typically expressed as fold changes. Careful consideration to the choice of reference genes is required and normalization against a single gene is not considered acceptable. The reference genes selected must be shown to be consistently expressed in the tissues or cells under investigation and not vary in response to experimental treatment (Vandesomple *et al.*, 2002).

Reference interval

This is defined as the normal range of values of the analyte being investigated from the lower reference limit to the upper reference limit (Sloan, 2007). Typically for qualitative assays where the nucleic acid target is always absent in healthy individuals, the reference value is negative (or not detected) and evaluation is unnecessary (Burd, 2010). For quantitative assays, the reference interval is usually reported as below a particular quantitative value (e.g. *X* copies/ml), often this will be the LOD of the assay. In some cases the reference interval may be different from the LOD, particularly in the case of viruses that exhibit a latent life cycle, such as CMV or EBV, where a cut-off, distinguishing asymptomatic infection from active disease has to be established.

Reproducibility

This is a measure of the test's ability to provide consistent results, demonstrated by estimates of precision. This is best achieved through inter-laboratory proficiency testing. A minimum of three participating laboratories should be included in the trial, each of which receives aliquots of the same set of samples for testing. The panel can consist of a subset of that used in the comparative study of precision and accuracy. A minimum of 20 samples (including negative specimens) should be tested providing additional information on the assay's robustness.

Robustness

This parameter is a measure of the assay's ability to withstand slight variations in reagents and machine-to-machine differences while still providing the expected results. Robustness is most usefully measured by monitoring the performance of an internal reaction control included in each run of the assay.

Sensitivity and specificity

These two fundamental characteristics of an assay can cause confusion if used imprecisely when describing analytical and diagnostic (or clinical) performance. The analytical sensitivity of an assay defines its ability to reliably detect very low levels of target molecule in a biological specimen. Analytical sensitivity is commonly referred to as the limit of detection (LOD), which is the lowest concentration of an analyte consistently detected (n=95%) in the specimens tested (Saah and Hoover, 1997; Burd, 2010). Analytical sensitivity is empirically derived by testing serial (log) dilutions of an extracted specimen with a known concentration of the target molecule. Clinical sensitivity is defined as the percentage of individuals with a given disorder correctly identified as positive by the assay (Saah and Hoover, 1997).

Analytical specificity is a measure of an assay's ability to exclusively identify the intended gene-target of the pathogen and not components of related or different pathogens likely to be found in the type of specimen under investigation. Pathogens likely to cause similar symptoms should also be included in the specificity analysis. Clinical specificity is a measure of the proportion of previously confirmed negative specimens that test negative with the new assay (Saah and Hoover, 1997; OIE, 2010).

Verification and validation

The terms verification and validation are often used interchangeably in the literature when describing the process of determining the clinical utility of an assay. The ISO define verification as 'confirmation through the provision of objective evidence, that specified requirements have been fulfilled' (ISO, 2007) and validation as 'confirmation through the provision of objective evidence, that the requirements for a specific intended use or application have been fulfilled' (ISO, 2007). The fact that the ISO has gone to the trouble of providing separate definitions for the two terms indicates that they should be used in a specific way. However, the distinction between these two definitions is not immediately clear. In the context of this chapter the term 'verification' is used to define the process of establishing whether the individual components of an assay meet the analytical performance requirements established at the start of the development process. 'Validation' refers to the process of ensuring that the completed assay conforms to the users' needs, requirements, and/or specifications under defined operating conditions. Validation is an ongoing process and, therefore, encompasses both quality control and quality assurance plans, put in place at the start of verification to ensure that the assay continues to meet the original design specifications (Sloan, 2007). There are numerous aspects of an assay that need to be continuously monitored throughout its use in the laboratory. Micro-organisms, particularly viruses, often mutate. This means that the efficiency of the PCR needs to be monitored for potential false-negative results, as this may be the first sign that the primers and/or probe need to be up-dated and re-validated. Manufacturers continually develop new buffers, enzymes and extraction kits, these must be assessed and if adopted, re-verified, and the assay validated once more. New types of probe chemistry are also constantly being introduced and these, too, need testing to maintain or improve the assay's performance.

The validation process

The process begins with the development of a validation plan and involves preliminary decisions based on the clinical need for the assay (Budowle, *et al.*, 2008). This is defined at the start of the process because the way the assay will be used (epidemiological studies, infection control, screening etc.) can affect the choice of commercial assay or the fundamental design of a laboratory-developed test (Peeling *et al.*, 2008). As discussed previously, there are differences between the levels of experimental validation laboratories are expected to perform when introducing commercial or in-house assays. Nevertheless, as a minimum, the laboratory introducing a commercial test must establish that the manufacturer's performance claims can be reproduced. Once the performance characteristics of an assay have been met, whether commercial or laboratory-designed, the validation exercise must continue on a daily basis (see Fig. 10.1). This involves continually monitoring the levels of internal and external positive controls to ensure the validation status of the assay is maintained (Budowle *et al.* 2008; Raymaekers *et al.*, 2009; OIE, 2010).

Preliminary considerations

The initial step is to define clearly the purpose of the assay; all the subsequent steps in the validation process are guided by this decision (OIE, 2010). The following three variables that can affect an assay's performance must be considered at this stage:

1 The sample type and the host–pathogen interactions, which determine whether a qualitative or quantitative assay is required.
2 The assay system: the biological, technical and operator-related factors, which affect the assay's ability to detect the target in the specific sample type.
3 The result: does the result accurately predict the status of an individual or population in regard to the analyte detected?

The sample type (e.g. tissue, whole blood, CSF) in which the analyte is present may contain endogenous inhibitors that can significantly affect the enzymatic activity of the polymerase in the PCR. Therefore, the extraction process needs to be evaluated and necessary alternatives considered. Quantification assays are useful in the management of patients, where pathogen load can be monitored during therapy (Madej *et al.*, 2010). Some viral pathogens, such as EBV and CMV, can establish a primary low-level, latent infection and subsequently become reactivated, switching to a productive lytic infection. In such cases it is necessary to quantify the virus, rather than use a qualitative assay, to distinguish between latent infection and lytic disease (Jebbink *et al.*, 2003). There may also be non-host factors, such as contamination, stability of the nucleic acid and post extraction cold-storage facilities (particularly for RNA targets), that may affect the subsequent performance of the assay. Another significant factor is the availability of sufficient numbers of well-characterized positive control samples to enable the validation experiments to be carried out.

Further points to consider relate to instrumentation choice (Will the reaction chemistry work on the thermal cycler?), reagent choice and staff resources and training. The possibility of reagents, or consumables as a source of contamination, either from the target analyte, or one closely related, is a significant problem, particularly for bacterial and pan-fungal PCRs targeting ribosomal genes (Millar *et al.*, 2002; Newsome *et al.*, 2004).

The factors that can affect an assay's ability to predict accurately the infection status of the host are related to its diagnostic or clinical sensitivity and specificity, together with the prevalence of the disease in the local population. Since sensitivity and specificity measure will be related to the reference panel tested, careful consideration must be given to this aspect of the validation plan. Because of the potential difficulties in assessing the data, it is important, where possible, to use well-characterized specimens or reference samples with known target values (Burd, 2010). In order to maintain objectivity, the method chosen to resolve discrepant results must be established as part of the validation plan before testing begins. Specific factors that could influence sensitivity and specificity, for example, sequence variation amongst different viral subtypes, as found in HIV-1, must also be taken into account at the planning and design stage of the assay. Arguably the most critical attribute of an assay is its ability to accurately determine the positive or negative status in a sample for the analyte under investigation. The positive and negative predictive values are heavily influenced by the prevalence of the disease in a given population. If accurate and up-to-date reference intervals are not obtained, the interpretation of positive and negative results may be difficult (OIE, 2010).

A quality assurance plan must be established for the assay, this is normally an addition to the laboratory's existing scheme. Consideration must also be given to the availability of external QA reagents; this can be a problem with new assays targeting rare pathogens, where the established providers of QA panels are unlikely to have a suitable reagent. Under these circumstances, the developing laboratory may need to consider working with the providers to develop a suitable panel. Once the requirement for a test has been established, the next decision to be made concerns whether to use a commercial assay or develop an in-house test. In the USA the CLIA 88 rules, updated in 2003, specify that laboratories using FDA-cleared assays must verify that the manufacturer's performance specifications for accuracy, precision, reportable range and reference intervals can be replicated by their personnel when following the procedure (Sloan, 2007; Raymaekers *et al.*, 2009; Burd, 2010). This is the approach typically adopted by many laboratories and should be established as a formal protocol, incorporated into the laboratory's validation plan and QA paperwork. It is suggested that the following aspects of the assay should be tested (Sloan, 2007):

- accuracy/trueness (qualitative)
- accuracy/trueness (quantitative)
- precision
- reportable range (quantitative)
- reference range (normal values).

For modified commercial assays the following should also be tested:

- analytical sensitivity (LOD)
- analytical specificity.

The validation plan

Preparation of the validation plan starts by defining the criteria used to evaluate the performance of the method (Budowle *et al.*, 2008). Although each new assay may present its own particular challenges, the basic criteria for validation, such as specificity, sensitivity and reproducibility will apply in most cases and any specific differences can be incorporated without difficulty. The first steps concern decisions based around the clinical need for the test, i.e. the scope, purpose and application of the assay and whether a commercial or in-house design will be required. It is useful at this stage to consider the validation documentation that will be submitted for inclusion in the laboratory standard operating procedure (SOP) and/or for publication. For all in-house developments it is strongly recommended that the comprehensive guidelines as laid down in the MIQE guidelines are followed (Bustin *et al.*, 2009) and described in Chapter 11.

The individual components of the initial verification study, performed prior to validation will now be considered.

Analytical verification

Reference materials and sample numbers

There are a number of factors related to specimen type that need to be considered. Selection, preparation and storage of samples are crucial to the function of an assay (OIE, 2010). The first question that arises when developing in-house assays, particularly for rare and emerging infectious disease pathogens, is the availability of samples. If insufficient samples are available in the laboratory developing the test, are they obtainable elsewhere? If they are not, it may be necessary to construct test samples by spiking various concentrations of the analyte into a suitable matrix. Consideration needs to be given to potential inhibitory substances, likely to be found in the type of specimens routinely tested. Therefore, paired control specimens should be prepared by adding low concentrations of the analyte, with and without the known inhibitors, to suitable negative samples (Burd, 2010). However, it must be recognized that although it may be the only approach initially available, such artificially constructed samples will not necessarily have the same properties as 'natural' samples. Therefore, when sufficient numbers of genuine samples do become available, the assay and extraction methods may need to be re-assessed. Where samples are available, their type is normally dictated by the pathogen to be analysed, although more than one specimen-type may be used. For example, respiratory pathogens can be extracted from nasopharyngeal aspirates, bronchoalveolar lavages or nose and throat swabs. A literature search should always be carried out to determine the range of disease-specific specimens to be assessed (Raymaekers *et al.*, 2009). The different types of specimen may each require different extraction techniques and alternative methods may be available. In such cases a comparative study should be conducted to determine the most efficient extraction method for each of the sample types used in the assay.

The number of samples required during the verification stages will vary depending on the specific parameters to be measured. For example, when determining the reportable range, 7–11 concentrations of the analyte across the expected range should be tested (Burd, 2010). More details of sample numbers will be given in each of the following specific sections. In general, there is no set figure for the number of specimens required for validation, in our laboratory, we typically use a 100 sample-set for initial evaluations, which comprises 80 positive and 20 negative samples. Sloan (2007) recommends that 50 positive and 50 negative samples should be tested over at least 10 days for both commercial and in-house developed qualitative tests. Ideally the number of samples should be determined by statistical analyses, where the sample size needed to detect a significant difference is determined by the standard deviation from the difference of the means of paired samples used in the test under development and the gold standard comparator (Burd, 2010).

When assessing a commercial assay the extraction process will also need to be verified. In most cases the manufacturer's protocol will include details on the type of extraction process to use, although occasionally the assay will need to be validated with a number of different extraction methods. The choice of technique in many laboratories is determined by the type of equipment available. Whatever approach is taken, a suitable sample size must be used to confirm that the assay performs satisfactorily.

Template, primers and probes

The first choice that needs to be considered is whether to use sequence-specific probes or non-specific labels to detect the analyte. Dyes such as SYBR green 1 have been used since the inception of real-time PCR and can be useful, particularly when exploiting melt-curve analyses to differentiate between viral subtypes or resistance mutations. However, since intercalating dyes bind to all dsDNA species in the reaction, their use is somewhat limited in diagnostic PCR, where the additional security provided by sequence-specific probes is preferred (Espy, *et al.*, 2006; Gunson *et al.*, 2006).

A thorough literature review is first carried out to identify a suitable locus to be amplified. At this point it may be necessary to consider using degenerate primers and possibly probes to cover sequence variants that occur in different strains of the pathogen to be detected. Alternatively, if an assay is required to detect specific sequence variants, such as the viral subtypes HSV-1 and HSV-2, it may be necessary to design conserved primers and sequence-specific probes. Whichever type of assay is being worked-up, careful and thorough sequence searches with BLAST (Basic Alignment Search Tool; http://www.ncbi.nlm.nih.gov/) will be required to ensure as many variants as possible are included in the assay design. A local database can be constructed and aligned using one (or more) of the many sequence alignment programmes available, such as CLUSTALW (http://www.ebi.ac.uk/Tools/msa/clustalw2/). Having a suitable alignment greatly aids the process of primer and probe selection. Optimal primer and probe function is critical to successful PCR (Raymaekers *et al.*, 2009). There a number of software packages available to assist in primer and probe design, such as Primer Express (Applied Biosystems) and OLIGO (http://www.oligo.net/) although many researchers prefer to design 'longhand'. Whichever method is chosen a set of rules must be consistently applied to ensure optimal performance. The following is an amalgam of our (and many others) experiences over the years:

- Avoid G+C rich regions in the proposed amplicon sequence because they are difficult to amplify.
- Avoid homopolymeric (repetitive sequences of A+T and/or G+C) stretches within the amplicon.
- Select shorter amplicons, less than 300–400 nucleotides which are more efficiently amplified and less likely to include secondary structures, which are difficult to melt and inhibit polymerases. These structures can arise from a number of sequence features, such as self-complementary bases, which may cause the template to fold into hairpin-like structures and sequential runs of guanines, which can fold the template into tetraplex structures.
- Primers should be 18–24 nucleotides long to provide specificity with a G+C content of 40–60% to provide a sufficiently high and discriminating annealing temperature. It is usually unnecessary to design primers longer than 24 nucleotides; in general, the longer the primer, the more inefficient the annealing, with fewer templates primed at each step, resulting in a significant decrease in amplicon production.
- Both forward and reverse primers should be thermally balanced, i.e. they should have the same calculated meting temperature (T_m). There have been many attempts over the years to provide the most accurate method of calculating primer and probe T_m values, the best of which take into account the nearest-neighbour effects on the thermodynamic properties of the overall oligonucleotide. We have found the web-based Oligo Calc (rhttp://www.basic.northwestern.edu/biotools/oligocalc.html) to be consistently useful in this respect and it also has useful secondary structure and primer-dimer formation prediction functions. Ultimately, the only way of establishing the actual (and most efficient) T_m of each set of primers and probes is by experimentation. This is because none of the theoretical methods will provide a correct, absolute, value for T_m values; however, the relative values between primers and probes they do provide is necessary for a correct, balanced design.
- The annealing temperature of a primer is generally approximately 5°C lower than its T_m and should be in the range 50–60°C. The most efficient temperature for *Taq* polymerase to extend the primer is at 72°C and most three-step programmes will incorporate this as a third temperature step. However, it is possible to run a two-step programme, with annealing and extension both occurring at 60°C, since the widely used TaqMan hydrolysis probes elongate equally efficiently at this temperature. This approach has the added advantages of shorter run times and by subjecting the heat-labile components in the reaction to less thermal inactivation, improving the overall sensitivity of the assay.

During the PCR, the primer–template combination is stabilized as the polymerase extends the primer. However, because the probe is not extended during this process, it is not stabilized (Raymaekers *et al.*, 2009). Therefore, the probe (both hydrolysis and hybridization types) should have a higher Tm than the primers, typically 5–10°C to ensure stable binding through annealing and extension phases of the reaction. To achieve higher Tms, probes are often longer than the primers (up to 35 nucleotides); however, probes longer than this can result in a loss in specificity. If there are significant constraints on the probe-binding region that prevent the optimal Tm being achieved using conventional probes, the use of minor groove-binding or locked nucleic acid (LNA) probes should be considered. LNA nucleosides are ribonucleotide analogues that have a methylene bridge between the 2′-O and 4′-C of the ribose ring. The resulting sugar moiety is locked in the 3′-endo conformation that preorganizes the LNA nucleotide monomers for high-affinity hybridization (Singh *et al.*, 1998; Kaur *et al.*, 2006). The substitution of DNA with LNA monomers allows for shorter probe lengths whilst maintaining or increasing the thermal stability of the probe towards its complementary DNA target. This provides a more flexible approach to probe design for problematic target sequences, while maintaining accurate gene quantification and allelic discrimination.

The typical G+C content for a probe ranges from 30–80%, preferably with a greater number of Cs than Gs. Guanosine residues are capable of quenching fluorescence and by reducing their number in the probe a greater change in normalized fluorescence (ΔRn) will be produced. This enables the results to be more easily interpreted, particularly when low-copy numbers of target are present and the signal would otherwise be difficult to distinguish from the background fluorescence (Gunson *et al.*, 2006; Raymaekers *et al.*, 2009). As for the primers, homopolymeric runs of single bases should be avoided in the probe design. For optimal efficiency of hybridization (TaqMan) probes, the 5′-terminal of the probe carrying the reporter dye should be as close as possible to the 3′-end of the forward primer. Avoid placing a guanosine residue at the extreme 5′ end of the probe adjacent to the reporter dye because this may lead to spontaneous quenching of the fluorophore.

- As with the selected template, homopolymeric tracts must be avoided in both primers and probes to avoid mispriming and subsequent non-specific amplification/detection. There is an absolute requirement for 3′-end fidelity in the sequence of the primers, usually the first three nucleotides (at least) must be a perfect match to ensure specific amplification and this must be considered when the requirement for degenerate oligonucleotides arises. There is some evidence for designing the so-called 3′-prime 'clamp' (i.e. two to three G or C residues at the 3′ end) in each primer, to ensure enhanced stability in the annealing process, but this needs careful consideration to ensure no secondary structure could form by inclusion of the 'clamp'.
- Avoid any regions of potential self-annealing within the primers and probe; complementary sequences that could anneal between primers and probe must also be avoided.
- Particular care must be taken when designing primer-sets for multiplex assays, the same rules apply; however, the primer and probe mixes are necessarily more complex and the potential for unwanted interactions greater.
- Once the primers and probes have been selected and checked for the parameters described above, they must be checked for specificity by BLAST analysis. This *in silico* verification is the first basic check for specificity. Any matches with other sequences, particularly from other pathogens likely to be encountered in the sample types under investigation, will necessitate redesign. It must be stressed that this preliminary specificity check against the databases is not conclusive evidence. The assay itself must be tested against the range of pathogens likely to be encountered in the sample types under investigation.

Reverse transcription PCR

The ability to detect and analyse mRNA transcripts and other RNA species in a sample is critical in many experimental studies. It is also critical in detecting infectious pathogens, such as RNA viruses and in other pathogens, where differential gene-expression can be used to distinguish between infection and colonization (Klugman *et al.*, 2008; Meade *et al.*, 2009). Reverse transcription-PCR can be carried out as a two-step reaction, where the reverse transcription is carried out separately and an aliquot is subsequently transferred to the PCR. More usually, in infectious disease assays, a one-step reaction is employed, where the reverse transcription and PCR occur in the same tube on the thermocycler.

There are advantages and disadvantages with both techniques, often the two-step reaction is considered the more sensitive method, while the one-step is quicker and more suited to the high-throughput diagnostic laboratory's work-flow. Furthermore, by reducing the number of steps that require opening tubes, it also reduces the potential for contamination. The choice of priming the reverse transcription step needs to be considered; typically the downstream, anti-sense primer is used. However, many workers find that random hexamers, or oligo-dT are more efficient. Often the choice is down to a sensitivity/efficiency balance between the hexamers ability to bind and hence copy all the RNA species in the sample and the sensitivity provided by the sequence-specific binding of the downstream primer. As with all aspects of the PCR, the most effective way of determining whether to use a one- or two-step reaction and the optimum primer-type is by experimentation. It must be stressed that the RT step can lead to significant variation in final PCR result, particularly in qPCR assays. Therefore, during the initial verification stage of the development, it is important to provide details of the amount of RNA transcribed into cDNA, priming strategy and enzyme(s) used, together with the timing of the RT step (Bustin *et al.*, 2009).

Choice of the quantification standards

There are a number of strategies available for quantifying by a real-time PCR assay, which can be either absolute or relative. In most diagnostic assays absolute quantification is chosen. A set of separately amplified, log-diluted, quantification standards (i.e. with known concentrations) are used to form a standard curve (see Fig. 10.2a,b). This curve is derived from the C_t values obtained during exponential amplification of the standards plotted against their log concentrations. The copy number of any sample is estimated from its C_t and the standard curve. Standards can be presented in any units appropriate to the analyte or specimen type, most frequently they are given in copy numbers/ml or copy number/cell. Standards must be reliable, stable and accurately quantified; often they are constructed by cloning the amplicon. This method has the advantage of providing a large amount of consistent and easily reproducible material; however, it is unlikely that a circular plasmid will be amplified with the same efficiencies as the target analyte due to its rapid renaturation. Therefore, the plasmid must be linearized with a suitable restriction enzyme, diluted in a negative sample and analysed to determine the conditions that provide the same amplification performance as the native target.

Relative quantification assays are normally used in determining the fold-changes in gene expression levels, relative to a standard curve, based on serial dilutions of a calibrator with arbitrary units. Relative changes in expression can also be determined by the $\Delta\Delta C_t$ method, which has the advantage of dispensing with the need for quantification standards. To compensate for target concentration variations in the different samples, normalization against one or more reference genes is required.

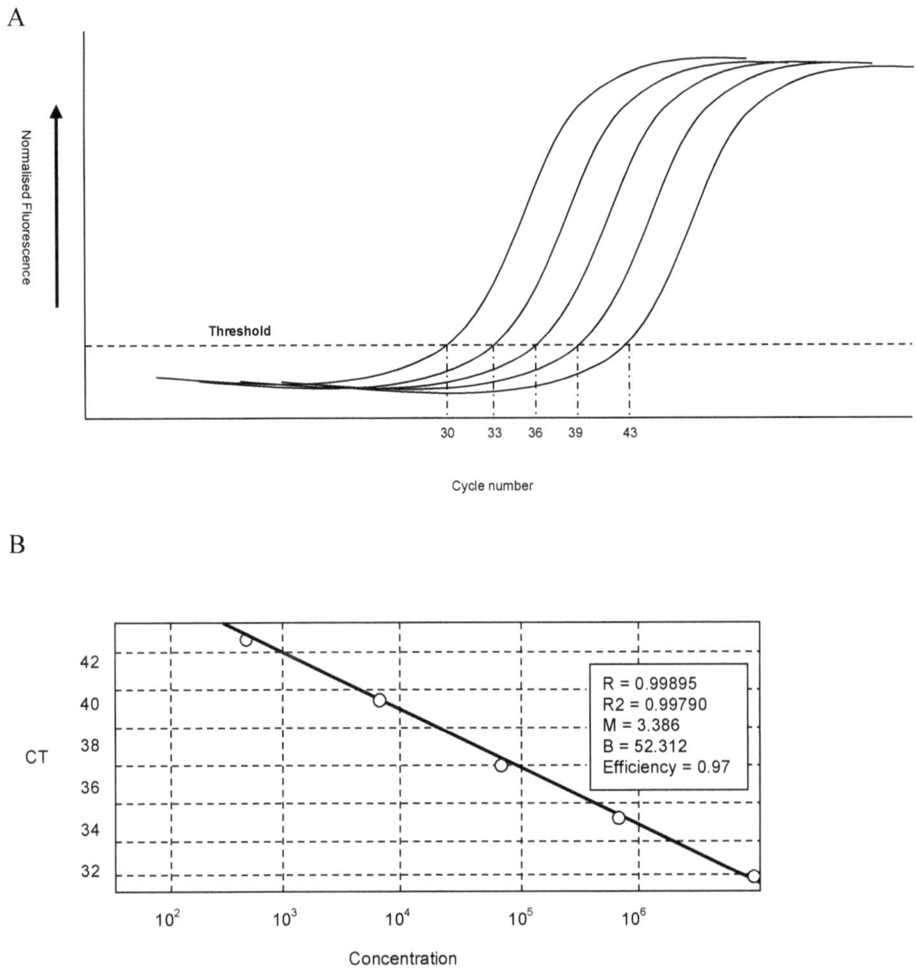

Figure 10.2 (a) Typical amplification curves from a series of five 10-fold dilutions based on output from a Rotor-Gene 6000. (b) Standard curve based on the amplification plot shown in (a). The table within the figure shows the statistical data derived from the curve.

Internal and extraction controls

Each time the diagnostic assay is run it must include a suitable number of positive controls (PC) and negative, no template controls (NTC). Furthermore, the level of amplification of the positive control must be routinely monitored and the C_t values recorded as part of the assays ongoing quality assurance. In our laboratory we run a single positive control for each specific pathogen in each assay and a NTC after every eight samples, also ensuring that a NTC is the last sample on the run. This is to monitor all the handling steps in the process, placing one or two NTCs at the start of the set-up could give a false impression of any contamination problems. The PC can be produced from extracted clinical samples, or if available, from commercial sources and should be diluted to be reproducibly amplified at the lowest detectable level (typically a C_t of 30). An internal PCR control (IC) is becoming an absolute requirement for any diagnostic assay. Nucleic acid from a suitable non-target pathogen can be spiked into each extracted specimen prior to the PCR and amplified and detected with a separate set of primers and probe. By spiking the samples, the success (or failure) of the reaction can be demonstrated, together with an indication of the presence of inhibitors carried over from the sample during extraction. This is done by comparing the C_t values of the amplified internal control in NTCs with the C_t value in any negative specimen. A difference, typically greater than three C_t values (i.e. approximately equivalent to one log) indicates inhibition and a potentially false-negative result. It is necessary to establish that the primers and probes used in the IC reaction do not interfere with the amplification of the target. Similarly, the control template itself must be titrated to ensure that its amplification does not compete with the target reaction.

Ideally, an extraction control (EC), using a non-related virus or bacterium should be used to monitor extraction efficiency, one of the most important steps in the whole process. Using this approach, the EC is spiked into the sample, in this case before extraction. This has the advantage over using 'naked' nucleic acid as an EC, because it controls for the critical lysis step as well as the subsequent steps in the extraction process. Phocine herpes virus is a useful EC for assays detecting viral DNA pathogens and for RNA viruses, MS2 phage or mengovirus can be used. The use of an RNA template in RT-PCR controls for both the reverse transcription step and the PCR.

Westgard rules

For each run the C_t values of controls can be plotted on Levy-Jennings or Shewhart charts to monitor the test system for shifts in performance. Such visual presentation of the control data is often very helpful in spotting variations in the operation of the assay. Westgard rules can also be applied to determine when a corrective action should be taken to prevent test failure. Westgard multirule analysis is recommended, particularly when using two or more control measures per run (see Table 10.1). This approach reduces the number of false rejections and improves the laboratory's capability of detecting analytical errors. It will also give some indication of the type of analytical error that caused the run to be rejected, assisting in problem-solving (Westgard *et al.*, 1981; Burd, 2010). The use of Westgard rules and Shewhart charts must form part of the on-going monitoring of the assay's performance to maintain its validated status.

Initial optimization

Once the basic parameters of the assay have been established, the assay must be experimentally optimized. The first step is to confirm that the amplicon product is what is expected, to do this it must be analysed by sequencing and BLAST analysis. Due to the typically short amplicon lengths in real-time PCR, direct sequencing of the product may not be possible and the fragment will need to be cloned into a suitable plasmid vector.

Optimal primer and probe concentrations must be verified, together with the PCR programme itself. To do this, a sufficient supply of a well-characterized, stable positive control must be used throughout this verification process. This will provide an invariant baseline from which the effects of changes to the various components can be more easily assessed. The formation of primer dimers and the degree of non-specific amplification can be checked by running the assay, initially with SYBR green and analysing the melting

Table 10.1 The basic Westgard rules

A	1_{2SD}	If one control measure exceeds the mean ±2SD, the control values from the previous run should be considered to rule out a trend
B	2_{2SD}	This rule detects systematic error. The rule is violated when two consecutive control values exceed the same mean ±2SD limit
C	4_{1SD}	This rule detects systematic error. The rule is violated when four consecutive control values exceed the same (mean +1SD or mean −1SD) limit. The run need not be rejected if this rule is violated but should trigger recalibration or equipment maintenance
D	1_{3SD}	This rule detects random error. Violation of this rule may also point to systematic error. The assay run is rejected if one control value exceeds the mean ±3SD
E	R_{4SD}	This is a range rule and it detects random error. The rule is violated when the difference in the SD between two control values exceeds four SDs (i.e. −2SD and +2SD)
F	10_X	This rule detects systematic error. The rule is violated when the last ten consecutive values are on the same side of the mean. Its violation often indicates the deterioration of assay reagents

curve. However, more information is usually provided by combining the investigation with analytical agarose-gel electrophoresis, which will also provide direct evidence of the size of the amplicon.

To optimize primer and probe concentrations, ideally two log-dilutions of the positive control template are used in a chequerboard layout of varying primer/probe combinations. This chequerboard is used throughout the optimization of the conditions employed, so that effect of each variation can be assessed correctly. The optimal primer and probe concentrations and thermal cycling conditions are those that provide the lowest C_t values, the greatest ΔRn and a consistent C_t difference of approximately three cycles between the two log-diluted samples (Raymaekers *et al.*, 2009).

Normalization

Expression analyses with reverse transcription PCRs is not normally a technique used in diagnostic PCR. However, the range of diseases and their associated pathogens being diagnosed by RT-PCR is increasing. In some cases, e.g. with fungal species, there is a need to distinguish between carriage, environmental contamination and infection. Therefore, assays capable of detecting and quantifying differentially expressed genes specific to the disease condition are needed. This requires carefully selected reference genes to control for variations in extraction yield, reverse-transcription and amplification efficiencies, so that comparisons across different mRNA concentrations can be made (Bustin *et al.*, 2009). A detailed literature search for suitable reference gene mRNA-targets must be made and the selected candidates experimentally tested for stable expression in both the diseased and non-diseased sample range and show strong correlation with total mRNA levels. Ideally, at least two reference genes should be employed for normalization; however, a single gene can be used, if sufficient evidence can be provided showing its invariant expression (Bustin *et al.*, 2009).

Analytical specificity and sensitivity

Analytical specificity is a measure of the assay's ability to detect the target it was designed to and not cross-react with other analytes in the sample. Careful *in silico* analysis during the design of the primers and probes is crucial to the preliminary stages of verification. However, diagnostic assays operate in a genetically complex mixture and no amount of computer verification can predict the actual outcome of the reaction. Specificity can only be demonstrated either by spiking samples with a range of different pathogens prior to extraction, or adding the extracted nucleic acid from these pathogens to the extracted sample under investigation. All reactions should be analysed by agarose gel electrophoresis. This is to ensure that amplification has not occurred in any of the samples expected to be negative. In some cases primers may form close, but imperfect matches, on non-target templates, which will be amplified, but may not be detected by the probe. This will compromise the sensitivity of the assay and must be eliminated, either by re-evaluating the PCR conditions (typically raising the annealing temperature), or by redesigning the primers.

Analytical sensitivity is defined as an assay's ability to detect very low levels of a given analyte in a biological specimen. It is synonymous with the assay's LOD, i.e. the lowest concentration of the analyte consistently detected in ≥95% of samples tested with acceptable precision. The LOD must be determined for both qualitative and qPCR assays, where the values are usually given as copies/ml or copies/μg input nucleic acid in qPCRs. The LOD can also usefully provide information on the value at which the low level PC, used to monitor the consistency of the assay, should be set. There are two main approaches used in determining the LOD of an assay, statistically, by calculating the point at which the fluorescence from the target can be distinguished from background levels and empirically, by testing serial dilutions of the analyte. The empirical method is usually selected for diagnostic assays, using a set of log-diluted controls to initially determine the LOD (Burd, 2010). It is important to include a NTC in the LOD study to ensure that the PCR does not generate a signal that could interfere with true low-level positive signals from a sample (Burd, 2010).

There is no hard and fast rule as to how many samples to use; however, the general view in the literature is that 20 measurements, at, above and below the LOD, as determined by a preliminary dilution study should be employed (Burd, 2010). The Clinical and Laboratory Standards Institute (CLSI) also recommend laboratories to carry out 20 measurements to verify the manufacturer's claims with a commercially developed assay. However, CLSI also suggests a minimum of 60 data-points (12 separate measurements from each of five samples) are required from a manufacturer to establish the LOD. This seems a reasonable figure to adopt as a standard approach for both in-house and commercially produced assays and will provide the necessary intra-assay variation measurements. This measurement should also be carried out on a second thermocycler to establish machine-to-machine variation. The samples used in the study must be from a well-characterized source, such as patient samples, tested by an accepted method (extracted and diluted in a suitable matrix), a suitable cell line, or proficiency panels (Burd, 2010). Repeated testing and analysis in this way is required because of the inherent sensitivity of the PCR and random sampling errors when the target template is at low concentrations. Sampling errors are caused by nucleic acid aggregation during repeated freeze–thaw cycles, together with pipetting and dilution errors (Chandler *et al.*, 1998). Repeated testing at very low concentrations of the analyte invariably results in a positive result on one occasion and a negative on another. The 'Monte Carlo effect' describes difficulties with the reproducible amplification and quantification of low abundance templates (below 1000 copies) where the true abundance is less likely to be reflected in the subsequent amplified product (Karrer *et al.*, 1995).

Estimation of the LOD is usually carried out by probit analysis, a type of regression used to analyse binomial response variables, where a sigmoid dose–response curve is transformed into a straight line that can be analysed by regression either through least squares or maximum likelihood. For PCR assays the lowest concentration of analyte that can be detected with a stated probability can be determined by plotting the data from positive replicate results versus the analyte concentration (Sloan, 2007).

Inhibition study

The factors that inhibit or prevent the amplification of nucleic acids by PCR can be present in the samples and subsequent extracts from a number of sources. This type of inhibition is distinct from that caused by cross-reactions with other nucleic acids that may also be present in the sample. Inhibitors typically act at one or more of three essential points in the reaction in the following ways: (i) interference with the cell lysis necessary for extraction of nucleic acids; (ii) interference by degrading the nucleic acid or inhibiting its capture; and (iii) inhibiting polymerase activity during amplification of the target DNA (Wilson, 1997). Considerable thought needs to be given to the method of extraction because inhibitory substances are a potential cause of false-negative results or inaccurate quantification. Different sample types present different problems due to the range of endogenous inhibitors found, for example in faecal samples, complex polysaccharides, breakdown products from haemoglobin, and bile acids can all be present (Wilson, 1997; Monteiro *et al.*, 1997; Jacobson *et al.*, 2009). Other potential inhibitors include metabolites resulting from pathological conditions such as diabetes mellitus and homeostatic hepatitis, or from medications used in treatment (Burd, 2010).

Exogenous inhibitors can be introduced during the handling of the specimens and include hand cream, powder from protective gloves, pollen and serum separators (Wilson, 1997; Burd, 2010). While the majority of inhibitory substances are usually removed during extraction, some of the reagents used in the process, such as ethanol, EDTA, SDS and guanidine hydrochloride, also inhibit the PCR, therefore, care needs to be exercised and the extraction protocols followed precisely. Manufacturers continue to develop their PCR reagents specifically to overcome many of the common endogenous inhibitors found in clinical specimens and it is recommended that a number of different commercial amplification kits are evaluated for their performance in this respect.

As discussed previously, inhibitors can be detected using an IAC. If the IAC fails to amplify or exhibits suppression below the acceptable threshold, the amplification of the intended target sequence may also be inhibited. To avoid reducing the sensitivity of the assays, the IAC should be used at the lowest concentrations that can be reproducibly amplified, to minimize any competition between its amplification and that of the target pathogen. In some assays, a positive specimen may outcompete the IAC, which will fail to amplify; however, in such cases a positive result with a negative internal control is considered valid. IACs cannot differentiate between inhibition and amplification failure due to other potential reaction problems, such as thermocycler well failure, failure to add enzyme or other reagents, etc., but will highlight the problem for further investigation. Various constructs of IAC are used in molecular assays and include homologous internal, heterologous external or heterologous internal controls. Further details on the types of control and their uses can be found elsewhere in this book.

The IAC must demonstrate inhibition by the substance expected to be found in the sample, for which an inhibition study needs to be carried out. The study must first identify what types of inhibitors are likely to be present in the sample types used in the assay. Details of inhibitors and strategies to overcome them can be found through literature searches, e.g. useful reviews can be found in Rådström *et al.* (2004) and Wilson (1997). The IAC can be tested in negative samples with and without the interfering substance(s) in parallel and the effect on the C_t values monitored. This test must be repeated with the full assay (including the pathogen detection primers and probes) using paired samples, with and without the target analyte to validate the assay. Once interference is found, of either or both the IC and target, samples can be serially diluted to test the limit at which the target will amplify. In most cases dilution of inhibited samples provides a straightforward method of enabling amplification (Wilson, 1997). This information can be used in preparing both the SOP and testing-algorithms for routine use of the assay. The test must also be carried out with each sample type likely to be analysed with the assay and may require a number of different extraction methods to be evaluated (Monteiro *et al.*, 1997; Jacobson *et al.*, 2009). The assay must then be tested against clinical samples; this is because artificially constructed inhibition control samples may not accurately reflect the assays performance in routine use. There is no recommended figure for the number of samples to be used for the inhibition study, although power analyses using standard deviation of the differences between the means of paired samples can be used to determine the sample size required (Burd, 2010).

PCR efficiency

The efficiency of the PCR can have a significant impact on the robustness and precision of the assay. There are a number of factors that contribute to the efficiency of the reaction; the primer design is of paramount importance, together with the cycling conditions and the reagents used in the reaction mixture. PCR efficiency is particularly important in assays where relative quantification strategies are used, e.g. reporting concentrations of mRNA for target genes relative to those of reference genes, where both templates must be amplified with equal efficiencies (Bustin *et al.*, 2009). PCR amplification efficiency is typically established using calibration curves and it makes sense to use the dilution series (covering five to six orders of magnitude) from the LOD analyses for this purpose (see Fig. 10.2a and b). The standard or calibration curve is constructed by plotting the $\log_{10}$ of the starting quantity of template against the C_t value obtained during amplification of each dilution of the template. The equation of the linear regression line, together with Pearson's correlation coefficient (r) and the coefficient of determination (r^2) are used to determine amplification efficiency. Amplification efficiency itself is determined from the slope of the log-linear portion of the calibration curve and is given by equation 10.1:

$$E = 10^{(-1/\text{slope})} \tag{10.1}$$

Ideally the amount of template will double during each round of exponential amplification, this translates to a reaction efficiency of 2, therefore, using equation 10.1: $2 = 10^{(-1/\text{slope})}$ which gives a slope for the standard curve of −3.32. This ideal figure of 3.32 also represents the difference in cycle number for each log dilution in the series and is characteristic of ten-fold changes in template concentration for real-time PCR. The efficiency can also be expressed as a percentage of the template amplified in each cycle using equation 10.2:

$$\% \text{ Efficiency} = (\text{E} - 1) \times 100\% \tag{10.2}$$

In the ideal example given above, % Efficiency $= (2-1) \times 100\% = 100\%$.

The theoretical maximum efficiency (100%) indicates that the amount of product doubles with each cycle. Ideally, the experimental measurement should be carried out in triplicate on at least two thermal cyclers and the CIs or SEs of the means of estimated PCR efficiencies included in the validation documentation. In practice acceptable assays should be expected to achieve efficiencies of 90–105% (97% for the example in Fig. 10.2). Reaction efficiencies lower than this may be caused by suboptimal reaction conditions or poor primer design. Reaction efficiencies of more than 100% could be due to measurement errors, e.g. in preparation of the dilution series, or co-amplification of primer-dimers, and/or other non-specific products. Improving specificity will improve the sensitivity and increase the dynamic range of the assay, by reducing competition between the specific and non-specific amplification products. The presence of inhibitors in the reaction may change the overall amplification efficiency and therefore the value of template copy numbers determined using external standards (interpolation of standard curve). A sample which is diluted prior to amplification may therefore have a different value from an undiluted one when the value is interpolated from the same standard curve. Indeed the presence of inhibitors in a template that is diluted for use as standards may produce a curve that suggests a greater or lesser efficiency value (by the slope method described) due to co-dilution of the inhibitory substance with template down the standard range.

Linear dynamic range or reportable range

The linear dynamic range of an assay can be conveniently determined during the LOD analysis by extending the range of the dilution series. The dilution series is often set at the range one would expect the analyte to be found in clinical specimens, although this may not always be known, especially for a novel pathogen and so initial testing can be carried out from 0 up to 10^7 or higher copies of target. If using a generic assay to detect a range of pathogens, e.g. a panfungal PCR, each of the species the assay is deigned to detect must be tested. Similarly, for multiplex assays, the LOD must be determined individually for each target analyte in the assay. Additionally, testing must include the individual targets in the multiplex at high and low concentrations to ensure that high levels of one target do not out-compete amplification and detection of the other targets in the sample. The reportable range is defined as the lowest and highest results (in suitable units of concentration) reliably detected in the assay (Sloan, 2007). The linearity of the range can be established by calibration curves, where an r^2 value (coefficient of determination) of 1.00 indicates a perfect fit of the data points. Generally, in qPCR assays an $r^2$of not less than 0.99 is considered acceptable, full details of the correlation coefficients and preferably the confidence intervals for each data-point should be provided throughout the linear dynamic range (Bustin *et al.*, 2009). The dynamic range should cover at least three and preferably five orders of magnitude. The linearity and reportable range should be carried out on at least five log-dilutions of the target nucleic acid extracted from an appropriate sample type [serum, urine, nasopharyngeal aspirate (NPA), etc.] in triplicate, ideally on two different thermocyclers. Quantitative results from the test can only be reported when they fall within the linear range of the assay. A reportable range is not applicable to qualitative assays, since the result is simply positive or negative (below the LOD). However, a log-diluted panel should be tested, because the information derived is useful in assessing the clinical utility of the test and the results (as C_t values) reported as part of the supporting documentation for the assay.

Clinical validation

Precision and accuracy

Accuracy experiments are designed to measure the random error or imprecision of an assay over a pre-determined period of time (Sloan, 2007). This can be achieved by multiple measurements of an aliquot derived from a homogeneous sample over a set period of time (Burd, 2010). For qualitative assays random error can be established by testing the PC and NC material in triplicate over a period of 10 days on two different thermocyclers. For qPCR assays, estimates can be determined from the quantification standards and two positive controls at the lower and upper ends of the assay's reportable range as described previously.

The term accuracy, or systematic error, as a measure of an assay's performance with a comparative or reference method, has been replaced by 'trueness'. Accuracy now refers to the closeness in agreement between a single measurement and the true value of the analyte under investigation (ISO, 2003). Trueness is determined by analysing the average value obtained from a series of measurements with the new assay and the true value of the analyte (if an international standard is available) or a reference method (if a standard is not available). Experiments to determine trueness must be carried out using the full test system, i.e. from receiving the sample and its preparation through to the result (Sloan, 2007). There are two approaches to evaluating trueness (Burd, 2010):

1. A comparison of methods study, where samples are tested in parallel with an appropriate gold standard method.
2. A recovery study, where proficiency samples, or other verified sample types are compared with the assay under development.

Although either of these methods is acceptable, a split-sample, comparative study using clinical samples is preferred. However, in some cases, e.g. when developing an assay for a novel pathogen, a suitable gold standard assay may not available and a recovery study will have to suffice.

The gold standard comparative study

The comparative study is the cornerstone of the validation exercise. These experiments determine many of the defining characteristics of the assay, such as positive and negative predictive values. Therefore, the estimates of diagnostic sensitivity and specificity must be as precise as possible. Classical estimates of sensitivity and specificity are derived from empirical comparisons of the index test (the assay being developed) and the established gold standard, which is the same type of test procedure and has an assumed sensitivity and specificity of 100% (de Clare Bronsvoort *et al.*, 2010). That is, by definition the true gold standard assay is an error-free diagnostic method; however, in practice the gold standard rarely (if ever) exists. Consequently, an earlier PCR assay, or a culture-based method must be used for comparison.

Once the method has been chosen a suitable set of samples must be selected for parallel testing. Specimens for the study should be representative of the infectious condition and population being investigated for the planned use of the test and should also be suitably distributed across both age-range and gender. Where only limited samples are available, it may be necessary to use archived positive and negative specimens, or cultured material spiked into suitable negative clinical samples. It may also be necessary to carry out a prospective analysis and use clinical data to resolve any discrepant results. Other sources of suitable samples may be other clinical/research laboratories or commercial standards, quality control materials or proficiency panels (Sloan, 2007; Burd, 2010). The number of samples can be determined by the laboratory using statistical calculations, based on the confidence intervals and levels considered acceptable. However, the sample numbers are more often selected for more pragmatic reasons, such as cost and feasibility and, therefore, less than the statistically optimal number will be tested. The consensus seems to be that 100 samples should be used, 50 positive and 50 negative and they should be tested over a period of 5–10 days. In our laboratory we routinely test 80 positive and 20 negative samples in triplicate over a period of 5 days for both qualitative and quantitative assays.

The difficulty comes in interpreting the results of a study where the comparative test is not a 'perfect' gold standard, (i.e. the 'alloyed standard') since it is unlikely that the two methods will yield the same result. This is particularly so if the comparative method is cell culture, which is thought to be 100% specific, but provides less than optimal sensitivity for many pathogens (Hadgu *et al.*, 2005). The strength of the PCR lies in its ability to provide a positive result from very low levels of analyte (theoretically a single DNA copy) from which it is hypothesized that the PCR has extremely good sensitivity (Hadgu *et al.*, 2005). Thus a newly developed PCR assay may produce many more positive results from a test-panel than the reference method, leading to a misleading estimate of relative sensitivity. The researcher is then left with the problem of deciding whether these are true- or false-positives. In some cases, basic experimental analysis can help resolve some of these difficulties. A thorough and detailed assessment involving consideration of the sample derivation, clinical history and sequencing can help to determine the true status of any discrepant samples (OIE, 2010). Another approach is to use discrepant analysis, in use since the 1990s, to determine the true outcome of conflicting results between a new test and an imperfect gold standard. In discrepant analysis the discordant results are resolved by a third test, such as another PCR directed to a different gene-target of the pathogen and the results from this 'resolver' test are used to assign the final results. Typically, only the discrepant results are analysed. While the resolver provides additional information, obviously it cannot provide information on samples that are not tested, which are usually the majority. There are a number of problems with discrepant analysis, including the introduction of bias in favour of the new test, miss-calling false-positives due to carryover-contamination and potentially poorly validated comparative or reference assays. There are a very few special circumstances where the decision to use discrepant analysis can be justified. For example, where there is a 'perfect' test that can be used as a resolver and a random sample of concordant results is also verified by the same resolver test using the correction method proposed by Begg and Greenes (Begg and Greenes, 1983; Rutjes *et al.*, 2007). Because of these issues, discrepant analysis is now considered an unsuitable method for estimating sensitivity and specificity (Miller, 1998; Rutjes *et al.*, 2007; Hadgu and Sternberg 2009) and a recent FDA document considers that 'discrepant resolution does not solve the bias problem; it is just a more complicated wrong solution' (FDA, 2003).

Bias can be avoided in the statistical analyses by making use of several scientifically and statistically valid approaches to test validation. Where the true disease status of the samples is difficult to evaluate, or the diagnostic specificity and sensitivity are not precisely known, the agreement between the new test and comparator test can be expressed as the kappa value (Conraths and Schares, 2006; Burd, 2010). The kappa statistic is a generic term for a number of similar measures of agreement applied to categorical data and is a measure of the proportion of agreement between results beyond chance. Typically, it is used in assessing the degree to which two or more raters, (i.e. tests) examining the same data, (i.e. specimens) agree in assigning the data to categories (positive or negative). A kappa value of 1.00 indicates perfect agreement; a value of 0.00 indicates no agreement above that expected by chance, and a kappa value of −1.00 indicates complete disagreement. There is some debate as to what constitutes good agreement, but generally, kappa statistics above 0.80 are considered almost perfect (Conraths and Schares, 2006; Burd, 2010). The kappa value is calculated from the results in the standard 2×2 contingency table (Table 10.2), used to record the results from the comparative test.

$$\mathrm{OP} = \frac{\mathrm{TP+TN}}{n}$$

$$\mathrm{EP} = \left[\frac{\mathrm{TP+FP}}{n}\right] \times \left[\frac{\mathrm{TP+FN}}{n}\right] \times \left[\frac{\mathrm{FN+TN}}{n}\right] \times \left[\frac{\mathrm{FP+TN}}{n}\right]$$

$$\mathrm{K} = \frac{\mathrm{OP}-\mathrm{EP}}{1-\mathrm{EP}}$$

where OP = observed proportion of agreement, EP = expected proportion of agreement, TP = true positive, FP = false-positive, TN = true negative and FN = false-negative. Diagnostic sensitivity and specificity (DSe and DSp) are also calculated from the results in the contingency table.

Two further important probabilities can be calculated from the data derived from the trueness study, the positive and negative predictive values (PPV and NPV). The PPV is a

Table 10.2 2×2 contingency table

	Gold standard		
New Test	Positive	Negative	Totals
Positive	TP	FP	TP+FP
Negative	FN	TN	FN+TN
Totals	TP+FN	FP+TN	*n*

FN, false-negative; FP, false-positive; TN, true negative; TP, true positive.

measure of a positive result from the test to truly predict the presence of disease/infection, while the NPV is the probability that a negative result accurately indicate a non-diseased/uninfected status.

The study without a gold standard

In many cases, particularly for novel pathogens, a comparative method will not be available, under these circumstances the gold standard may be a diagnosis determined by accepted clinical methods (CLSI, 2002). It may also be necessary to incorporate results from more than one method to provide the comparative results (Conraths and Schares, 2006; Mattocks *et al.*, 2010). Whatever the approach chosen, estimating the performance characteristics of a new assay without a true comparative test is a challenging task (Hadgu and Sternberg, 2009). Typically, agreement is measured as the overall percentage or fraction of samples that have the same result (i.e. both either positive or negative). The difficulty with these simple agreement measures is that they do not take agreement by chance into account (Dendukuri *et al.*, 2004; Hadgu *et al.*, 2005; Burd, 2010). However, over the last 10–20 years, there has been progress in developing statistical solutions to these challenges. A number of models have been proposed, based on latent class analysis, log-linear modelling and other techniques, reviewed by Hui and colleagues (1998), Hadgu (1998), Hadgu and Qu (1998) and Hadgu and colleagues (2005).

Latent class analysis involves multiple imperfect tests that are used to construct a gold standard. Such models are based on the concept that the observed results of different, imperfect, tests for the same disease are influenced by a common but unobserved (latent) variable, i.e. the true disease status. Increasing the number of such imperfect tests in a test panel also increases knowledge of the latent disease status (Hadgu *et al.*, 2005). When only two tests are involved in the study (the new test and the non-gold standard), the results alone cannot be used to determine the outcome of the analysis, because there are too many unknown quantities. This is elegantly referred to by statisticians as a non-identifiable problem and the solution can only be found through a Bayesian approach (Bayes, 1763; Hadgu *et al.*, 2005). In the Bayesian approach, the unknown parameters are treated as random variables, with each following a probability distribution. Data already available on each parameter (prior probability distribution) are combined with data after testing (observed data) to produce the posterior probability distribution for each parameter. This is basically an update of the prior knowledge of each parameter, using the observed data or results after the test. This posterior distribution is used to determine Bayesian confidence intervals of, for example, sensitivity and specificity. The prior distributions are usually determined from literature searches or expert opinion. To solve this problem, distributions are required on at least two variables, usually the sensitivities and specificities of the non-gold standard test (Hadgu *et al.*, 2005). Further information can be found in an excellent review of statistical approaches associated with non-gold standard comparative tests (Rutjes *et al.*, 2007).

The PPV and NPV depend on the DSp, DSe and the disease prevalence and for various reasons, the samples selected for the trueness study are often in a proportion that does not reflect the true prevalence of infection in the population targeted by the assay being developed (Conraths and Schares, 2006; OIE, 2010). In areas of low disease prevalence (i.e. lower than the prevalence used in the trueness study) the PPV of a test will be low, due to the proportion of false-positive to total samples tested remaining relatively constant, although the assay has not been altered (Conraths and Schares, 2006). Consequently, an estimate of disease prevalence in the target population is necessary for the estimation of PPVs and NPVs. It is useful to provide an interpretative statement with the test results indicating the anticipated positive and negative predictive values covering the range of disease prevalence expected in the target population (OIE, 2009).

The sample panel used for comparative analysis, or a subset from it, can also be used for proficiency testing. The ultimate evidence of an assay's fitness for purpose is its successful integration into other laboratories. For this purpose, aliquots of the reagents used in the precision and accuracy studies during validation should be sent to a minimum of three laboratories willing to participate in proficiency testing. This will also provide additional data on the assay's reproducibility and robustness.

Reference intervals

The reference interval or range is usually the last parameter to be established because it is not actually used to determine whether an assay is acceptable (Burd, 2010). A reference interval for an assay is defined as the normal range of values for the analyte in individuals who do not have the disease or infection being analysed by the assay and is more usually associated with biochemical tests. However, the reference range is a useful guide to clinicians, providing objective information that can be used as a guide to patient management. For many infectious diseases the analyte will not be present in a healthy population (i.e. the normal population) and for qualitative assays the reference range will be reported as negative or not detected and a reference interval study will not be required. For quantitative assays, the reference interval is usually reported as below the LOD or a set level of quantification. However, for some analytes, the reference interval may differ from the LOD and a clinical decision will need to be taken on where the limit should be set. This is particularly relevant to infections caused by herpesviruses such as CMV and EBV, which can establish life-long latent infections, where a limit may need to be established to distinguish between asymptomatic carriage and active disease.

If a reference interval needs to be determined (e.g. in the case of a new and emerging or re-emerging infectious pathogen) a study is required. CLSI recommend that the best method to establish a reference interval is to collect a minimum of 120 samples from 'well qualified reference individuals'. However, it is recognized that conforming to this recommendation is difficult and so it is rarely carried out in practice (CLSI, 2000). An acceptable alternative is to verify an established reference interval using as few as 20 samples from individuals who represent the healthy population. If no more than two specimens fall outside the reference limits the value is verified (CLSI, 2000; Sloan, 2007). Commercial assays are likely to have been developed using an entirely different patient population from that served by the laboratory and therefore, it is necessary to verify the manufacturer's figures (Sloan, 2007; Burd, 2010).

Maintaining the validated status and re-validation

Validation is not a process that leads to a single, final end-point. It is an on-going process carried out after verification experiments have demonstrated that an assay performs to the laboratory's required standards. A validated assay must be monitored consistently for repeatability through the performance of the run controls to evaluate any potential changes in the assay's precision and accuracy (Conraths and Schares, 2006; Mattocks *et al.*, 2010;

OIE, 2010). All reagents should be assigned batch numbers and new batches tested with the controls and a suitable number of positive and negative samples from a recent, previous run. This is to ensure the new reagents perform with the same efficiencies as the previous batch. If more than one thermal cycler is used in the laboratory, each machine should be referenced and each test carried out on it identified. This will enable the laboratory to monitor the performance of each machine and if required, take the appropriate corrective action. The results of the PCR controls and quantification standards should be plotted daily on Shewhart charts and assessed by Westgard rules to ensure that the reaction efficiencies are maintained.

Participation in regular IQA and EQA schemes is a fundamental requirement for maintaining an assay's validated status. The utility of molecular tests, particularly qPCR assays for patient management, relies on the ability to relate new results to those obtained previously, or to clinical practice guidelines. Ideally, to ensure reliable, accurate and reproducible results, assays should be assessed against externally produced reference standards (Madej *et al.*, 2010). The purpose of EQA testing is not solely to monitor the performance of the assay, but assess all aspects of the test procedure, including the extraction method and reporting accuracy. If an external panel is not available, alternatives can include blind sample testing, sample exchange with other laboratories, or clinical note reviews (Raymaekers *et al.*, 2009). There is undoubtedly a need for more collaboration between laboratories and the organizations responsible for producing QA schemes. This is particularly important where a group is developing an assay directed at a rare or new and emerging pathogen. There are a number of recent publications highlighting this successful approach (Niedrig *et al.*, 2006; Templeton *et al.*, 2006; Domingo *et al.*, 2010; de Pagter *et al.*, 2010).

The frequency with which laboratories participate in EQA schemes varies considerably. A recent survey carried out in Europe showed that microbiology laboratories participated in between one and 12 rounds of proficiency testing per year, with the majority testing four times. The number of samples per round varied from one (31%) to more than 20 (0.5%), with the majority of schemes supplying between one and three samples per round (83%) (Thomas, 2009). The OIE (2009) recommends inter-laboratory proficiency testing should be conducted at least twice a year and this should be considered the minimum level of participation. The EQA panels for proficiency testing should comprise a suitable number of samples covering the full dynamic range of the assay, not just high, medium and low levels of target, as is often the case (Conraths and Schares, 2006).

As more and more assays are developed for new and/or emerging infections, EQA panels may not be widely available. In such cases the laboratory that developed the assay can make use of the samples used for assessing diagnostic sensitivity and specificity. The panel, or a subset of the samples from it, can be used in inter-laboratory proficiency testing, or ring-trials as part of the validation process and wider acceptance of the assay. The originating laboratory would then receive the results back from the participating laboratories, analyse the data and report results back (Conraths and Schares, 2006). Ideally, if successful, the new assay will replace any existing assays and itself become the accepted, standard analytical method and can also be used to further develop reference reagents for national and international QA schemes, to ensure its continued validated status (OIE, 2010).

Over time it is likely that an assay will require some degree of re-validation, either for purely technical reasons, or because of changes in the nature of the analyte detected.

Technical modifications continually arise as manufacturers develop and improve their reagents for extracting nucleic acids, the PCR and instrumentation. Technical modifications, such as the assessment of a new probe chemistry or the transfer of an assay to full or semi-automated instrumentation, normally only requires a methods comparison study. In this way any changes to the diagnostic sensitivity and specificity can be assessed quickly and accurately and if acceptable, the updated assay introduced with the minimum delay.

Regular monitoring of the PCR amplification efficiency together with associated clinical information will give an early warning of any changes in the sequence of the amplicon detected in an assay. Point mutations are common in the genomes of many infectious microorganisms, particularly RNA viruses. New viral lineages may also be introduced into the population being tested due to travel from abroad, which continues to increase. Although one of the major aims of the Globalization process is focused on reducing disease world-wide, an inescapable consequence is increased international trade and travel, leading to increasing exposures to disease-causing agents, formerly confined to remote areas. Under such circumstances, significant modifications to the primers and probe may be required, necessitating re-verification and re-validation of the assay. Amplicon sequencing should be part of the validation plan and an integral part of routine trouble-shooting algorithms, since it is capable of resolving errors due to non-specific binding of primers and probe (OIE, 2010).

Conclusion

The proposals presented in this chapter are intended as a practical guide to help inform and standardize both the process of assay validation and the accompanying data that supports the validation. These concepts represent a synthesis of ideas, based on work in our laboratory and from many of the excellent publications that have appeared over the recent years. Some flexibility in the application of these guidelines is expected and necessary; not all assays are the same and for some infectious diseases reference materials and comparative tests may not be available. The introduction of the validation guidelines should not prevent innovation, in fact by applying common sense to the process and by using the appropriate experimental methods and statistical tests, properly validated assays will be far more readily accepted.

References

Apfalter, P., Reischl, U., and Hammerschlag, M.R. (2005). In-house nucleic acid amplification assays in research: how much quality control is needed before one can rely upon the results? J. C. Micro. 43, 5835–5841.

Bayes, T. (1763). An essay towards solving a problem in the doctrine of chances. Phil. Trans. 53, 370–418.

Begg, C.B., and Greenes, R.A. (1983). Assessment of diagnostic tests when disease verification is subject to selection bias. Biometrics 39, 207–215.

Bossuyt, P.M., Reitsma, J.B., Bruns, D.E., Gatsonis, C.A., Glasziou, P.P., Irwig, L.M., Moher, D., Rennie, D., de Vet, H.C.W., and Lijmer, J.G. (2003). The STARD statement for reporting studies of diagnostic accuracy: explanation and elaboration. Clin. Chem. 49, 7–18.

Budowle, B., Schutzer, S.E., Morse, S.A., Martinez, K.F., Chakraborty, R., Marrone, B.L., Messenger, S.L., Murch, R.S., Jackson, P.J., Williamson, P., *et al.* (2008). Criteria for validation of methods in microbial forensics. Appl. Environ. Microbiol. 74, 5599–5607.

Burd, E.M. (2010). Validation of laboratory-developed molecular assays for infectious diseases. Clin. Microbiol. Rev. 23, 550–576.

Bustin, S.A., Benes, V., Garson, J.A., Hellemans, J., Huggett, J., Kubista, Mueller, R., Nolan, T., Pfaffl, M.W., Gregory, L., *et al.* (2009). The MIQE guidelines: minimum information for publication of quantitative real-time PCR experiments. Clin. Chem. 55, 611–622.

Chandler, D.P., Wagnon, C.A., and Bolton Jr, H. (1998). Reverse Transcriptase (RT) Inhibition of PCR at low concentrations of Template and its implications for quantitative RT-PCR. Appl. Environ. Microbiol. *64*, 669–677.

Chen, Y.A., Lin, C.C., Wang, C.D., Wu, H.-B., and Hwang, P.I. (2007). An optimized procedure greatly improves EST vector contamination removal. BMC Genomics *8*, 416–427.

de Clare Bronsvoort, B.M., von Wissmann, B., Fèvre, E.M., Handel, I.G., Picozzi, K., and Welburn, S.C. (2010). No gold standard estimation of the sensitivity and specificity of two molecular diagnostic protocols for *Trypanosoma brucei* spp. in western Kenya. PLoS One *5*, e8628.

Clinical and Laboratory Standards Institute. (2000). Defining, establishing and verifying reference intervals in the clinical laboratory; approved guideline. CLSI document C28-A3.

Clinical and Laboratory Standards Institute. (2002). Method comparison and bias estimation using patient samples. Approved guideline, 2nd ed. CLSI document EP9-A2.

Cohen, P. (2010). The changing regulatory environment of *in vitro* diagnostics: implications for sponsors, manufacturers and laboratory users. Aust. J. Med. Sci. *31*, 9–15.

Conraths, F.J., and Schares, G. (2006). Validation of molecular-diagnostic techniques in the parasitological laboratory. Vet. Parasitol. *136*, 91–98.

Dendukuri, N., Rahme, E., Bˇlisle, P., and Joseph, L. (2004). Bayesian sample size determination for prevalence and diagnostic test studies in the absence of a gold. standard test. Biometrics *60*, 388–397.

Domingo, C., Niedrig, M., Teichmann, A., Kaiser, M., Rumer, L., Jarman, G.G., and Donoso-Mantke, O. (2010). 2nd International external quality control assessment for the molecular diagnosis of dengue infections. PLoS Negl. Trop. Dis. *4*, e833.

Drosten, C., Günther, S., Preiser, W., van der Werf, S., Brodt, H.-R., Becker, Rabenau, S., H., Panning, M., Kolesnikova, L., Fouchier, R.A., *et al.* (2003). Identification of a Novel Coronavirus in Patients with Severe Acute Respiratory Syndrome. New Engl. J. Med. *348*, 1967–1976.

Drosten, C., Chiu, L.-L., Panning, M., Leong, H.N., Preiser, W., Tam, J.S., Gunther, S., Kramme, S., Emmerich, P., Ng, W.L., *et al.* (2004). Evaluation of advanced reverse transcriptase-PCR assays and an alternative PCR target region for detection of severe acute respiratory syndrome-associated coronavirus. J. Clin. Microbiol. *42*, 2043–2047.

Ellis, J., Iturriza, M., Allen, R., Bermingham, A., Brown, K., Gray, J., and Brown, D. (2009). Evaluation of four real-time PCR assays for detection of influenza A (H1N1) v viruses. Eurosurveillance *14*, 1–3.

Espy, M.J., Uhl, J.R., Sloan, L.M., Buckwalter, S.P., Jones, M.F., Vetter, E.A., Yao, J.D.C., Wengenack, N.L., Rosenblatt, J.E., Cockerill, I.I.I., F.R., *et al.* (2006). Real-time PCR in clinical microbiology: applications for routine laboratory testing. Clin. Microbiol. Rev. *19*, 165–256.

Food and Drug Administration. (2003). Statistical guidelines on reporting results from studies evaluating diagnostic tests (FDA, Washington, DC).

Gerriets, J.E., Greiner, T.C., and Gebhart, C.L. (2008). Implementation of a T4 extraction control for molecular assays of cerebrospinal fluid and stool specimens. J. Mol. Diagn. *10*, 28–32.

Gunson, R.N., Collins, T.C., and Carman, W.F. (2006). Practical experience of high throughput real time PCR in the routine diagnostic virology setting. J. Clin. Virol. *35*, 355–367.

Hadgu, A. (1998). Bias in the evaluation of DNA-amplification tests for detecting *Chlamydia trachomatis.* Stat. Med. *17*, 1064–1066.

Hadgu, A., and Qu, Y. (1998). A biomedical application of latent class models with random effects. J. R. Stat. Soc. Ser. C *47*, 603–616.

Hadgu, A., and Sternberg, M. (2009). Reproducibility and specificity concerns associated with nucleic acid amplification tests for detecting *Chlamydia trachomatis*. Eur. J. Clin. Microbiol. Infect. Dis. *28*, 9–15.

Hadgu, A., Dendukuri, N., and Hilden, J. (2005). Evaluation of nucleic acid amplification tests in the absence of a perfect gold-Standard test – a review of the statistical and epidemiologic issues. Epidemiology *16*, 604–612.

Hartley, J.L., and Rashtchian, A. (1993). Dealing with contamination: enzymatic control of carryover contamination in PCR. Genome Res. *3*, S10–S14.

Hauk, W.W., Koch, W., Abernathy, D., and Williams, R.L. (2007). Making sense of trueness, precision, accuracy and uncertainty. Pharmacopeial Forum *34*, 838–842.

Higuchi, R., Dollinger, G., Walsh, P.S., and Griffiths, R. (1992). Simultaneous amplification and detection of specific DNA sequences. Nat. Biotechnol. *10*, 413–417.

Higuchi, R., Fockler, C., Dollinger, G., and Watson, R. (1993). Kinetic PCR analysis: real-time monitoring of DNA amplification reactions. Nat. Biotechnol. *11*, 1026–1030.

Hui, S.L., and Zhou, X.H. (1998). Evaluation of diagnostic tests without gold standards. Stat. Methods Med. Res. *7*, 354–370.

International Organization for Standardization. (2003). *In vitro* diagnostic medical devices-measurement of quantities in biological samples. Metrological traceability of values assigned to calibrators and control materials. ISO 17511. International Organization for Standardization, Geneva, Switzerland.

International Organization for Standardization. (2007). Medical laboratories-particular requirements for quality and competence. ISO 15189, 2nd ed. International Organization for Standardization, Geneva, Switzerland.

Jacobson, M., Norling, B., Gunnarson, A., and Aspan, A. (2009). Flotation – a new method to circumvent PCR inhibitors in the diagnosis of *Lawsonia intracellularis*. Int. J. Microbiol. *2009*, 410945.

Jebbink, J., Bai, X., Rogers, B.B., Dawson, D.B., Scheuermann, R.H., and Domiati-Saad, R. (2003). Development of real-time PCR assays for the quantitative detection of Epstein–Barr virus and cytomegalovirus, comparison of TaqMan probes, and molecular beacons. J. Mol. Diag. *5*, 15–20.

Karrer, E.E., Lincoln, J.E., Hogenhout, S., Bennett, A.B., Bostock, R.M., Martineau, B., Lucas, W.J., Gilchrist, D.G., and Alexander, D. (1995). *In situ* isolation of mRNA from individual plant cells: Creation of cell-specific cDNA libraries. Proc. Natl. Acad. Sci. U.S.A. *92*, 3814–3818.

Kaur, H., Arora, A., Wengel, J., and Maiti, S. (2006). Thermodynamic, counterion, and hydration effects for the incorporation of locked nucleic acid nucleotides into DNA duplexes. Biochem. *45*, 7347–7355.

Klugman, K.P., Mahdi, S.A., and Albrich, C. (2008). Novel approaches to the identification of *Streptococcus pneumoniae* as the cause of community-acquired pneumonia. CID *47*, S202–S206.

Lefever, S., Hellemans, J., Pattyn, F., Przybylski, D.R., Taylor, C., Geurts, R., Untergasser, A., and Vandesompele, J. (On behalf of the R.D.M.L consortium.) (2009). RDML: structured language and reporting guidelines for real-time quantitative PCR data. Nucleic Acids Res. *37*, 2065–2069.

Longo, M.C., Beminger, M.S., and James, L., and Hartley, J.L. (1990). Use of uracil DNA glycosylase to control carry-over contamination in polymerase chain reactions. Gene *93*, 125–128.

Madej, R.M., Davis, J., Holden, M.J., Kwang, S., Labourier, E., and Schneider, G.J. (2010). International standards and reference materials for quantitative molecular infectious disease testing. J. Mol. Diagn. *12*, 1–11.

Mahony, J.B., Petrich, A., Louie, L., Song, X., Chong, S., Smieja, M., Chernesky, M., Loeb, M., and Richardson, S. (2004). Performance and cost evaluation of one commercial and six in-house conventional and real-time reverse transcription-PCR assays for detection of severe acute respiratory syndrome coronavirus. J. Clin. Micro. *42*, 1471–1476.

Marra, M.A., Jones, S.J.M., Astell, C.R., Holt, R.A., Brooks-Wilson, A., Butterfield, Y.S., Khattra, J., Asano, J.K., Barber, S.A., Chan, S.Y., *et. al.* (2003). The genome sequence of the SARS-associated coronavirus. Science *300*, 1399–1404.

Mattocks, C.J., Morris, M.A., Matthijs, G., Swinnen, E., Corveleyn, A., Dequeker, E., Müller, C.R., Pratt, V., and Wallace, A. (2010). A standardized framework for the validation and verification of clinical molecular genetic tests. Eur. J. Human Genet. *18*, 1276–1288.

Meade, K.G., Narciandi, F., Cahalane, S., Reiman, C., Allan, B., and O'Farrelly, C. (2009). Comparative *in vivo* infection yield insights on early host immune response to Campylobacter in chickens. Immunogenetics *61*, 101–110.

Miffin, T.E. (1995). Setting up a PCR laboratory. In PCR: Essential Data, Newton, C.R., ed. (John Wiley and Sons Inc., New York), pp. 4–5.

Millar, B.C., Xu, J., and Moore, J.E. (2002). Risk assessment models and contamination management: implications for broad-range ribosomal DNA PCR as a diagnostic tool in medical bacteriology. J. Clin. Micro. *40*, 1575–1580.

Miller, W.C. (1998). Bias in discrepant analysis: when two wrongs don't make a right. J. Clin. Epidemiol. *51*, 219–231.

Monteiro, L., Bonnemaison, D., Vekris, A., Petry, K.G., Bonnet, J., Vidal, R., Cabrita, J., and Mégraud, F. (1997). Complex polysaccharides as PCR inhibitors in feces: *Helicobacter pylori* model. J. Clin. Micro. *35*, 995–998.

Mullis. K., Faloona, F., Scharf, S., Saiki. R.K., Horn, G.T., and Erlich, H.A. (1986). Specific enzymatic amplification of DNA *in vitro*: the polymerase chain reaction. Cold Spring Harbour Symp. Quant. Biol. *51*, 63–273.

Murphy, J., and Bustin, S.A. (2009). Reliability of real-time reverse-transcription PCR in clinical diagnostics: gold standard or substandard? Expert Rev. Mol. Diagn. *9*, 187–196.

Newsome, T., Li, B.-J., Zou, N., and Lo, S.-C. (2004). Presence of bacterial phage-like DNA sequences in commercial *Taq* DNA polymerase reagents. J. Clin. Micro. *42*, 2264–2267.

Niedrig, M., Meyer, H., Panning, P., and Drosten, C. (2006). Follow-up on diagnostic proficiency of laboratories equipped to perform orthopoxvirus detection and quantification by PCR: the second international external quality assurance study. J. Clin. Micro. *44*, 1283–1287.

Office International des Epizooties. (2009). Principles of validation of diagnostic assays for infectious diseases. In: OIE Manual of Diagnostic Tests and Vaccines for Terrestrial Animals (Office International des Epizooties, Paris), pp. 1–14.

Office International des Epizooties. (2010). Principles of validation of diagnostic assays for infectious diseases. In: OIE Manual of Diagnostic Tests and Vaccines for Terrestrial Animals (Office International des Epizooties, Paris), pp. 1–18.

de Pagter, P.J., Schuurman, R., de Vos, N.M., Mackay, W., and van Loon, A.M. (2010). Multicenter external quality assessment of molecular methods for detection of human herpesvirus. J. Clin. Micro. *48*, 2536–2540.

Peeling, W.R., Smith, P.G., and Bossuyt, P.M.M. (2008). A guide for diagnostic evaluations. Nat. Rev. Micro. S2–S6.

Price, C.P., and Christenson, R.H. (2008). Evaluating New Diagnostic Technologies: perspectives in the UK and US. Clin. Chem. *54*, 1421–1423.

Prince, A.M., and Andrus, L. (1992). PCR: how to kill unwanted DNA. Biotechniques *12*, 358–360.

Rådström, P., Knutsson, R., Wolffs, P., Lövenklev, M., and Löfström, C. (2004). Pre-PCR processing: strategies to generate PCR-compatible samples. Mol. Biotechnol. *26*, 133–146.

Raymaekers, M., Smets, R., Maes, B., and Cartuyvels, R. (2009). Checklist for optimization and validation of real-time PCR assays. J. Clin. Lab. Anal. *23*, 145–151.

R.C.P. (2006). Evaluating and introducing new diagnostic tests: the need for a national strategy. Report from the short life working group on the evaluation and introduction of new tests. Royal College of Pathologists. Available at: http://www.rcpath.org/Resources/RCPath/Migrated%20Resources/Documents/E/Evaluating_and_introducing_new_diagnostic_tests.pdf.

Rolfe, K.J., Parmar, S., Mururi, D., Wreghitt, T.G., Jalal, H., Zhang, H., and Curran, M..D. (2007). An internally controlled, one-step, real-time RT-PCR assay for norovirus detection and genogrouping. J. Clin. Virol. *39*, 318–321.

Rutjes, A.W.S., Reitsma, J.B., Coomarasamy, A., Khan, K.S., and Bossuyt, P.M.M. (2007). Evaluation of diagnostic tests when there is no gold standard. A review of methods. Health Technology Assessment *11*, 1–69.

Saah, A.J., and Hoover, D.R. (1997). 'Sensitivity' and 'specificity' reconsidered: the meaning of these terms in analytical and diagnostic settings. Ann. Int. Med. *126*, 91–94.

Saiki, R.K., Scharf, S., Faloona, F., Mullis, K.B., Horn, G.T., Erlich, H.A., and Arnheim, N. (1985). Enzymatic amplification of beta-globin genomic sequences and restriction site analysis for diagnosis of sickle cell anemia. Science *230*, 1350–1354.

Singh, S.K., Nielsen, P., Koshkin, A.A., and Wengel, J. (1998). LNA (locked nucleic acids): synthesis and high-affinity nucleic acid recognition. Chem. Commun. *4*, 455–456.

Sisti, D., Guescini, M., Rocchi1, M.B.L., Tibollo, P., D'Atri1, M., and Stocchi, V. (2010). Shape based kinetic outlier detection in real-time PCR. BMC Bioinformatics *11*, 1–12.

Sloan, L.M. (2007). Real-time PCR in clinical microbiology: verification, validation, and contamination control. Clin. Microbiol. Newsletter *29*, 87–95.

Templeton, K.E., Forde, C.B., van Loonc, A.M., Claas, E.C.J.,. Niesters, H.G.M., Wallace, P., and Carman, W.F. (2006). A multi-centre pilot proficiency programme to assess the quality of molecular detection of respiratory viruses. J. Clin. Virol. *35*, 51–58.

The T.D.R Diagnostics Evaluation Expert Panel. (2010). Evaluation of diagnostic tests for infectious diseases: general principles. Nat. Rev. Microbiol. *December*, S17–S29.

Thomas, A. (2009). External quality assessment in laboratory medicine: is there a rationale to determine frequency of surveys?. Accred. Qual. Assur. *14*, 439–444.

Vandesomple, J., De Preter, K., Pattyn, F., Poppe, B., Van Roy, N., De Paepe, A., and Spelman, F. (2002). Accurate normalization of real-time quantitative RT-PCR data by geometric averaging of multiple internal control genes. Genome Biol. *3*, 1–12.

Weile, J., and Knabbe, C. (2009). Current applications and future trends of molecular diagnostics in clinical bacteriology. Anal. Bioanal. Chem. *394*, 731–742.

Westgard, J.O., Barry, P.L., and Hunt, M.R. (1981). A multi-rule shewhart chart for quality control in clinical chemistry. Clin. Chem. *27*, 493–501.

Wilson, I., G. (1997). Inhibition and facilitation of nucleic acid amplification. Appl. Environ. Microbiol. *63*, 3741–3751.

MIQE: Guidelines for the Design and Publication of a Reliable Real-time PCR Assay

11

Jim Huggett, Tania Nolan and Stephen A. Bustin

Abstract

The capacity to amplify and detect trace amounts of nucleic acids has made the polymerase chain reaction (PCR) the most formidable molecular technology in use today. Its versatility and scope was further broadened first with the development of reverse transcription (RT)-PCR, which opened up the entire RNA field to thorough exploration and then, most conspicuously, with its evolution into real-time quantitative PCR (qPCR). Speed, simplicity, specificity, wide linear dynamic range, multiplexing and high-throughput potential, reduced contamination risk, simplified detection and data analysis procedures as well as availability of increasingly affordable instrumentation and reduced reagent cost have made qPCR the molecular method of choice when quantifying nucleic acids. Detection of pathogens, SNP analyses and quantification of RNA, even real-time analysis of gene expression *in vivo* have become routine applications and constant enhancements of chemistries, enzymes, mastermixes and instruments continue to extend the scope of qPCR technology by promising added benefits such as extremely short assay times measured in minutes, low reagent usage and exceptionally rapid heating/cooling rates. The whole process is driven by the insatiable demand for ever-more specific, sensitive, convenient and cost-effective protocols.

However, it has also become clear that variable pre-assay conditions, poor assay design and incorrect data analysis have resulted in the regular publication of data that are inconsistent, inaccurate and often simply wrong. The problem is exacerbated by a lack of transparency of reporting, with the details of technical information wholly inadequate for the purpose of assessing the validity of reported qPCR data. This has serious consequences for basic research, reducing the potential for translating findings into valuable applications and potentially devastating implications for clinical practice. In response, guidelines proposing a minimum standard for the provision of information for qPCR experiments ('MIQE') have been launched. These aim to establish a standard for accurate and reliable qPCR experimental design as well as recommendations to ensure comprehensive reporting of technical detail, indispensable conditions for the maturing of qPCR into a robust, accurate and reliable nucleic acid quantification technology.

Introduction

Quantitative real-time PCR (qPCR) has become a ubiquitous, mainstream technology widely referred to as the gold standard for nucleic acid detection and quantification in basic research as well as for clinical applications (Bustin, 2010a). Awkwardly, the remarkable transparency and practical simplicity of qPCR has been replaced by numerous conceptual, reagent, protocol and analysis alternatives that continue to yield quantitative results, but require a detailed understanding of the methodology as well as the methods involved to allow an assessment of the data validity. In its simplest form, qPCR is extensively used as a low throughput method for the detection or quantification of limited numbers of individual SNPs or RNAs in relatively few samples or for corroborating fluorescent microarray analyses, as it complements their relatively limited sensitivity, and dynamic range. However, its power and versatility have propelled a growing enthusiasm towards more complex applications. Novel chemistries, both fluorescence- and non-fluorescence-based, fast cyclers with rapid ramping and cooling capability and able to process thousands of assays at the same time, and fast reagents allowing minimal cycling times have combined with advances in automation technology to convert qPCR into a very high-throughput technology generating results in very short time. One of the most remarkable technical developments is the integration of qPCR with nanotechnology. The emergence of high-throughput nanolitre qPCR (Morrison, 1995) and microfluidic digital PCR (Zhang and Xing, 2007) is coupled to advances such as the integration of DNA purification, concentration and qPCR into a real time micro-PCR chip (Min *et al.*, 2011) and the development of innovative, portable PCR instruments (Qiu *et al.*, 2011). Amongst many novel applications, qPCR is used for the rapid and reliable detection of biological threat agents (Christensen *et al.*, 2006), custom-made commercial detection systems are available for the detection of clinical pathogens (Rossney *et al.*, 2008; Novak-Weekley *et al.*, 2010; Zidaric *et al.*, 2011), genetically modified organisms (Dorries *et al.*, 2010) and a combination of nanotechnology and electrochemical qPCR (Yeung *et al.*, 2008) has resulted in the development of hand-held devices that allow local and immediate assays for point-of-care testing applications (Won *et al.*, 2011). This has channelled the technology from a research setting, where, in general, qPCR assays are conceived, designed, performed and controlled by experienced researchers into a wider arena, with high volume 'black box' assays performed as rapidly as possible by less proficient staff.

Current qPCR assay design and reporting are flawed

Unsurprisingly, the ubiquity of the assay itself, combined with abundant choices of methodology, methods and applications, has given rise to numerous protocol alternatives with the potential to produce conflicting data (Bustin and Nolan, 2004; Bustin, 2008a,b; Murphy and Bustin, 2009). Any latent tendency towards inconsistency is exacerbated by the significant probability of variability associated with the numerous steps that make up a qPCR assay (Bustin, 2010a) (Fig. 11.1). Pre-analysis variability derives from issues such as poorly defined sample selection or handling, patchy nucleic acid quality, inconsistent use of controls, poor assay design as well as non-existent optimization and validation. Inconsistencies with post-qPCR data processing are caused by a lack of quality control of the post-qPCR data, especially unsuitable methods of normalization, misguided data analysis procedures and challenges associated with applying the correct statistical methods (Hellemans *et al.*, 2007).

For RT-qPCR, in particular, these problems are critical and have been well documented over many years (Bustin, 2000, 2002, 2004; Bustin and Nolan, 2004; Bustin *et al.*, 2005; Nolan *et al.*, 2006a). Poorly designed, validated and executed assays may, and indeed do,

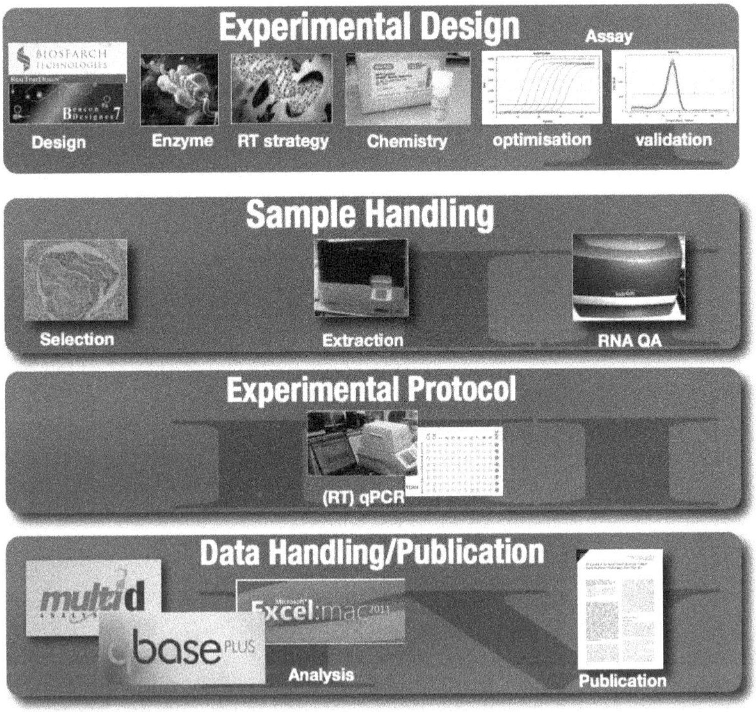

Figure 11.1 RT-qPCR assay. A complete RT-qPCR assay consists of four key steps: (1) Experimental design which includes in silico and empirical validation of each assay. (2) Consistent sample handling and quality assessment. (3) The RT-qPCR steps themselves. (4) Data analysis and publication.

generate a wealth of measurements, but the results can be highly variable and thus inaccurate, ultimately reducing the precision of the measurement or, worse, introducing biases leading to statistically positive but incorrect results. Hence the persistent appeal of developing a set of standard and objective quality control measures (Bustin and Mueller, 2005, 2006; Bustin, 2006, 2008a,b,d, 2010; Murphy *et al.*, 2007; Murphy and Bustin, 2009). The most reliable means of judging the plausibility of a body of work is to study the 'Materials and Methods' section for the relevant information, particularly that relating to the latter aspect of the experimental protocol. A recent survey confirms that this section is avidly read by researchers (Bustin, 2010b), but a second survey shows that the detail of the information provided is often wholly inadequate (Huggett and Bustin, 2011). Unfortunately, not only are many assays designed poorly but also many publications utilizing qPCR technology provide insufficient information to allow a reader to assess the assay and evaluate the validity of conclusions derived from the qPCR data (Bustin, 2010b).

Common omissions that could easily be remedied include simple things like the accession numbers for mRNA or genomic sequences or specifics of the reverse transcription step (Huggett and Bustin, 2011). The latter is especially important, since cDNA priming method and choice of RT have long been known to have a significant impact on results (Stahlberg *et al.*, 2004a,b). Even if the claim is that the experimental protocol recommended by the manufacturer was followed, experience shows that individual researchers frequently introduce subtle, yet consequential variations. Several other pieces of key information are often omitted:

1 Detailed sample information: It remains a remarkable feature of many publications utilizing qPCR technology that very basic information with respect to the samples under investigation is not published. This is of particular importance when considering gene expression analyses from tissue biopsies, where sample selection, acquisition, handling and storage can significantly affect quantification results. Since it should be taken for granted that the researchers have gathered as much information as possible about their samples, the release of that information should add no additional burden to the exacting task of gathering data for publication. Furthermore, it is also necessary to provide details of sample processing procedures, since samples pass through numerous preparative steps prior to the qPCR assay, every one of which can introduce additional variability (Hammerle-Fickinger *et al.*, 2009; Tichopad *et al.*, 2009).
2 Sample quality: it is accepted that RNA integrity and quality assessment is essential for reliable quantification of RNA using RT-qPCR assays (Perez-Novo *et al.*, 2005; Fleige and Pfaffl, 2006; Fleige *et al.*, 2006) and a recent, comprehensive study demonstrates that compromised RNA quality has a measurable impact on the significance of differential expression of prognostic marker genes between cancer patient risk groups, and on risk classification performance using a multigene signature (Vermeulen *et al.*, 2011). The availability of microfluidics-based devices for nucleic acid quality assessment allows automated, rapid and standardized quality assessment of very small amounts of total RNA with quality metrics such as the RIN (Agilent), RQI (BioRad) or SDV (lab901) to represent the level of degradation in a sample. However, it must be remembered that assessing the integrity of rRNA does not necessarily equate with assessing mRNA integrity, and specifically did not in the study by Vermeulen *et al.* (2011), hence the suggestion of introducing a 3′:5′ mRNA-specific integrity assessment (Nolan *et al.*, 2006a) as a more suitable alternative. Importantly, inhibition of reverse-transcription or PCR should be checked by dilution of the sample or use of a universal inhibition assay such as SPUD (Nolan *et al.*, 2006b). Yet very few publications mention the term RNA quality (Huggett and Bustin, 2011).
3 Normalization: normalization strategies are aimed at selecting stable and minimally regulated reference genes and identifying the optimal number of reference genes for the calculation of a normalization factor. This reduces technical variability and increases confidence in data by controlling for experimental error introduced during the multistage process required to extract and process the RNA. Different strategies are not mutually exclusive and the general recommendation is to match sample size, RNA quantity and protocol around three validated internal controls for the final normalization (Huggett *et al.*, 2005). Certainly, there is abundant published evidence that the use of a single reference gene to show small changes in target copy numbers generates unreliable data (Tricarico *et al.*, 2002; Dheda *et al.*, 2004). However, most papers normalize target gene copy numbers against single, invalidated reference genes, even though there are several published methods clearly detailing methods for selection of appropriate reference genes sets (Andersen *et al.*, 2004; Pfaffl *et al.*, 2004; Vandesompele *et al.*, 2002). When this is combined with the reporting of small changes (e.g. differences of less than threefold) it is impossible to conclude whether findings are

differences in expression of the gene of interest, the references gene or a combination of both; consequently statistically significant measurement bias can be introduced leading to incorrect findings (Dheda *et al.*, 2005).

4 PCR efficiency: relative expression levels of mRNA of genes of interest (GOI) are frequently reported relative to those of one or, ideally, several reference genes using the comparative *Cq* method (Livak and Schmittgen, 2001). This involves normalizing the expression of GOI against the internal reference gene(s) *within* two samples, one of which acts as control (ΔCq), followed by a comparison of the normalized difference in expression *between* the two samples ($\Delta\Delta Cq$). In order to be reliable, this requires constant doubling of amplicon with each cycle for each of the targets, or equal amplification efficiency at the very least. However, actual efficiency can range between 60% and 100%, depending on assay quality, sample and target characteristics, reagents as well as instrument variability. Hence small differences between GOI and reference genes can lead to huge differences in relative expression ratios and generate distorted results. Again it is remarkable that not only is this a well-known fact, but that there are a number of efficiency-corrected relative quantification models that provide an efficient and reliable means of quantifying nucleic acids (Pfaffl, 2001; Peirson *et al.*, 2003). They depend on the calculation of PCR efficiency for every target and incorporate this information into relative expression algorithms. Popular software packages such as Genex (Bergkvist *et al.*, 2010), qBase (Hellemans *et al.*, 2007) and REST (Relative Expression Software Tool) (Pfaffl *et al.*, 2002) all make use of such models.

Since all of this information is essential to assay performance, omitting it in a publication makes it impossible to judge the technical quality of published data.

One of the most egregious examples of the enormous implications for the health and lives of individuals that result from inappropriate use of this technology is provided by the use of RT-qPCR data that purported to demonstrate the presence of measles virus (MV) RNA in children with developmental disorders (Uhlmann *et al.*, 2002). It provided sustenance to the controversy surrounding the triple measles mumps and rubella (MMR) virus vaccine, as the data were interpreted as providing evidence for a link between MMR, gut pathology and autism. However, a detailed analysis of the raw data underlying that report carried out by one of the authors (SAB) acting as an expert witness to the UK High Court and the US Vaccine Court, revealed that these data were obtained amongst a catalogue of mistakes, inaccuracies and inappropriate analysis methods as well as contamination and poor assay performance (Bustin, 2008c). A reanalysis of the data concluded that the assay had been detecting DNA and since MV is an RNA-only virus, the RT-qPCR data had been erroneously interpreted, a conclusion confirmed elsewhere (Afzal *et al.*, 2006; D'Souza *et al.*, 2006; Hornig *et al.*, 2008).

MIQE

A growing consensus has been developing around the need to improve published information with relevant experimental detail that covers every aspect important to the qPCR assay itself, as well as issues relating to pre- and post-assay parameters. Specifically, it became clear that there is a requirement for a set of recommendations that can be used by journal reviewers, who need to be able to evaluate the reliability of the experimental protocols and ensure the inclusion of all essential information in the final publication. This need was addressed by the 'Minimum Information for Publication of Quantitative Real-Time PCR Experiments' (MIQE) guidelines (Bustin *et al.*, 2009), with a recent amendment clarifying the disclosure of primer sequences (Bustin *et al.*, 2011). The last year has seen a rapid expansion in the number of researchers aware of the existence of these guidelines (Fig. 11.2) as well as an increasing number of citations of the original publication in the per-reviewed literature (Fig. 11.3). There even is an iPAD/iPOD/iPhone as well as an Android app available from the iTunes store.

The ultimate goal of the MIQE guidelines is to make all technical aspects of the qPCR assay sufficiently transparent, so that the reader of a publication can take technical excellence for granted and focus on the biological relevance of the data. MIQE is made up from nine sections that list 85 parameters that constitute the minimum information required to allow potential reproduction as well as unambiguous quality assessment of a qPCR-based experiment. The nine key areas are 'Experimental design, sample, nucleic acids, reverse transcription, target, primers and probes, assay details, PCR cycling and data analysis'. At first sight, these look overwhelming, arduous and over-exacting. In practice, it is clear that most, if not all of these parameters describe information that would be obtained as a matter of course during the experimental design, optimization and validation stages. Importantly, there is a clear hierarchy with some parameters, labelled 'E' (essential) in the published guidelines, indispensable for an adequate description of the qPCR assay, whereas other components, labelled 'D' (desirable) more peripheral, yet constituting an effective foundation for the realization of best practice protocols. There is increasing recognition that the MIQE guidelines provide the basis for much-needed standardization as well as encourage the publication of essential information that should be accessible to reviewer and reader. Of course, these parameters are not set in stone and are open for discussion; indeed, a core set of sections is being implemented by the BMC group of open-access journals (Bustin *et al.*, 2010).

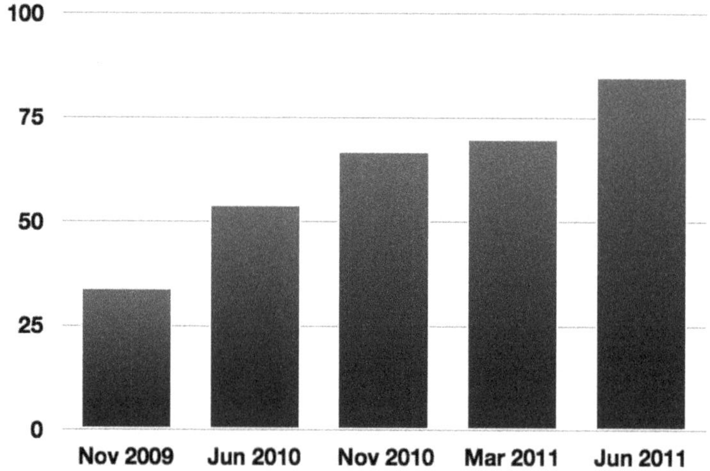

Figure 11.2 MIQE awareness. Attendees of qPCR meetings (2009: qPCR USA, San Francisco; 2010: Gothenburg and qPCR USA; 2011: Freising and Prague) were asked whether they had heard of the MIQE guidelines. The graph shows the percentage of 'yes' responses.

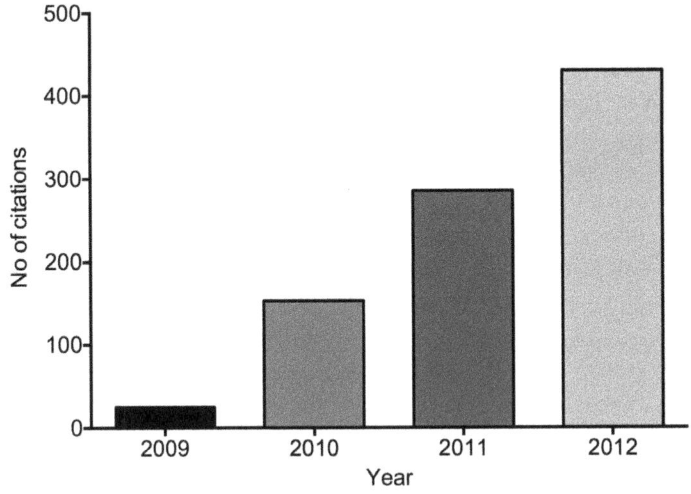

Figure 11.3 Citations of the MIQE paper in the peer-reviewed literature. The number of citations were obtained from Web of Knowledge (www.webofknowledge.com) and stood at 453 at the end of November 2011.

The most controversial aspects of the original MIQE guidelines related to the proposal that publications must divulge the sequences of any primers used and especially should also report the sequences of any probes. The rationale behind releasing the primer sequences was straightforward: an experiment cannot be reproduced if one of the principal reagents is unavailable. Lack of access to a probe sequence, on the other hand, does not preclude analysis of the specificity, efficiency and sensitivity of an assay; however, for completeness' sake it is but a small step to take for most researchers. Many commercial qPCR assays are not supplied with the primer/probe sequences, since most vendors consider this commercially sensitive information; usually there are also no details provided on empirical validation of each individual assay. Publications utilizing such assays could not satisfy the original MIQE requirements, placing limits on a universal acceptance of MIQE. Consequently, an amendment of the original guidelines now requires either primer sequences or a clearly defined amplicon context sequence. This guidance was issued based on the assessment that in the absence of full primer sequence disclosure it is possible to achieve an adequate level of transparency, but only if there is an appropriate level of background information and disclosure of validation results on the qPCR assay. Consequently, if primer sequences are not disclosed, a MIQE-compliant publication should institute the same validation criteria used for assays reporting primer/probe sequences. Specifically, when reporting a precise fold-change for a transcript it remains an essential requirement that the PCR efficiency, analytical sensitivity and specificity of each individual assay be determined. This information should be verified by the investigator for the actual assay that is being reported using the conditions and personnel in their laboratory and not extrapolated from commercial assays validated by the vendors.

The implementation of appropriate statistical methodologies for data handling and processing is an essential complement to any improvements introduced to the practical workflow. Obviously, there are a large number of statistical tools that can be used to address and minimize the variability discussed above and amongst many, specific studies have been published that look at the identification and handling of outliers and precision associated with calibration curves (Burns *et al.*, 2005), the relative merits of obtaining *Cq*s from the threshold method or sigmoidal functions (Rutledge and Cote, 2003; Rutledge, 2004; Rutledge and Stewart, 2008) and limits of detection modelled from sample replication and *Cq* values (Burns and Valdivia, 2008). Furthermore, tools have become available that allow management and analysis of qPCR data (Muller *et al.*, 2002; Jin *et al.*, 2006; Hellemans *et al.*, 2007; Gallup and Ackermann, 2008; Ritz and Spiess, 2008). However, since there are no guidelines or universally accepted standards for data handling and interpretation, the use of multiple statistical tools adds to variability and discordance. Hence MIQE guidelines propose the disclosure of key statistical features associated with a qPCR assay, as well as reporting of what software was used to analyse results.

Appropriate statistical modelling and analysis for the interpretation of qPCR data is of particular importance for clinical applications, where false-positive or negative results can have serious implications. A systematic evaluation of the various qPCR data analysis methods has shown that they differ substantially in their performance (Karlen *et al.*, 2007); hence MIQE guidelines specify the importance of providing detailed information on the methods of data analysis and confidence estimation, especially identifying the statistical methods used to evaluate variances. Since expanding sample size can increase the power of a statistical test, technical repeats are a much favoured and reported sign of qPCR virility. Whilst these do help reduce measurement error, technical repeats simply provide a commentary on the researchers', or their robots' ability to pipette accurately. Technical replicates are needed when optimizing a PCR reaction to ensure that the assay is optimum. However the high precision of an optimal qPCR assay (typically less than 10% coefficient of variation) means that when this is replicated, far from increasing the reliability of results, it can distort the statistics of determining confidence in experimental data. Furthermore if technical replicates are to be included when conducting RT-qPCR it is essential that the more variable reverse transcription reaction is replicated (a process that is automatic with one step RT-qPCR), as a frequent conceptual error is to perform a single reverse transcription and replicate the PCR step on the cDNA sample. This action has surely led to numerous publications of very precise yet totally biased results.

Appropriate biological replication is essential if findings are meant to be valid in the context of a conceptually large population from which the subjects were sampled, rather than only for the particular individuals considered in the experiment (Mehta *et al.*, 2004). Since biological variability is larger than technical variation, increasing biological replication usually translates into more effective gains in power. However, increasing sample size generally leads to added cost and increased time for performing the experiments. In addition, some biological replication cannot be increased, e.g. when comparing large numbers of healthy individuals with a limited number of patients with a particular disease.

It is important to note that the need for appropriate biological replication and other aspects of the MIQE recommendations do not reflect the deliberations of a handful of purists frequenting the whitest of ivory towers. MIQE matters because it is more important that findings truly reflect what is being investigated than it is to publish a Cell or Nature paper. When, as is often the case, high impact publications are the sole aim of a study a poor design can lead to the embarrassing process of a retraction.

Conclusions

qPCR and RT-qPCR are powerful technologies that have enabled many of the advances made in our understanding of basic biological and disease processes; both are also increasingly used for clinical diagnostic purposes. However, the combination of ease of use and lack of rigorous standards of practice has resulted in widespread misinterpretation of data and consequent publication of erroneous conclusions. Any solution to the challenge of how to make PCR-based assays more reliable requires both an appreciation and an understanding of numerous attributes that include biological concepts, statistics, mathematical modelling, technical know-how and a willingness to share this knowledge. This range of fundamental variables must be addressed by guidelines that permit a shift of focus from questions regarding the technological relevance underlying a publication's conclusion to the actual biological or diagnostic issues being addressed. MIQE constitutes a reference framework for communication within the research community, instrument and reagent manufacturers and publishers that promises to deliver guidelines that promote transparency of experiments and confidence in results and conclusions that advance, rather than impede our knowledge.

References

Afzal, M.A., Ozoemena, L.C., O'Hare, A., Kidger, K.A., Bentley, M.L., and Minor, P.D. (2006). Absence of detectable measles virus genome sequence in blood of autistic children who have had their MMR vaccination during the routine childhood immunization schedule of UK. J. Med. Virol. *78*, 623–630.

Andersen, C.L., Jensen, J.L., and Orntoft, T.F. (2004). Normalization of real-time quantitative reverse transcription-PCR data: a model-based variance estimation approach to identify genes suited for normalization, applied to bladder and colon cancer data sets. Cancer Res. *64*, 5245–5250.

Bergkvist, A., Rusnakova, V., Sindelka, R., Garda, J.M., Sjogreen, B., Lindh, D., Forootan, A., and Kubista, M. (2010). Gene expression profiling – clusters of possibilities. Methods *50*, 323–335.

Burns, M.J., and Valdivia, H. (2008). Modelling the limit of detection in real-time quantitative PCR. Eur. Food Res. Technol. *226*, 1513–1524.

Burns, M.J., Nixon, G.J., Foy, C.A., and Harris, N. (2005). Standardisation of data from real-time quantitative PCR methods – evaluation of outliers and comparison of calibration curves. BMC Biotechnol. *5*, 31.

Bustin, S.A. (2000). Absolute quantification of mRNA using real-time reverse transcription polymerase chain reaction assays. J. Mol. Endocrinol. *25*, 169–193.

Bustin, S.A. (2002). Quantification of mRNA using real-time reverse transcription PCR (RT-PCR): trends and problems. J. Mol. Endocrinol. *29*, 23–39.

Bustin, S.A. (2004). Meaningful quantification of mRNA using real-time PCR. In PCR Technology Current Innovations, Griffin, H.G., and Griffin, A., ed. (CRC Press Boca, Raton, FL), pp. 225–233.

Bustin, S.A. (2006). Nucleic acid quantification and disease outcome prediction in colorectal cancer. Personalized Med. *3*, 207–216.

Bustin, S. (2008a). Molecular medicine, gene-expression profiling and molecular diagnostics: putting the cart before the horse. Biomarkers Med. *2*, 201–207.

Bustin, S.A. (2008b). Real-time polymerase chain reaction – towards a more reliable, accurate and relevant assay. Eur. Pharm. Rev. *6*, 19–27.

Bustin, S.A. (2008c). Real-time quantitative PCR-opportunities and pitfalls. Eur. Pharm. Rev. *4*, 18–23.

Bustin, S.A. (2010a). Why the need for qPCR publication guidelines? – The case for MIQE. Methods *50*, 217–226.

Bustin, S.A. (2010b). Developments in real-time PCR research and molecular diagnostics. Expert Rev. Mol. Diagn. *10*, 713–715.

Bustin, S.A., and Mueller, R. (2005). Real-time reverse transcription PCR (qRT-PCR) and its potential use in clinical diagnosis. Clin. Sci. (Lond.) *109*, 365–379.

Bustin, S.A., and Mueller, R. (2006). Real-time reverse transcription PCR and the detection of occult disease in colorectal cancer. Mol. Aspects Med. *27*, 192–223.

Bustin, S.A., and Nolan, T. (2004). Pitfalls of quantitative real-time reverse-transcription polymerase chain reaction. J. Biomol. Tech. *15*, 155–166.

Bustin, S.A., Benes, V., Nolan, T., and Pfaffl, M.W. (2005). Quantitative real-time RT-PCR – a perspective. J. Mol. Endocrinol. *34*, 597–601.

Bustin, S.A., Benes, V., Garson, J.A., Hellemans, J., Huggett, J., Kubista, M., Mueller, R., Nolan, T., Pfaffl, M.W., Shipley, G.L., *et al.* (2009). The MIQE guidelines: minimum information for publication of quantitative real-time PCR experiments. Clin. Chem. 55, 611–622.

Bustin, S.A., Beaulieu, J.F., Huggett, J., Jaggi, R., Kibenge, F.S., Olsvik, P.A., Penning, L.C., and Toegel, S. (2010). MIQE precis: Practical implementation of minimum standard guidelines for fluorescence-based quantitative real-time PCR experiments. BMC Mol. Biol. *11*, 74.

Bustin, S.A., Benes, V., Garson, J.A., Hellemans, J., Huggett, J., Kubista, M., Mueller, R., Nolan, T., Pfaffl, M.W., Shipley, G.L., *et al.* (2011). Primer sequence disclosure: a clarification of the MIQE guidelines. Clin. Chem. 57, 919–921.

Christensen, D.R., Hartman, L.J., Loveless, B.M., Frye, M.S., Shipley, M.A., Bridge, D.L., Richards, M.J., Kaplan, R.S., Garrison, J., Baldwin, C.D., *et al.* (2006). Detection of biological threat agents by real-time PCR: comparison of assay performance on the R.A.P.I.D., the LightCycler, and the Smart Cycler platforms. Clin. Chem. 52, 141–145.

D'Souza, Y., Fombonne, E., and Ward, B.J. (2006). No evidence of persisting measles virus in peripheral blood mononuclear cells from children with autism spectrum disorder. Pediatrics *118*, 1664–1675.

Dheda, K., Huggett, J.F., Bustin, S.A., Johnson, M.A., Rook, G., and Zumla, A. (2004). Validation of housekeeping genes for normalizing RNA expression in real-time PCR. Biotechniques 37, 112–119.

Dheda, K., Huggett, J.F., Chang, J.S., Kim, L.U., Bustin, S.A., Johnson, M.A., Rook, G.A., and Zumla, A. (2005). The implications of using an inappropriate reference gene for real-time reverse transcription PCR data normalization. Anal. Biochem. *344*, 141–143.

Dorries, H.H., Remus, I., Gronewald, A., Gronewald, C., and Berghof-Jager, K. (2010). Development of a qualitative, multiplex real-time PCR kit for screening of genetically modified organisms (GMOs). Anal. Bioanal. Chem. *396*, 2043–2054.

Fleige, S., and Pfaffl, M.W. (2006). RNA integrity and the effect on the real-time qRT-PCR performance. Mol. Aspects Med. 27, 126–139.

Fleige, S., Walf, V., Huch, S., Prgomet, C., Sehm, J., and Pfaffl, M.W. (2006). Comparison of relative mRNA quantification models and the impact of RNA integrity in quantitative real-time RT-PCR. Biotechnol. Lett. *28*, 1601–1613.

Gallup, J.M., and Ackermann, M.R. (2008). The 'PREXCEL-Q Method' for qPCR. Int. J. Biomed. Sci. *4*, 273–293.

Hammerle-Fickinger, A., Riedmaier, I., Becker, C., Meyer, H.H., Pfaffl, M.W., and Ulbrich, S.E. (2009). Validation of extraction methods for total RNA and miRNA from bovine blood prior to quantitative gene expression analyses. Biotechnol. Lett. 32, 35–44.

Hellemans, J., Mortier, G., De Paepe, A., Speleman, F., and Vandesompele, J. (2007). qBase relative quantification framework and software for management and automated analysis of real-time quantitative PCR data. Genome Biol. *8*, R19.

Hornig, M., Briese, T., Buie, T., Bauman, M.L., Lauwers, G., Siemetzki, U., Hummel, K., Rota, P.A., Bellini, W.J., O'Leary, J.J., *et al.* (2008). Lack of association between measles virus vaccine and autism with enteropathy: a case-control study. PLoS One 3, e3140.

Huggett, J., and Bustin, S.A. (2011). Standardisation and reporting for nucleic acid quantification. Accredit. Qual. Assur. *16*, 399–405.

Huggett, J., Dheda, K., Bustin, S., and Zumla, A. (2005). Real-time RT-PCR normalisation; strategies and considerations. Genes Immun. *6*, 279–284.

Jin, N., He, K., and Liu, L. (2006). qPCR-DAMS: a database tool to analyze, manage, and store both relative and absolute quantitative real-time PCR data. Physiol. Genomics 25, 525–527.

Karlen, Y., McNair, A., Perseguers, S., Mazza, C., and Mermod, N. (2007). Statistical significance of quantitative PCR. BMC Bioinformatics *8*, 131.

Livak, K.J., and Schmittgen, T.D. (2001). Analysis of relative gene expression data using real-time quantitative PCR and the 2(-Delta Delta C(T)) Method. Methods 25, 402–408.

Mehta, T., Tanik, M., and Allison, D.B. (2004). Towards sound epistemological foundations of statistical methods for high-dimensional biology. Nat. Genet. *36*, 943–947.

Min, J., Kim, J.H., Lee, Y., Namkoong, K., Im, H.C., Kim, H.N., Kim, H.Y., Huh, N., and Kim, Y.R. (2011). Functional integration of DNA purification and concentration into a real time micro-PCR chip. Lab. Chip *11*, 259–265.

Morrison, D.K. (1995). Mechanisms regulating Raf-1 activity in signal transduction pathways. Mol. Reprod. Dev. 42, 507–514.

Muller, P.Y., Janovjak, H., Miserez, A.R., and Dobbie, Z. (2002). Processing of gene expression data generated by quantitative real-time RT-PCR. Biotechniques 32, 1372–1379.

Murphy, J., and Bustin, S.A. (2009). Reliability of real-time reverse-transcription PCR in clinical diagnostics: gold standard or substandard? Expert Rev. Mol. Diagn. 9, 187–197.

Murphy, J., Dorudi, S., and Bustin, S.A. (2007). Molecular staging of colorectal cancer: new paradigm or waste of time? Expert Opin. Med. Diagn. *1*, 31–45.

Nolan, T., Hands, R.E., and Bustin, S.A. (2006a). Quantification of mRNA using real-time RT-PCR. Nat. Protocols *1*, 1559–1582.

Nolan, T., Hands, R.E., Ogunkolade, B.W., and Bustin, S.A. (2006b). SPUD: a qPCR assay for the detection of inhibitors in nucleic acid preparations. Anal. Biochem. *351*, 308–310.

Novak-Weekley, S.M., Marlowe, E.M., Miller, J.M., Cumpio, J., Nomura, J.H., Vance, P.H., and Weissfeld, A. (2010). *Clostridium difficile* testing in the clinical laboratory by use of multiple testing algorithms. J. Clin. Microbiol. *48*, 889–893.

Peirson, S.N., Butler, J.N., and Foster, R.G. (2003). Experimental validation of novel and conventional approaches to quantitative real-time PCR data analysis. Nucleic Acids Res. *31*, e73.

Perez-Novo, C.A., Claeys, C., Speleman, F., Van Cauwenberge, P., Bachert, C., and Vandesompele, J. (2005). Impact of RNA quality on reference gene expression stability. Biotechniques 39, 52, 54, 56.

Pfaffl, M.W. (2001). A new mathematical model for relative quantification in real-time RT-PCR. Nucleic Acids Res. *29*, E45.

Pfaffl, M.W., Horgan, G.W., and Dempfle, L. (2002). Relative expression software tool (REST) for group-wise comparison and statistical analysis of relative expression results in real-time PCR. Nucleic Acids Res. *30*, e36.

Pfaffl, M.W., Tichopad, A., Prgomet, C., and Neuvians, T.P. (2004). Determination of stable housekeeping genes, differentially regulated target genes and sample integrity: BestKeeper – Excel-based tool using pair-wise correlations. Biotechnol. Lett. *26*, 509–515.

Qiu, X., Chen, D., Liu, C., Mauk, M.G., Kientz, T., and Bau, H.H. (2011). A portable, integrated analyzer for microfluidic-based molecular analysis. Biomed. Microdevices 13, 809–817.

Ritz, C., and Spiess, A.N. (2008). qpcR: an R package for sigmoidal model selection in quantitative real-time polymerase chain reaction analysis. Bioinformatics 24, 1549–1551.

Rossney, A.S., Herra, C.M., Brennan, G.I., Morgan, P.M., and O'Connell, B. (2008). Evaluation of the Xpert methicillin-resistant Staphylococcus aureus (MRSA) assay using the GeneXpert real-time PCR platform for rapid detection of MRSA from screening specimens. J. Clin. Microbiol. *46*, 3285–3290.

Rutledge, R.G. (2004). Sigmoidal curve-fitting redefines quantitative real-time PCR with the prospective of developing automated high-throughput applications. Nucleic Acids Res. 32, e178.

Rutledge, R.G., and Cote, C. (2003). Mathematics of quantitative kinetic PCR and the application of standard curves. Nucleic Acids Res. *31*, e93.

Rutledge, R.G., and Stewart, D. (2008). A kinetic-based sigmoidal model for the polymerase chain reaction and its application to high-capacity absolute quantitative real-time PCR. BMC Biotechnol. *8*, 47.

Stahlberg, A., Hakansson, J., Xian, X., Semb, H., and Kubista, M. (2004a). Properties of the reverse transcription reaction in mRNA quantification. Clin. Chem. *50*, 509–515.

Stahlberg, A., Kubista, M., and Pfaffl, M. (2004b). Comparison of reverse transcriptases in gene expression analysis. Clin. Chem. *50*, 1678–1680.

Tichopad, A., Kitchen, R., Riedmaier, I., Becker, C., Stahlberg, A., and Kubista, M. (2009). Design and optimization of reverse-transcription quantitative PCR experiments. Clin. Chem. *55*, 1816–1823.

Tricarico, C., Pinzani, P., Bianchi, S., Paglierani, M., Distante, V., Pazzagli, M., Bustin, S.A., and Orlando, C. (2002). Quantitative real-time reverse transcription polymerase chain reaction: normalization to rRNA or single housekeeping genes is inappropriate for human tissue biopsies. Anal. Biochem. *309*, 293–300.

Uhlmann, V., Martin, C.M., Sheils, O., Pilkington, L., Silva, I., Killalea, A., Murch, S.B., Walker-Smith, J., Thomson, M., Wakefield, A.J., *et al.* (2002). Potential viral pathogenic mechanism for new variant inflammatory bowel disease. Mol. Pathol. *55*, 84–90.

Vandesompele, J., De Preter, K., Pattyn, F., Poppe, B., Van Roy, N., De Paepe, A., and Speleman, F. (2002). Accurate normalization of real-time quantitative RT-PCR data by geometric averaging of multiple internal control genes. Genome Biol. 3, 0034.1–0034.11.

Vermeulen, J., De Preter, K., Lefever, S., Nuytens, J., De Vloed, F., Derveaux, S., Hellemans, J., Speleman, F., and Vandesompele, J. (2011). Measurable impact of RNA quality on gene expression results from quantitative PCR. Nucleic Acids Res. *39*, e63.

Won, B.Y., Shin, S., Baek, S., Jung, Y.L., Li, T., Shin, S.C., Cho, D.Y., Lee, S.B., and Park, H.G. (2011). Investigation of the signaling mechanism and verification of the performance of an electrochemical real-time PCR system based on the interaction of methylene blue with DNA. Analyst *136*, 1573–1579.

Yeung, S.S., Lee, T.M., and Hsing, I.M. (2008). Electrochemistry-based real-time PCR on a microchip. Anal. Chem. *80*, 363–368.

Zhang, C., and Xing, D. (2007). Miniaturized PCR chips for nucleic acid amplification and analysis: latest advances and future trends. Nucleic Acids Res. *35*, 4223–4237.

Zidaric, V., Kotnik Kevorkijan, B., Oresic, N., Janezic, S., and Rupnik, M. (2011). Comparison of two commercial molecular tests for the detection of Clostridium difficile in the routine diagnostic laboratory. J. Med. Microbiol. *60*, 1131–1136.

12 Management Aspects of Real-time PCR-based Assay Development, Validation, Verification and Implementation

Jacob Moran-Gilad and Nick Saunders

Abstract

There are significant risks associated with the introduction of new diagnostic assays based on real-time PCR. A consistent approach to the management of the development, validation, verification and implementation of such assays is essential to meet good practice. Adoption of a strategic framework which is followed rigorously by the team is important to ensure that the project is successful. The project core team must have clearly defined roles and produce adequate documentation that can be assessed by a separate review team. Project planning should include aspects such as setting clear objectives, identification of materials and the resources required to complete the project. Of particular importance for real-time PCR diagnostics is the inclusion of adequate positive controls. Tools for the selection of appropriate positive controls are presented and discussed here as a key aspect of diagnostic PCR project management. Following the introduction of assays to practice it is vital to maintain the standard operating procedure to ensure that it is followed consistently and so that any necessary changes are documented and adequately validated. The users of diagnostic assays must be aware of the contents of the project dossiers or have other means to verify their provenance and performance. Project documentation should be maintained to ensure that the quality of the validation data is strengthened over time.

Background

The introduction of new or modified diagnostic tests, either through development in-house, or by adoption of methods developed elsewhere (whether available commercially or not) is frequently undertaken within modern clinical laboratories. There are significant risks associated with this process and therefore there is a need to adopt a consistent 'validation framework' within which assay development, validation, verification and implementation are managed. Whilst guidance on assay validation and verification has been issued by various authorities and is included in this volume, the scope of such guidance has usually been limited to technical and scientific aspects of the process. Many important issues concerning management of the process have been given insufficient attention.

A well designed management approach to diagnostic assay validation is essential to ensure good practice. No less important, a managed process will help to ensure fitness for purpose, efficiency in the use of resources, quality assurance, attention to risk management, compliance with regulatory requirements such as the EC IVDD directive and guarantee that competing commercial rights and constraints are adequately considered.

To make sure of the best outcomes it is preferable that a strategic validation framework is adopted and systematically followed by the project managers and scientists engaged in assay validation. Here we will attempt to highlight the various issues that should be addressed, within the envelope of such a management approach, during the development of a diagnostic assay relying on real-time PCR.

When is a validation framework essential?

The short answer to this question is 'always'. The term 'validation' is often used very loosely and can cover a variety of different processes but is essentially a quality assurance process establishing evidence that provides a high degree of assurance that a product, service, or system accomplishes its intended requirements. Validation is an evidence-based process that requires proper planning to ensure that newly developed assays comply with laboratory standard systems and meet clinical governance and risk management requirements. There are a number of stages in this process including planning and inception, assay development and optimization, assay validation, roll out and verification, and finally implementation.

Scientists, especially those who are experienced with development work, may perceive the management aspects of assay validation as onerous and bureaucratic. Therefore, it is important that the organizational culture emphasizes the importance of this process by highlighting its outputs and creating appropriate feedback mechanisms. Although the validation framework should be all-encompassing, ground rules defining the main settings to which it applies and specifying which of its components should be used in different situations are required.

The following are typical settings where use of a validation framework is appropriate:

- development of a novel diagnostic assay;
- adoption of a method developed elsewhere;
- adoption of a commercial method;
- application of an assay to a different clinical or laboratory application; and
- change or modification of an assay beyond its original specifications.

Producing adequate documentation

The assay validation process entails the production of adequate documentation. Such documentation should include all the information analysed during the various stages of the project including all crude data. Additionally, any pre-existing evidence such as published papers, external or internal quality assessment (EQA or IQA) results, the results from comparisons to other tests and other documented studies that provide evaluation data, should be included. Once satisfactory validation has been completed it is essential to provide a standard operating procedure (SOP) according to which the assay is performed.

Project managers should investigate the requirements for particular documentation to satisfy organizational, local, national and international regulations as applicable.

Setting up the project core team

The following describes a model team that may be recognizable within many organizational structures. A Project Manager, most appropriately an individual at Team Leader, Unit Head or Laboratory Director level, should be designated. This person has overall responsibility for the completion of the project and for signing off the completed validation file and SOP. The Project Manager may either be at Board level or report directly to a Board level Director. This arrangement provides the project with a high level champion which is especially important for the development of novel assays or projects that demand significant resource.

A Project Leader is assigned with responsibility for the project work including the laboratory activities needed for development of the test and its validation. The Project Leader is also responsible for the data analysis, compilation of the validation file, report writing, presentation of data to review meetings, writing the SOP and training staff to carry-out the new SOP. For small or medium scale projects or in organizations of small size the Project Leader may also be the Project Manager.

All project personnel should have appropriate knowledge and experience. It is desirable that records of qualifications be reviewed periodically.

Establishment of a project review team

The drivers for development of new assays are gaps in capability or capacity, or opportunities for improved service presented by new knowledge and technology.

The first stages in the development process are to secure high level management agreement for the work to proceed and approval of any necessary funding.

The aim of the planning phase is to produce a clear, agreed project plan. A suitable panel ('review team') should be established to examine all aspects of the plan. This panel will also review the progress of development work, the validation plan, the evidence that the test is fit for purpose and plans for post-deployment monitoring of test performance. It is advisable that at least one member of the panel should have experience of molecular assay development and validation work within the particular institutional or organizational framework. This person will be capable of providing advice to project team members on conforming to institutional standards.

An example of a suitable review team for an assay designed to detect a human pathogen would be two or three scientists (Clinical or Biomedical scientists in the context of UK pathology services), one or two medical microbiologists and the local Quality representative. In other situations the team members should originate from relevant laboratory scientific or, where relevant, clinical disciplines. It is highly desirable that at least one of the members of the team should be an end user of the assay as this will be useful in providing input to design and validation parameters to ensure utility. It is recommended that consideration is given to the inclusion of a statistician in the review team. Alternatively, a statistician should be consulted to advise on the validation study design. The project manager or project leader but not both, unless both roles are held by a single person, should be members of the review team.

Planning and inception of the project

The review team may have the following responsibilities in relation to planning and inception:

- To ensure that the purpose and objectives of the new assay development are well defined. Specifically how it is supposed to improve or fill in gaps in the current repertoire of tests. This includes identification of the niche that the test is intended to fill and how it fits with currently available alternatives (e.g. for clinical tests the diagnostic need) and the intended users and stakeholders.
- To ensure that a risk register is compiled. The risk register should include risks associated with project success or failure and with implementation of the assay (e.g. users might use inappropriate specimen types or misinterpret the results). The register should specify design actions to be taken in mitigation of the risks.
- To ensure that laboratory safety issues associated with inherent features of the developed test and various stages of the development and validation process have been assessed.
- To ensure that equipment used in the assay development and validation exercise is maintained, serviced, calibrated and monitored as appropriate to ensure that it is suitable for use.
- Analysis of the business case including any procurement issues, the cost of performing the new assay, the price to be charged for the test taking into consideration any alternative assays that may be available. Consideration should also be given to any potential financial risks, for example, existing patents that may be associated with the proposed development.
- To perform commercial analysis including assessing opportunities for intellectual property protection and ownership, and the potential for commercialization. If the assay or reagents are to be supplied to other organizations the cost of fulfilling regulatory requirements (e.g. In Vitro Diagnostic Medical Device Directive, IVDMDD (98/79/EC)) should be considered.
- To ensure that the engagement of collaborative partners to provide expertise or share costs has been considered.
- To ensure that human resource aspects of the project, including secondment of personnel to the project, time-sharing or training issues are addressed.
- To ensure that the means for efficient project management have been established including the nomination of advisors and reviewers as necessary.
- Approval and sign-off of the detailed project plan.
- Approval and sign-off of the validation plan.
- Review of the validation study data and recommendation on whether the assay is suitable for deployment.
- Review of deployment and post deployment plans.
- To ensure that the project dossier is available for future scrutiny and is maintained.

Assay development plan

The objectives of assay development are to ensure the assay accurately and reliably measures only the sequence of interest in real (e.g. patient) samples, to ensure uniformity of assays and reagents over time and maintain full traceability of the results. Project objectives must be well defined (e.g. Development of a robust, convenient real-time PCR for the *mecA* gene of *Staphylococcus aureus* in extracts from blood, urine, tissue and broth cultures, capable of being performed by clinical laboratory staff with basic molecular biology training and producing results by mid-to-late afternoon on day of sample extraction.

The project plan is developed and implemented as illustrated in Fig. 12.1. The review panel and project lead should conduct a series of planning meetings, perform literature searches, appraise options and where possible consult with other centres that carry out the same or similar assays. The panel and project lead agree on the Assay Definition, which should include a name for the test and version control.

The development plan should include a description of the technological details of the assay including information on the platforms and chemistry to be used (Bustin *et al.*, 2009). The reagents to be employed and the assay protocol should be described. Where practical the plan should minimize the number of separate reagent additions by combining reagents. For PCR an ideal is to use only two reagent mixes that can be stored ready for use. Whenever possible kits of finished reagents assembled to run the assay in a diagnostic setting should be formatted so that it is cost effective to perform the test with batches of samples of the size generally encountered.

The development plan must detail the requirements for success and it may be useful to refer to the ICH Harmonized Tripartite Guideline (ICH Steering Committee, 1994). Factors to consider include the desired analytical specificity, analytical sensitivity, linearity, limit of

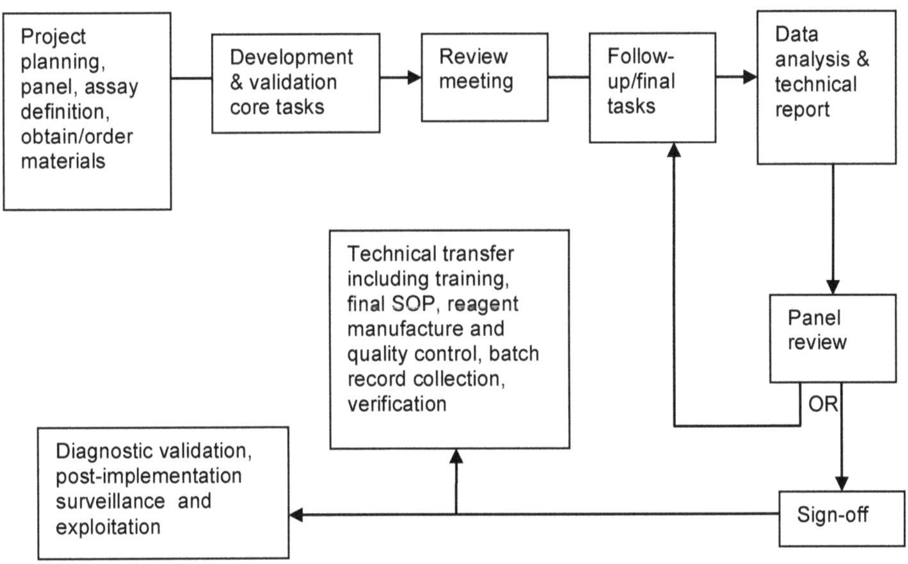

Figure 12.1 Assay development and validation plan. The flow chart illustrates the process. When the project plan has been agreed, the project lead will be responsible for performance of the laboratory tasks. A review panel meeting should then be held with the project lead presenting the data. Any follow up work and/or analysis, if necessary will be agreed, following which the project lead will generate the technical report. The technical report is then circulated to the review panel, who will then either sign-off the assay as ready for roll-out to routine use (technical transfer) or request further work is carried out before the assay can be signed off as suitable. Training of routine diagnostic staff is then undertaken.

detection, PCR efficiency, clinical sensitivity, clinical specificity and reproducibility. These factors will determine the future utility of the assay (Dimech *et al.*, 2004). For example, an assay that is expected to rule out a severe disease should have a high negative predictive value. Requirements for success should be determined with input from appropriate stakeholders, for example, those who would be expected to use or interpret the assay results.

The plan should identify any risk assessments and assessments related to hazardous materials (e.g. COSHH, BS EN ISO 14971) which need to reviewed or written.

Implementation of positive controls for risk management in real-time PCR assays

The implementation of positive controls for real-time PCR is dealt with in some depth here as this is a key issue in the design of assays with a diagnostic purpose. False-negatives in diagnostic real-time PCR may arise due to problems encountered at various stages of the diagnostic cycle. Although some of the problems are generic, also relating to other testing methods, some are specific to the design and use of diagnostic PCR. The problems of diagnostic PCR that are most intractable in terms of false-negative results are extraction failure and reaction inhibition. The probability of false-negative results can be decreased by good practice and by following relevant laboratory guidelines. The utilization of appropriate controls aids in identifying false-negative results and greatly reduces the risk of incorrect diagnosis and reporting. Consequently, it is essential to include a whole of process control (i.e. one capable of detecting both extraction failure and PCR inhibition).

Although internal controls are preferred, external controls (i.e. where reference materials are tested in a parallel PCR to that containing the sample) are useful and are simple to design and implement. The risk that false-negatives remain undetected, though greatly reduced, remains. However, for quantitative PCR it is advantageous that the control reaction cannot interfere with test accuracy.

Appropriate test controls when established may be used to monitor the performance of the quantitative assay over time. Commonly, the controls are spiked at a level that is reliably positive but not at a concentration significantly higher than a typical clinical sample in a typical real-time PCR reaction (e.g. as a guide the control should give a C_t value of 30). Test controls allow assay performance to be monitored over time and assessed by Westgard rules (Westgard and Barry, 1986).

Design and implementation of internal controls

The problems of diagnostic PCR that are most intractable in terms of false-negative results are extraction failure and reaction inhibition. Efficient extraction of target nucleic acid is dependent upon the nature of the micro-organism and extraction protocol used and is thus highly variable. The problem of reaction inhibition due to the presence of inhibitor in the specimen is unpredictable in that it may affect one or more sample within a batch even if no technical error has occurred. Extraction failure and reaction inhibition are far more common with clinical samples subject to primary diagnostic testing.

The application of appropriate controls that facilitates the identification of false-negative results, especially those related to extraction failure and/or reaction inhibition is thus generally recommended and should always be included in PCRs that are intended for clinical diagnosis affecting patient management or significant public health decisions (Clinical and Laboratory Standards Institute, 2006; Burd, 2010) and are implicitly or explicitly required in external standards (i.e. ISO, UKAS, CPA).

There are various forms of PCR controls for detection of false-negative results (Brightwell *et al.*, 1998; Pugnale *et al.*, 2006; Picard, F.J. *et al.*, 2009; Lange *et al.*, 2010). These are summarized in Table 12.1, which indicates the authors' view of the general level of assurance provided by each. It must be emphasized that the overall degree of assurance also depends on which stages of the PCR are under control (extraction, reaction and inhibition).

Different overall approaches to the provision of controls are summarized in Table 12.2. These approaches are graded according to the authors' perception of the level of overall assurance they confer, based on whether they control for extraction and/or inhibition and the robustness of the used methods. The level of assurance is an arbitrary scale. In general, methods utilizing robust controls (i.e. use of an internal control (IC) in multiplex assays) and addressing both extraction and inhibition were judged to confer a good level of assurance. Methods utilizing less robust controls (i.e. external controls in combination with IC) for inhibition and extraction were judged to confer an acceptable level of assurance. Methods not utilizing IC at all but only external controls were judged to confer low level of assurance. It should be noted that occasional assays were judged not to require controls for a false-negative result (e.g. typing assays detecting particular base sequences where a 'negative' necessarily implies a faulty assay that would be repeated).

External controls

External controls use reference material in a parallel PCR reaction. In its simplest form, external control utilizes one or few spiked vessels within a single run of a certain testing batch. Alternatively, external control can be achieved by running each sample in parallel reactions with one tube being the test and the other serving as a control. The use of external controls (parallel controls) has advantages in that the control PCR reaction cannot interfere with test accuracy and the control amplicon does not need to be distinguished from the test amplicon. For that reason external controls are very simple to design and implement. However, two main disadvantages exist as compared to IC. First, the number of PCR reactions that must be set up is doubled. Second, and more importantly, the potential for false-negatives to occur due to sample inhibition, though greatly reduced, remains.

Table 12.1 Categories of PCR controls for detection of false-negative results

Method	General level of assurance
Internal control using same amplimers (mimic controls)	Good
Internal control using different amplimers (non-mimic)	Good
Endogenous internal control (e.g. housekeeping gene)	Good
External control amplified in parallel paired reactions	Acceptable
Simple batch external control	Low

Table 12.2 Summary of control methods

Extraction control	Inhibition control	Degree of assurance
Control (phage coated RNA and/or DNA as appropriate) spiked into each sample	Extracted positive nucleic acid detected in same tube	Good
Human DNA indicative of good sample	Extract tested for intrinsic human template in the same tube	Good
Generic target (typing or gene variant assay)	Generic target detected in same tube	Good
Positive control cells extracted in separate tube	Positive target spiked into each sample detected in same tube	Good
No extraction control	Positive target spiked into each sample detected in same tube	Acceptable
Control DNA (e.g. plasmid) spiked into each sample	Extract tested for control template in a parallel tube	Acceptable
Human DNA indicative of good sample	Extract tested for intrinsic human template in a parallel tube	Acceptable
Extraction of nucleic acid tested by a parallel assay	Inhibition tested by a parallel assay	Low
Batch extraction control	Extract tested for spiked low copy template in a parallel tube	Low
No extraction control	Extract tested for spiked control template in a parallel tube	Low

Shaded cells represent external controls or no control and white cells represent internal controls.

Internal controls

ICs use multiplex PCR reactions. A second target molecule, the IC, that can be amplified with, but readily distinguished from, other products of the reaction is included in the reaction together with primers and probes as appropriate.

An ideal IC should provide control for all procedures, reagents and samples included in the PCR test and therefore would be present in all correctly taken samples, would be released from the same cellular compartment as the target sequence under the extraction conditions, would be non-competitively amplified using the same primers as the target sequence and also produce an amplicon of identical size and similar GC content, that can easily be distinguished from it. In practice this ideal is very difficult to achieve and consequently ICs generally represent a compromise though they are nevertheless valuable.

A simple IC for reaction inhibition might involve spiking the IC sequence into the PCR master mix at an appropriate concentration. IC targeting both extraction failure and reaction inhibition might involve spiking the IC sequence, or any form of target control nucleic acid that requires extraction, into the sample prior to the extraction phase. Spiking should involve a discrete quantity of exogenous nucleic acid into the sample at a known concentration.

Spiked ICs can be engineered to be a molecular mimic of the target sequence (see Chapter 2). The mimic sequences generally have priming sites identical to those of the target and intra-primer sequences ('stuffer') differing only to the extent needed to allow straightforward discrimination from the target. The degree of difference required between authentic sequence and mimic depends upon the method used for analysis. With gel electrophoresis, the IC amplicon must be sufficiently different in size from the target amplicon to allow them to be reliably distinguished. However, when a real-time probe system (e.g. TaqMan) is available, even a single base difference between IC and target may be sufficient. The disadvantage of a molecular mimic IC is competitive binding to primers and feedback inhibition due to amplicon annealing. Where this is likely to be a significant problem the best alternative is to use an IC with similar characteristics to the target (i.e. length, GC content, melting temperature) but that requires different primers and that is sufficiently dissimilar that no annealing of IC to target sequences can occur.

Delivery of the IC into the assay needs some consideration, especially when it must be added to a complex sample prior to extraction of the target nucleic acids. The addition of naked dsDNA that includes the IC sequence, in either linear, open circle or supercoiled form, to samples immediately prior to or post the addition of the 'lysis' buffer (i.e. any solution added to the sample to promote cell lysis or to strip proteins from nucleic acids or to prepare the nucleic acid for adsorption to a solid support) is not expected to be problematic and should result in non-preferential recovery of the IC sequence together with the target sequence. However, when it is necessary to add the IC as RNA (e.g. when the assay is designed to detect an RNA virus) it is usual to take additional steps to avoid ribonuclease action which may affect sequence integrity especially during longer term storage of the control material or to produce an RNA transcript that is resistant to RNase. One approach is to prepare the control RNA as 'armoured RNA' (i.e. RNA coated with bacteriophage MS2 coat protein) or to make use of products such as RNAlater™ which is a concentrated solution of ammonium sulfate.

It is essential to use an appropriate RNA or DNA internal control depending on target amplified. The internal control should be assessed for any adverse effect on the sensitivity of the assay. It should also be demonstrated that the internal control is similarly affected by the presence of potential inhibitory substances as the target organism. This can be achieved by spiking the PCR reaction with potential inhibitors such as haemoglobin or reagents such as guanidinium salts that may be carried over from the extraction process. The shift in crossing threshold values can be used to assess the relative impact of inhibition on the amplification of both the internal control and target organism.

A frequently used alternative that is appropriate for certain specimens is to use a host nucleic acid sequence as control. In this case all appropriately taken samples will be positive for the control. For the diagnosis of human infectious disease an important consideration is that the sample is adequate and this may entail that human cells be present. Consequently, it may be advantageous (or at least acceptable) to use a human gene as the IC in assays designed to detect infectious agents. One constraint of this approach is that the quantity of IC present depends upon the sample quality, which may be highly variable. It is therefore important to ensure that when large quantities of IC are present amplification of the target is not overwhelmed, whilst maintaining an adequate signal from the IC when lower amounts are present in the sample. Generally, on the basis that human DNA is present in large quantities in properly taken samples of relevant tissues, the best solution to this problem is a compromise in which the IC PCR is detuned to prevent the quantity of IC amplicon reaching the high levels that inhibit target amplification.

When a PCR assay is designed for detection of a particular type of a pathogen that is known to be present in a sample, suitable control may be inherent in the test. For example, in tests where an amplicon is sequenced the invariant bases, which are always present, provide a suitable IC. Alternatively, when the pathogen is known, conserved genes in its genome can be used for the purpose of IC in a reaction multiplexed with amplification of the test target sequence.

Risk assessment of controls

A risk assessment tool for the use of controls in real-time PCR assays is summarized in Table 12.3. The risk score presented is derived from the arbitrary assessment of the impact and consequences of having a false-negative result and the probability for having such a result. The impact of a false-negative is considered highest for specimens affecting patient care, either through primary or confirmatory diagnostics and lowest for typing for surveillance purposes. The probability of a false-negative is considered highest for primary specimens owing to the varying target density, influence of sampling method and risk of inefficient extraction and reaction inhibition. The probability is lower for secondary specimens due to the high target density and predictable extraction and inhibition and lowest for typing assays which are applied on specimens known to be positive for the target micro-organism. The degree of assurance conferred by controls informs the mitigation required to manage the risk.

Validation design

The purpose and objectives of the validation should be clearly defined. For example, the assessment may be designed to validate the performance of a new assay or may aim to demonstrate that a significantly modified assay or protocol gives acceptable results.

Table 12.3 Risk assessment tool for diagnostic real-time PCR assays

PCR application	Impact of a false-negative[a]	Probability of a false-negative[a]	Risk score	Minimal risk mitigation needed via controls
Clinical diagnosis from primary specimen	3	3	9	Good assurance
Clinical diagnosis from secondary specimen	3	2	6	Good assurance
Confirmatory testing (reference) of primary specimen	3	3	9	Good assurance
Confirmatory testing (reference) of secondary specimen	3	2	6	Good assurance
Treatment critical typing from primary specimen	2	2	4	Acceptable assurance
Treatment critical typing from secondary specimen	2	1	2	Acceptable assurance
Typing for surveillance from primary specimen	1	1	1	Low assurance
Typing for surveillance from secondary specimen	1	1	1	Low assurance

[a]1, low; 2, moderate; 3, high.

The following features should be included in the validation design:

- Identification of any training needs of workers involved in the validation to ensure each has a suitable level of competency. Ensure training records are up to date for procedures being carried out.
- Identification of standard or reference materials where available for use as controls and to facilitate assay standardization, comparison and stability testing.
- Identification of any kit or reagent to be used for comparison with the assay undergoing validation. Ideally this should be the 'gold standard' currently recommended for detection of the test analyte.
- For clinical assays, design a clinical validation study including the clinical context (e.g. surveillance, screening, clinical diagnosis), choice of comparators and ethical considerations. Choose the study population including type of patients, case definitions, inclusion and exclusion criteria and study settings.
- Identify the types (i.e. specimen, method of sampling, transport and processing) and numbers of samples to be tested for the validation. Consider the need to include known positives, known negatives, low and high positives and samples which are known or likely to be problematic (e.g. containing inhibitors or possibly cross-reactive markers) or representative of a particular population.
- Consider statistical techniques to select an appropriate sample size and avoid bias. It is essential to consider statistical requirements during project design to ensure that results are statistically significant.
- The validation design should avoid discrepant analysis bias. Some samples may give discordant results with the new test compared to a 'gold standard', if only these samples are retested bias is introduced as there is a probability that the second analysis will give concordant results for some of these samples. To avoid bias, at least the same number of concordant as discrepant samples should re-tested.
- Consider documenting an assessment of assay usability including method practicability, user feedback, barriers to implementation.
- Plan to review the validation study in a timely manner, minute meetings detailing attendees, consensus to changes to the study plan and all other agreed actions.
- Ensure that all SOPs related to the modified or new kits or reagents are current. It may be necessary to maintain new or revised SOPs as working drafts while their contents are being validated, ensuring COSHH and risk assessments are up to date. SOPs should be authorized as fully controlled documents when the validation study has been completed.
- In some instances assay validation may be difficult to achieve within a reasonable timescale. For example, an assay designed to diagnose a disease for which relevant clinical material is hard to obtain or rare (e.g. the viral haemorrhagic fevers). In these cases it may be desirable to use the assay due to its potential value even though full validation criteria are not yet met. Deficiencies in assay validation should be documented in the development dossier together with justification for each instance where the data are inadequate. In such cases a plan for completion of satisfactory validation should be outlined. For example, it may be possible to obtain samples post-implementation or it may be reasonable to place greater reliance on the use of carefully designed simulated specimens.

- The validation process should provide a clear guide to the usability of the assay, the meaning of various results produced by the assay in the context of reference ranges, the diagnostic or clinical applications of the assay and any limitations.
- In the context of real-time PCR for detection of pathogens analytical validation ensures that the test chemistry reliably detects the target while diagnostic validation ensures that the test detects the target in clinical specimens and is fit for use on test samples. The next step is then to integrate the test into the clinical routine as part of a pathway. The clinical pathway is initiated by the set of clinical indications leading to the patient being tested and includes the expected range of results and details of patient management depending on these results (i.e. to establish workflows so that the test is performed at an appropriate point in the diagnostic cycle, that performance is triggered by specific clinical questions and that the results are interpreted in a way the feeds back into patient management).
- The validation process should include shelf life studies for the key reagents. It is important to know how long a reagent remains active. Activity may fall off gradually and so still give an acceptable result with QC material but not be performing at its optimum.

The validation file and technical report

The technical report and validation file should summarize assay validation, results and recommendations. This task also includes the setting up of the multirule run QC criteria (Westgard *et al.*, 1986) using positive control results from runs conducted as per routine SOP, i.e. positive controls from the clinical accuracy runs should be sufficient.

For new assays, validation may be supported by the research and development carried out during the development of the procedures. Workbook records can be cross-referenced if appropriate in the validation report.

When the validation work has been completed the results should be analysed according to the plan outlined in the study design. Results are compared with the specification to determine whether or not the kit or reagent is suitable for use. Analysis should include comparison with results obtained using the 'gold standard' or alternative reagents. For commercial kits the analysis will include comparison of the actual data with sensitivity and specificity information provided by the manufacturer.

The validation data should be presented in a written report, so that the review team can make a decision whether or not the assay is fit for purpose (as agreed in the Assay Definition). A conclusion should be documented based on the results analysis. This should include a formal declaration that the assay is suitable for use which should be signed by the Project Leader and the Project Manager. At this stage supplemental documents should be prepared such as instructions for use, preparation and labelling of reagents, requirements for equipment maintenance, etc. The validation should be completed and signed off before any use of the kit or reagent for diagnostic or reference purposes.

Validation of assays already in use

Once satisfactory validation has been completed using a defined SOP it is essential that the assay is performed according to the SOP. Minor changes to the method that are immaterial to the performance specification of the assay should, as far as possible, be accommodated within the original SOP provided suitable evidence is available. For example, it may be better for the SOP to specify the use of distilled, deionised or molecular grade water rather than a particular brand of molecular grade water. Similarly, it may be reasonable to avoid specifying the particular model of PCR instrument to be used.

When changes to the protocol that fall outside an SOP must be made, revalidation is required. Revalidation will vary from a single run showing that C_t values are unchanged for a range of reference samples (e.g. as might be appropriate for a change to a buffer) to full revalidation (e.g. for use of an alternative sample preparation method or primer change). A guide to the level of revalidation that might be considered acceptable for a range of changes to the protocol appears in Table 12.4. It should be noted that in many situations the appropriate level of revalidation depends upon the judgement of the trained staff making the protocol changes. In all cases it is essential to document the reasoning involved which should be referred to in the updated SOP and kept in the validation file.

Where assays that have inadequate validation records are in clinical use it is good practice to provide a compliant validation file retrospectively wherever practical. Historical records should be reviewed and summarized to provide evidence that the assay is fit for purpose and a validation checklist should be used to cross-reference these documents. This advice

Table 12.4 Revalidation

	Protocol change					
Revalidation activity	Primers/ probes	Reaction mix components (ex-primers/ probes)	Extraction method	Control material	Instrument or conditions	Analysis
Analytical sensitivity	Yes	Yes	Yes	No	Yes	Yes
Diagnostic sensitivity	Yes	a	Yes	No	a	a
Analytical specificity	Yes	Limited	Limited	Limited	Limited	Limited
Diagnostic specificity	Yes	a	a	No	a	a
Efficiency (quantitative assays)	Yes	Yes	No	No	Yes	Yes
Linearity (quantitative assays)	Yes	a	Yes	No	a	a
Measurement range	Yes	a	a	No	a	a
Precision (quantitative assays)	Yes	a	Yes	No	a	a
Reproducibility (quantitative assays)	Yes	a	Yes	No	a	a
Analytical accuracy (quantitative assays)	Yes	a	Yes	Yes	a	a
Clinical accuracy (quantitative assays)	Yes	a	Yes	No	a	a
Reference intervals	Yes	a	a	No	a	a
Clinical validation	Yes	a	a	No	a	a
Shelf-life	Partial	Partial	No	Partial	No	No

[a]The level of revalidation required will depend upon professional judgement in the particular circumstances of the assay.

Table 12.5 Validation framework checklist

Name of assay	
Specific targets (pathogens and genes) of the assay	
Project manager	
Project leader	
Staff performing the hands-on development work	

Section	Subsection	Contributor	Deadline	Completion
Planning and inception	Project drivers – gap analysis			
	Setting purpose and objective for new assay			
	Sign-off process			
	Setting up a project risk register			
	Health and safety issues			
	Business analysis			
	Commercial analysis			
	Regulatory compliance			
	Human resources aspects			
	Project planning and initiation			
Assay development	Identification of targets			
	Setting benchmarks and performance measures			
	Options appraisal for technologies			
	Identification of gold standards			
	Identification of reference material and reagents			
	In silico design			
	Technical design			
	Optimization			
	Design of post-amplification analysis			
	Development of controls			
	Assessment of analytical performance			
	Preparation of specific SOPs (MIQE compliant)			

Validation	Design of validation study			
	Assessment of analytical performance			
	Choice of study population			
	Choice of sample type			
	Sample size calculation			
	Performance of clinical validation			
	Data analysis and interpretation			
	Assessment of usability			
	Reagent stability studies			
	Further optimization if required			
	Production of documentation			
Roll-out and verification	Review of needs and users			
	Planning of roll-out			
	Field verification studies			
	Provision of validation panels			
	Inter-laboratory comparison			
Implementation	Implementation in laboratory work-flows			
	Implementation in diagnostic pathways			
	Implementation in clinical pathways			
	Post-marketing assessment			
	Long-term data collection			
	Periodic QC/QA			
	Periodic proficiency testing			
	Production aspects of implementation			
	Periodic reassessment of fitness for purpose			

applies to both commercial and in-house assays. Further validation studies may not be required even when insufficient data are available to produce a compliant validation file. For example, it may be possible to implement appropriate data collection to provide supportive information. A conclusion based on the information obtained from the historical data should be documented.

A management checklist

The management requirements for the performance of adequate validation of diagnostic assays are considerable and this requires a systematic and meticulous approach in order that many of these aspects are not overlooked. A proposed validation framework checklist appears in Table 12.5 and could be used in order to guide project management.

Assay 'roll-out'

The collection of diagnostic validation data will necessarily involve limited technical 'roll-out' within the developing laboratory. Full adoption of the new assay into clinical use will include consideration of the reagent supply chain and arrangements for quality assessment, including EQA wherever appropriate. If the test is likely to meet the needs of a range of end-users 'roll-out' is likely to be a consideration. A roll-out plan should include the timeline, equipment, personnel and risk management proposals. If the assay is for clinical diagnostic purposes, any legal requirements that apply must be complied with.

Following roll-out an assay verification study and continuing quality assessment should be performed. This should be organized by the developing laboratory or a qualified third party (e.g. an EQA body). Verification should involve the testing of validation panels or be limited in scope to a few well characterized samples and standards. In some circumstances it may be desirable that validation materials are originated in laboratories other than that of the developer. Round-robins may be useful in ensuring continuing attention to quality.

Once the assay has been validated and put into use the gathering of information on its performance should be continued (Clinical and Laboratory Standards Institute, 2006). Details of any problems, e.g. poor performance, false-positives or negatives, should be documented. The assay should be reviewed periodically to ensure continued improvement and to highlight performance compared to alternative assays, whether commercial or in-house. Periodic reviews should include review of additional nucleotide sequence data that may be available, and relevant scientific literature.

References

Brightwell, G., Pearce, M., and Leslie, D. (1998). Development of internal controls for PCR detection of *Bacillus anthracis*. Mol. Cell. Probes *12*, 367–377.

Burd E.M. (2010). Validation of laboratory-developed molecular assays for infectious diseases. Clin. Microbiol. Rev. 23, 550–576.

Bustin, S.A., Brenes, V., Garson, J.A., Hellemans, J., Huggett, J., Kubista, M., Mueller, R., Nolan, T., Pfaffl, M.W., Shipley, G.L., *et al.* (2009). The MIQE guidelines: minimum information for publication of quantitative real-time OCR experiments. Clin. Chem. 55, 1–12.

Clinical and Laboratory Standards Institute. (2006). Molecular Diagnostic Methods for Infectious Diseases; Approved Guidelines, 2nd edition. Available at: http://www.clsi.org/source/orders/free/mm1-a2.pdf.

Dimech, W., Bowden, D.S., Brestovac, B., Byron, K., James, G., Jardine, D., Sloots, T., and Dax, E.M. (2004). Validation of assembled nucleic acid-based tests in diagnostic microbiology laboratories. Pathology *36*, 45–50.

I.C.H. Steering Committee. (1994). ICH Harmonised Tripartite Guideline: Text on validation of analytical procedures Q2A. Available at: http://www.ics.trieste.it/media/135980/df3560.pdf.

Kotnik Kevorkijan, N., Oresic, S., Janezic, and Rupnik, M. (2011). Comparison of two commercial molecular tests for the detection of *Clostridium difficile* in the routine diagnostic laboratory. J. Med. Microbiol. *60*, 1131–1136.

Lange, M., Neubauer, H., and Seyboldt, C. (2010). Development and validation of a multiplex real-time PCR for detection of *Clostridium chauvoei* and *Clostridium septicum*. Mol. Cell. Probes *24*, 204–210.

Picard, F.J., Gagnon, M., Bernier, M.R., Parham, N.J., Bastien, M., Boissinot, M., Peytavi, R., and Bergeron, M.G. (2009). Internal control for nucleic acid testing based on the use of purified *Bacillus atrophaeus* subsp. *globigii* spores. J. Clin. Microbiol. *47*, 751–757.

Pugnale, P., Latorre, P., Rossi, C., Crovatto, K., Pazienza, V., Gottardi, A.D., and Negro, F. (2006). Real-time multiplex PCR assay to quantify hepatitis C virus RNA in peripheral blood mononuclear cells. J. Virol. Methods *133*, 195–204.

Westgard, J.O., and Barry, P.L. (1986). Improving quality control by use of multirule control procedures. In Cost-effective Quality Control: Managing the Quality and Productivity of Analytical Processes (AACC Press, Washington, DC), pp. 92–117.

Index

A

B

N

O

P

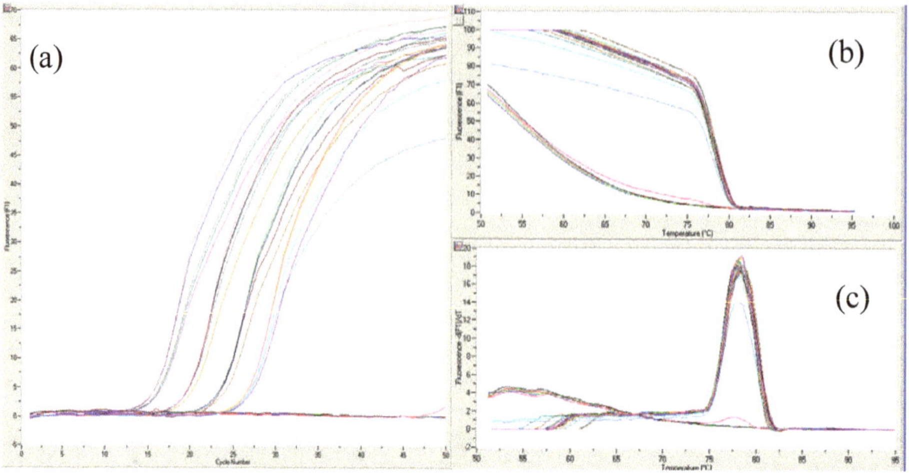

Figure 1.1 Amplification using generic binding dyes (minor-groove binding dye SYBR®Gold) on the LightCycler®. (a) The amplification plot shows 50 cycles (to completion) of amplification of four dilutions (four replicates) of a 10-fold dilution series and four no-template controls, one of which produces a low amplification signal. The reaction utilised anti-Taq antibody hot start and UNG carry-over protection and the data utilise the background subtraction algorithm. (b) Melting point analysis of the amplified products, and (c) the first negative differential of the fluorescence, with respect to Temperature, plotted against Temperature. From this plot it can observed that the signal in the positive no-template control has the same melting point as the specific amplification in positive samples, and therefore amplification is most likely a results of cross-contamination.

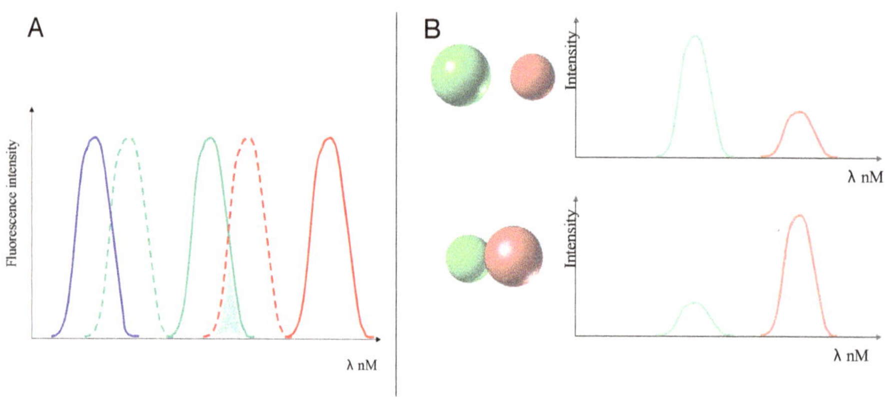

Figure 1.2 Fluorescence energy resonance transfer. (A) The emissions of a light source (blue line) falls within the excitation spectrum (green line dashed) of a fluorophore and cause this dye to fluoresce. The emissions (green solid) of this fluorophore fall within the excitation spectrum (red dashed line) of a second fluorophore that emits fluorescence at longer wavelengths (red solid). When the second fluorophore is close to the first on the same or neighbouring molecule, the energy that would be emitted by the first fluorophore (top of B) can be transferred to the second through a number of energy transfer mechanisms such that the second fluorophore emits this energy at its own emission wavelengths (bottom of B).

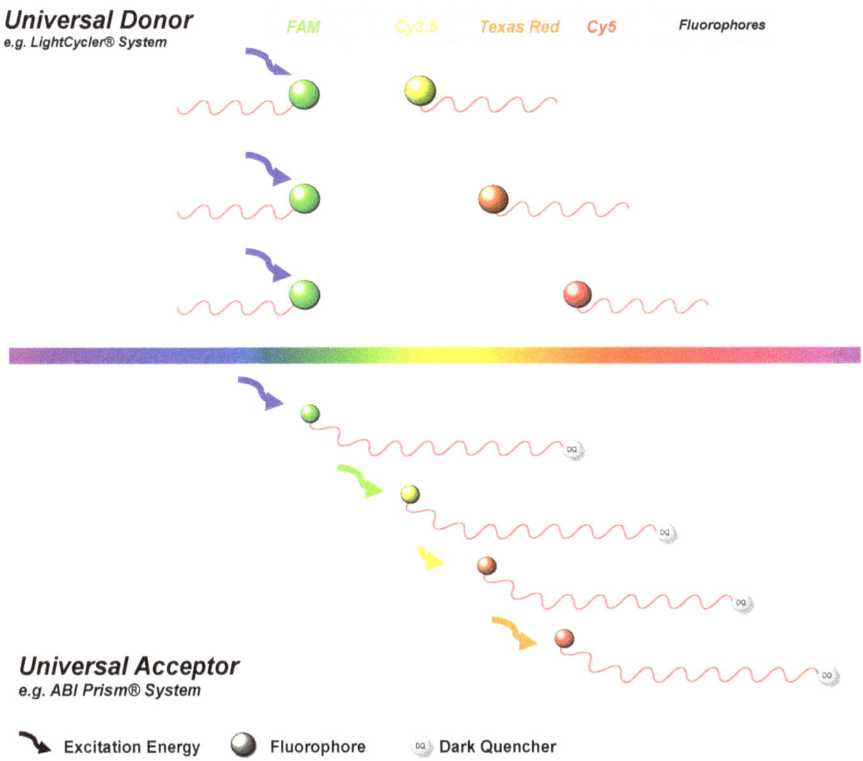

Figure 1.3 Schematic showing the two principal methods for multiplexing using FRET: (top) the universal donor on the LightCycler® and (bottom) the universal acceptor on the ABI 7700.

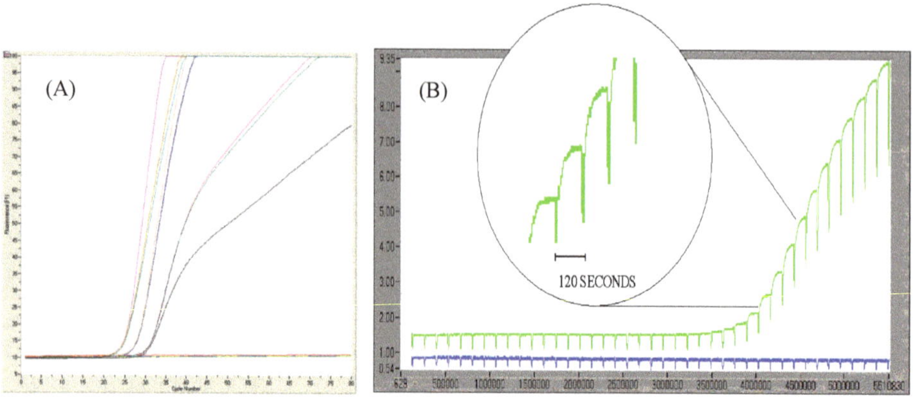

Figure 1.4 Amplification using the 5′ nuclease assay on the LightCycler®: (A) the amplification plot shows 80 cycles (in excess of completion) of thermal cycling of three dilutions (three replicates) of a 10-fold dilution series and three no-template controls on the Roche LightCycler®. The reaction utilised anti-Taq antibody hot start and UNG carry-over protection. The reaction used the primers and probes from the Applied Biosystems Human β-Actin kit which utilises a FAM-TAMRA internally quenched TaqMan® probe. The amplification plot shows raw data (no analysis/ background subtraction) and illustrates the excellent signal–noise ratio achievable with this chemistry. (B) Shows the same reaction amplified on the Idaho LightCycler® carried out by continuous monitoring of a positive and negative sample throughout all stages of amplification. The data shows that ~120 seconds are required each cycle to take the hydrolysis of bound probe to completion (plateau) each cycle.

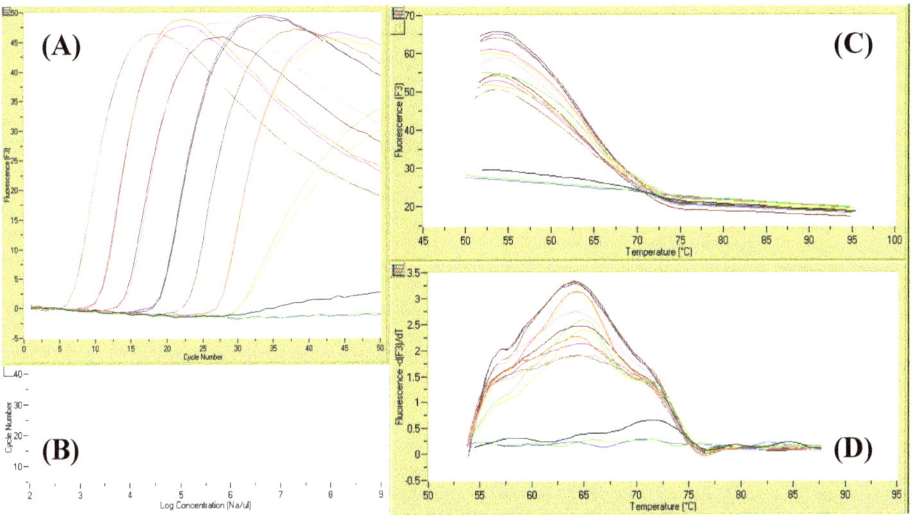

Figure 1.5 Amplification using dual hybridization probes on the LightCycler®: (A) the amplification plot shows 50 cycles of thermal cycling of replicates of eight dilutions of a 10-fold dilution series of a reference for the target, and no-template controls on the Roche LightCycler®. The amplification shows the data analysed using the background subtraction algorithm. The reaction utilised anti-Taq antibody hot-start and UNG carry-over protection. The reaction used the primers and probes for the Human β-globin gene using a 3′ labelled fluorescein donor probe and a 5′ Cy5.5 acceptor probe. The acceptor probe was blocked to prevent 3′ extension using an octendiol moiety. A 5′-3′ exonuclease Taq polymerase was used to illustrate that the 'hook' effect that is observed in linear hybridization probes as a result of probe displacement by product in later cycles. (B) The derived standard curve from the amplification plot. (C) The melt curve showing probe disassociation from product. (D) The first negative differential of fluorescence with respect to temperature showing the peak at which the maximum rate of probe dissociation occurs.

Table 1.1 Summary of fluorogenic reporting chemistries for real-time PCR. The schematics show the probe structure and mode of action

Method	Principle	Self-probing embodiment	**Principle** Probe structure	Nucleic acid interaction
DNA binding agent	Enhancement of fluorescence DNA binding into DNA duplex	N/A	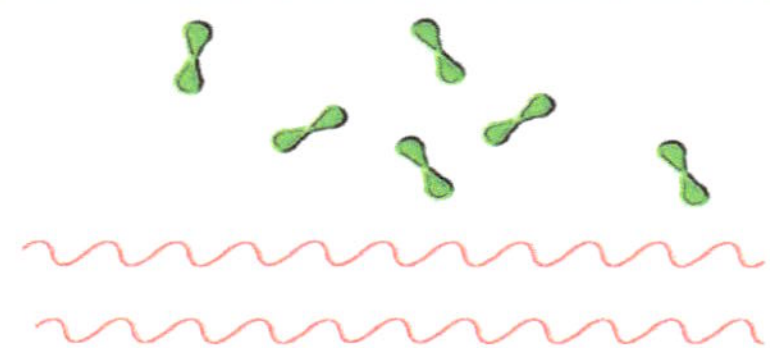	
5′ nuclease assay	Hydrolysis of probe by concomitant hydrolysis of probe during primer extension	Intrataq		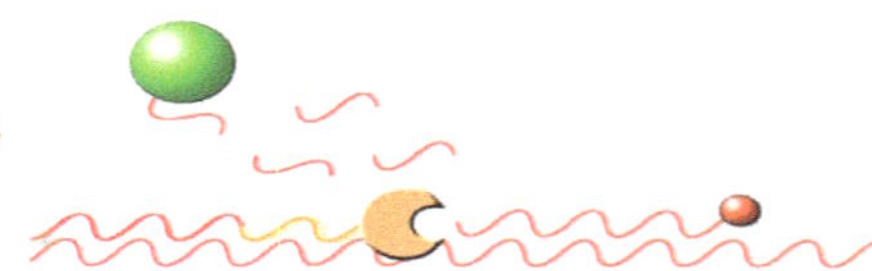
Dual hybridization probe (*cis*)	Hybridization of two linear probes to nascent PCR strand	Embodiment of Scorpions®	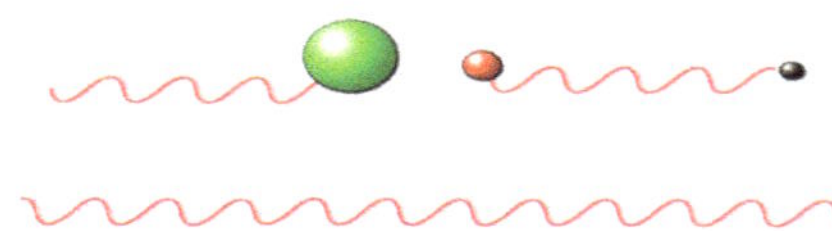	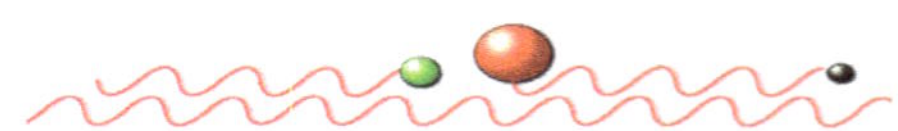
Dual hybridization probe (*trans*)	Hybridization of linear probe to nascent PCR strand generated by a labelled primer	Not reported	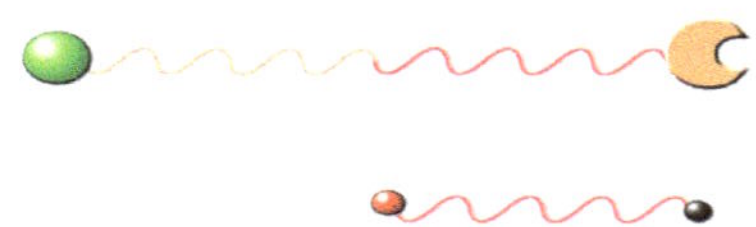	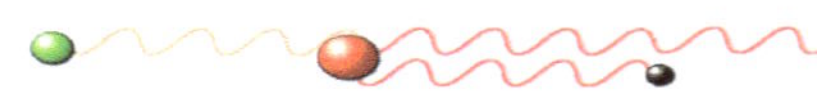

Table 1.1 (Continued)

Method	Principle	Self-probing embodiment	Principle: Probe structure	Principle: Nucleic acid interaction
ResonSense®	Hybridization of linear probe to nascent PCR strand. Energy is transferred through a DNA binding agent	Angler®	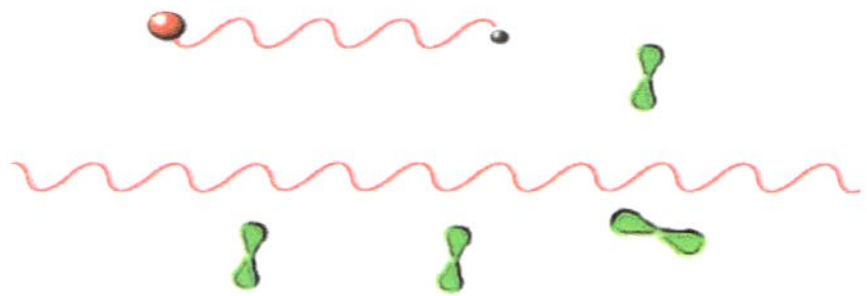	
Hybeacons®	Hybridization of linear probes.	Not reported	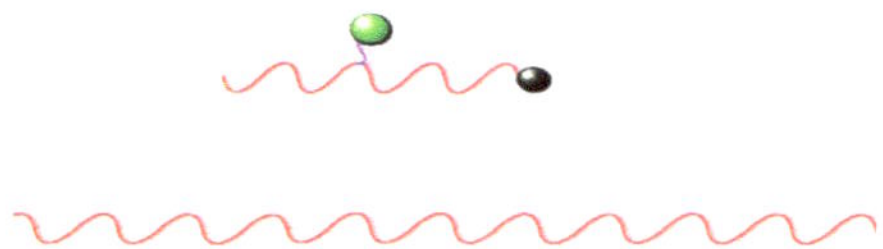	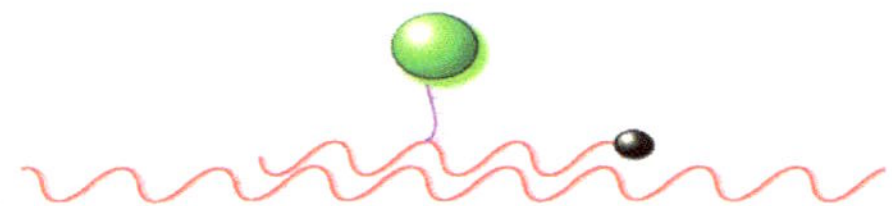
Eclipse®	Linear conformational change. Short probe is stabilized to target using a minor groove binding agent	Not reported	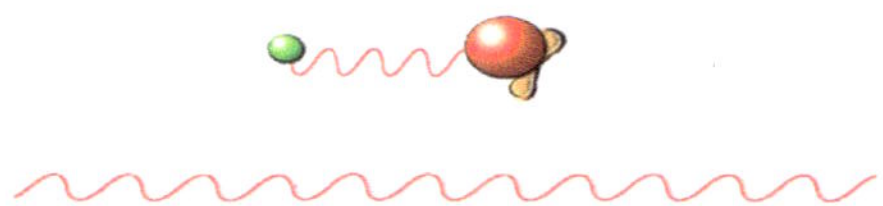	
Light-Up®	Fluorescent enhancement upon hybridization with DNA-binding agent covalently linked to probe	Not reported	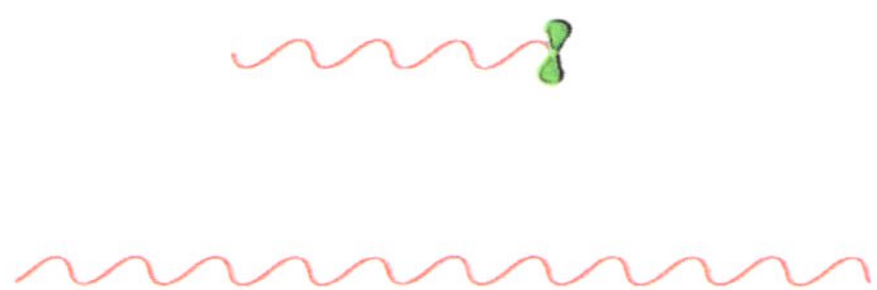	

Method	Principle	Self-probing embodiment	**Principle** Probe structure	Nucleic acid interaction
Molecular Beacons	Conformational change. Native probe stabilized with hairpin structure	Scorpions®	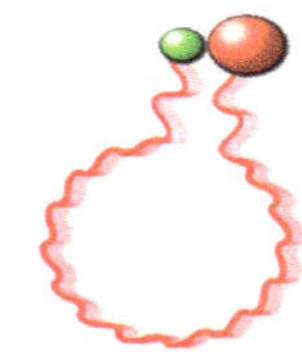	
Ying–Yang®	Competitive binding of linear probe	Duplex Scorpions®	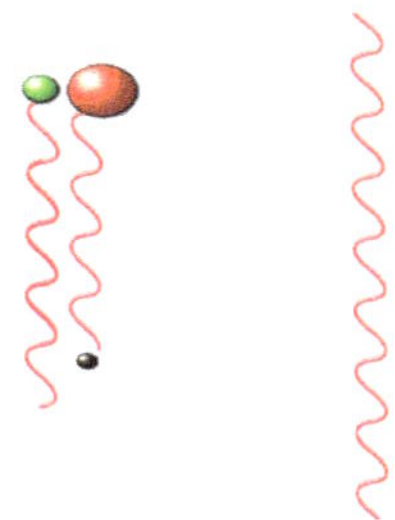	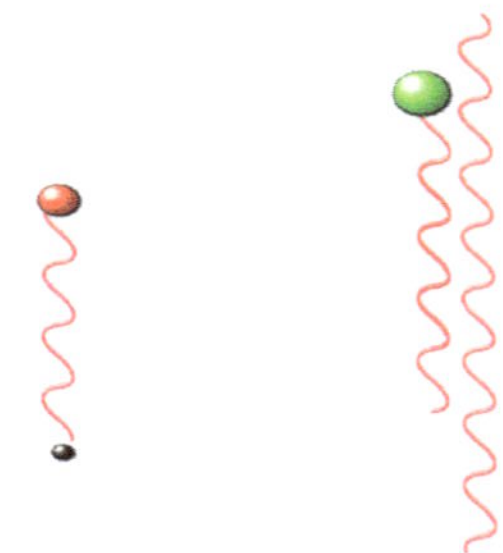

Key: DNA, donor dye, acceptor dye, DNA binding agent, minor groove binder, polymerase, 3' blocker.

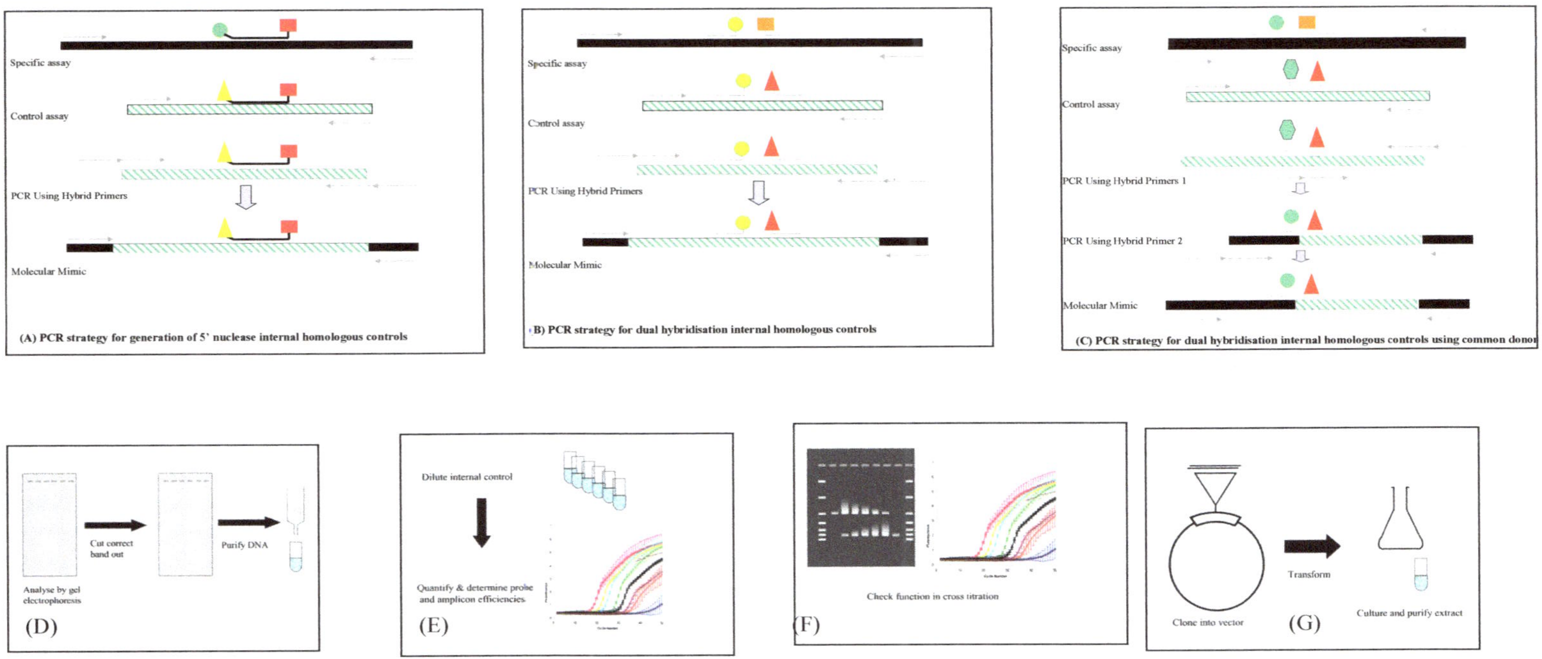

Figure 2.1 Schematic showing strategies for rapidly generating and evaluating competitive internal homologous controls by PCR. (A) For assays such as the 5′ nuclease where there is only one probe. (B) For dual hybridization assays using two additional probes. (C) Dual hybridization probes using one common probe for both the control and specific assay. (D) Amplify the IC using PCR and extract the correct band from the gel. (E) Serially dilute the product and determine the amplification and probe efficiencies (Fig. 2.2). (F) Cross-titre the IC with specific target to assess the efficacy of the IC in multiplex PCR before cloning into a suitable vector. (G) Transformation, culture, extraction and use.

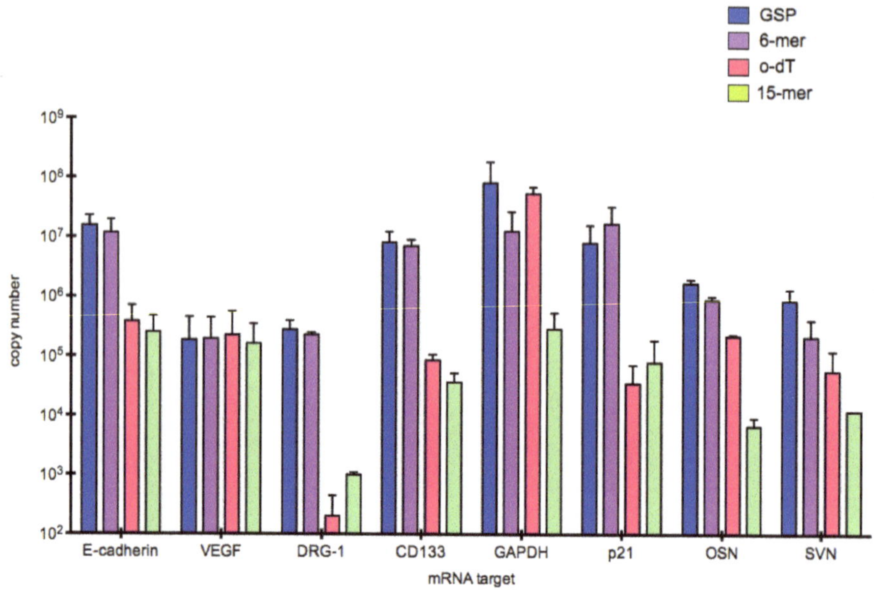

Figure 3.3 Choice of priming strategy affects RT-qPCR results. Four priming strategies were compared for the reverse transcription of eight target mRNAs. The cDNA from each was analysed using qPCR. In each case, gene-specific priming resulted in the most efficient reverse transcription, although the use of random hexamers generated very similar results. Reverse transcription from random 15-mers was consistently the least efficient priming method. Oligo-dT priming was less consistent, resulting in greater variability of quantification and relative quantity between targets.

www.ingramcontent.com/pod-product-compliance
Lightning Source LLC
LaVergne TN
LVHW051050210926
848486LV00001B/56